# Photophysics of Supramolecular Architectures

Edited by

## Paulpandian Muthu Mareeswaran

*Department of Industrial Chemistry*
*School of Chemical Sciences*
*Alagappa University, Karaikudi – 630 003*
*Tamilnadu, India*

## Palaniswamy Suresh

*Department of Natural Products Chemistry*
*School of Chemistry, Madurai Kamaraj University*
*Madurai – 625 021, Tamilnadu, India*

&

## Seenivasan Rajagopal

*Department of Physical Chemistry*
*School of Chemistry, Madurai Kamaraj University*
*Madurai – 625 021, Tamilnadu, India*

# Photophysics of Supramolecular Architectures

Editors: Paulpandian Muthu Mareeswaran, Palaniswamy Suresh and Seenivasan Rajagopal

ISBN (Online): 978-981-5049-19-0

ISBN (Print): 978-981-5049-20-6

ISBN (Paperback): 978-981-5049-21-3

need for a court order if at any point you breach any terms of this License Agreement. In no event will any delay or failure by Bentham Science Publishers in enforcing your compliance with this License Agreement constitute a waiver of any of its rights.

3. You acknowledge that you have read this License Agreement, and agree to be bound by its terms and conditions. To the extent that any other terms and conditions presented on any website of Bentham Science Publishers conflict with, or are inconsistent with, the terms and conditions set out in this License Agreement, you acknowledge that the terms and conditions set out in this License Agreement shall prevail.

**Bentham Science Publishers Pte. Ltd.**
80 Robinson Road #02-00
Singapore 068898
Singapore
Email: subscriptions@benthamscience.net

**BENTHAM
SCIENCE**

# CONTENTS

# FOREWORD

The introduction of surfactants and micelles to study the photophysics and photochemistry of molecules opened up a new approach to mimick the reactions taking place in nature. The opening of new horizons in molecular science after the introduction of surfactants brought a new approach to the field of molecular assemblies. The intense research in the fields of photochemistry and its applications in understanding the photosynthetic processes and solar energy conversion via chemical routes led to the growth of 'supramolecular chemistry', known as 'chemistry beyond the molecule'. The last four decades witnessed an explosion of research activities in the application of supramolecular assemblies in photophysics and photochemistry of molecules and the importance of supramolecular chemistry was understood after the award of 1987 Nobel Prize for Chemistry (Nobel Laureates: Donald J. Cram, Jean-Marie Lehn and Charles J. Pedersen) in this area of research.

The editors have chosen the interesting theme of 'Photophysics of Supramolecular Architectures' and presented a collection of important topics covering a wide variety of molecular assemblies and their applications, in particular, in the field of sensors and allied topics. The senior editor Prof. S. Rajagopal with his four decades of intense research in the fields of electron transfer reactions and photochemistry and his experienced co-editors are able to bring together a spectrum of scientists working in the areas of supramolecular systems and their applications. The authors of eight chapters put their efforts to update the knowledge gained in the fields of supramolecular chemistry and their applications in photophysics and photochemistry. The scientists and the young researchers working in the areas of photophysics and photochemistry of molecules in supramolecular assembly can immensely benefit from the book 'Photophysics of Supramolecular Architectures 2021' published by the editors. The editors have made an effort to cover a wide range of supramolecular systems from the conventionally known cyclodextrin systems to calixarene, resorcinarene crowns, pillararene, cucurbit[n]urils, cavitands, metallacycles and fluorescently labelled macromolecules and presented their variety of applications related to photophysics and photochemistry, in particular, the luminescent sensor systems for a wide range of analyte molecules. This book is a single source of collection of literature on the chosen topic and will help the young researchers to understand the field of research, up-to-date literature and design their future plan of action in the areas of supramolecular systems and their applications in photophysics and photochemistry.

**Dr. R. Ramaraj**
Department of Physical Chemistry
School of Chemistry
Madurai Kamaraj University
Madurai-625021, Tamilnadu
India

# PREFACE

Supramolecular architectures, the prevalent architectures in nature, are designed through a variety of non-bonding interactions like hydrogen bonding, $\pi$-$\pi$ staking, self-assembly, *etc*. From physics to biology, the functionalities of supramolecular architectures play an important role. For example, life is not possible without a DNA folding or protein self-assembly. This book mainly focuses on cavities containing supramolecular hosts and their photophysical properties by attaching luminescent molecules as guests. The host-guest chemistry is a widely established subject that can expand as an individual field of research with respect to the cavitand. The host-guest chemistry is envisaged as mimic for enzymatic catalysis. Also, they are used as drug delivery vehicles for the targeted payload delivery in biochemistry and biotechnology. The study of the interaction of guest molecules with light in the presence of host molecule opens a research opportunity to develop advanced research like optical tweezers.

The recent studies of photophysics of guest molecules with various cavitands like cylclodextrin, calixarene and their derivatives, cucurbiturils are highlighted in this book. The cyclodextrin complexes having different cavities and encapsulation of fluorescent guest molecules and applications of these systems are elaborately discussed in Chapter 1. Chapter 2 deals with the interaction of fluorescent guest molecules with calixarenes and their applications towards (as) sensors. The photophysical properties of coordinations complexes of calixarene-lanthanide systems are also discussed in this chapter.

Resorcinarenes, one of the important molecules in the calixarene family, receive importance for their hydroxyl group containing upper rim, which makes them suitable for catalysis applications. Chapter 3 mainly focuses on the upper rim modification at hydroxyl group to achieve crown ethers. Eventhough the crown ethers are separately known as supramolecules, this chapter discusses upper rim modified resorcinarene-crown and their applications using optical spectral techniques. Chapter 4 deals with the pillararenes, which are considered as young cavitand molecular system, reported only in 2008. This chapter focuses on the host-guest chemistry of pillararenes with fluorescent guest molecules. The self-assembly of pillarene derivative is also discussed to achieve sensor applications.

Chapter 5 is concerned with the molecular recognition of fluorescent guest molecules encapsulated cucurbiturils and their applications as sensors. The application of imaging and photodynamic therapy using cucurbiturils systems is also discussed. Chapter 6 deals with the control of photophysical properties and photochemical events of various cavities and capsules. Chapter 7 mainly focuses on the cavity containing rhenium(I) metallasupramolecules, i.e., metallacycles. The synthesis, photophysical properties and host-guest behavior of rhenium metallacycles ranging from simple to complex topologies are discussed in detail. Chapter 8 deals with the dynamics of macromolecules functionalized with fluorescent molecules. The macrocyclic environment acts as a host molecule and influences the optical properties of the functionalized fluorescent molecule. The folding and unfolding of macrocycles exhibit substantial variations in the fluorescent molecules.

This book strives to give collectively the applications of host-guest chemistry with recent applications. The recent advancements in the various host-guest systems will provide newer insights to readers into both conventional host molecules like cyclodextrin as well as young host molecules like pillararenes.

**Paulpandian Muthu Mareeswaran**
Department of Industrial Chemistry
School of Chemical Sciences
Alagappa University, Karaikudi – 630 003
Tamilnadu, India

**Palaniswamy Suresh**
Department of Natural Products Chemistry
School of Chemistry, Madurai Kamaraj University
Madurai – 625 021, Tamilnadu, India

&

**Seenivasan Rajagopal**
Department of Physical Chemistry
School of Chemistry, Madurai Kamaraj University
Madurai – 625 021, Tamilnadu, India

# List of Contributors

| | |
|---|---|
| **Bosco Christin Maria Arputham Ashwin** | Department of Chemistry, Pioneer Kumaraswamy College, Nagercoil 629 003, Tamilnadu, India |
| **E. Rajkumar** | Biomimetic and Biosensor Lab, Department of Chemistry, Madras Christian College (Autonomous), Affiliated to University of Madras, Chennai-600 059, Tamilnadu, India |
| **Jeyaraj Belinda Asha** | Supramolecular and Catalysis Lab, Dept. of Natural Products Chemistry, School of Chemistry, Madurai Kamaraj University, Madurai-625021, Tamilnadu, India |
| **Kandhasamy Durai Murugan** | Department of Bioelectronics and Biosensors, Alagappa University, Karaikudi-630003, Tamilnadu, India |
| **Liju R.** | Biomimetic and Biosensor Lab, Department of Chemistry, Madras Christian College (Autonomous), Affiliated to University of Madras, Chennai-600 059, Tamilnadu, India |
| **Murugesan Velayudham** | Department of Chemistry, Thiagarajar College of Engineering, Madurai-625015, Tamilnadu, India |
| **Palaniswamy Suresh** | Supramolecular and Catalysis Lab, Dept. of Natural Products Chemistry, School of Chemistry, Madurai Kamaraj University, Madurai-625021, Tamilnadu, India |
| **Pandi Muthirulan** | Department of Chemistry, Lekshmipuram College of Arts and Science, (Affiliated to MS University, Tirunelveli), Neyyoor-629802, Tamilnadu, India |
| **Paulpandian Muthu Mareeswaran** | Department of Industrial Chemistry, Alagappa University, Karaikudi 630 003, Tamilnadu, India |
| **Pounraj Thanasekaran** | Department of Chemistry, Pondicherry University, Kalapet, Puducherry - 605 014, India |
| **Selvaraj Devi** | P. G. Department of Chemistry, Cauvery College for Women, Tiruchirappalli-620018, Tamilnadu, India |
| **Somasundaram Anbu Anjugam Vandarkuzhali** | Department of Chemistry, Saveetha School of Engineering, Saveetha Institute of Medical and Technical Sciences, Saveetha University, Chennai-600005, Tamilnadu, India |
| **Vairaperumal Tharmaraj** | Environmental Science and Technology Research Group, Department of Chemical Engineering, SRM Institute of Science and Technology, Kattankulathur-603203, Tamilnadu, India |
| **Venkatesan Sethuraman** | Department of Industrial Chemistry, Alagappa University, Karaikudi 630 003, Tamilnadu, India |
| **Vijayanand Chandrasekaran** | Department of Chemistry, School of Advanced Sciences, Vellore Institute of Technology, Vellore-632014, Tamilnadu, India |

# CHAPTER 1

# Luminescent Cyclodextrin Systems and Their Applications

**Bosco Christin Maria Arputham Ashwin**[1], **Venkatesan Sethuraman**[2] and **Paulpandian Muthu Mareeswaran**[2,*]

[1] *Department of Chemistry, Pioneer Kumaraswamy College, Nagercoil 629 003, Tamilnadu, India*

[2] *Department of Industrial Chemistry, Alagappa University, Karaikudi 630 003, Tamilnadu, India*

**Abstract:** This chapter explains the most recent development on different luminophore tethered cyclodextrin (CD), a cyclic polysaccharide and these applications in distinct areas. The host-guest inclusion complexation studies of CD with different guest molecules using fluorescence techniques are discussed. The hybrid materials of CD in the detection of biological analytes, toxic compounds and *in-vivo* bio-imaging applications are discussed. The compatibility nature of CD leads to its usage in drug delivery and the controlled drug dosage using CDs is explained. The interesting usage of CDs in counterfeit recognition and tunable emission are emphasised. The dimers and self-assemblies of CDs utilized for the enhancement of photophysical properties are discussed in detail. The CD hybrid materials exhibited numerous usage in essential needs.

**Keywords:** Aggregation, Amino acids, Anion, Bioimaging, Biomolecule, Cyclodextrin, Chemosensor, Dimer, Drug, Dye, Emission, Fluorescence, Host-guest, Interaction, Lanthanoids, Luminescence, Macrocycle, Recognition, Nanocarrier, Nanoparticles.

## INTRODUCTION

Cyclodextrin (CD) is a cyclic polysaccharide derived from enzymatic hydrolysis of starch [1]. CD has three native forms, α-CD, β-CD and γ-CD containing six, seven and eight glucopyranose units, respectively. CD structure is a truncated cone with hydrophilic exterior and hydrophobic inner pocket, which makes it suitable for host guest inclusion complex formation [2]. The inner diameters of hydrophobic cavities of α, β, and γ-CDs are reported as 4.7–5.3, 6.0–6.5, and 7.5–8.3 Å, respectively [3]. Over the other supramolecular host molecules, CD

---

[*] **Corresponding author Paulpandian Muthu Mareeswaran**: Department of Industrial Chemistry, Alagappa University, Karaikudi 630 003, Tamilnadu, India; Tel: +919790963437; E-mails: muthumareeswaran@gmail.com and mareeswaran@alagappauniversity.ac.in

**Paulpandian Muthu Mareeswaran, Palaniswamy Suresh and Seenivasan Rajagopal (Eds.)**

attained wide interest and it was explored for its diverse functionalities in the therapeutic, biomedical, and food industries, as well as in biosensors/bio-imaging applications [4]. Despite CDs themselves being spectroscopically passive, by modification with appropriate chromophores, their derivatives are spectroscopically active [5]. Due to the non-luminescent nature of CDs, fluorescence spectral studies are prominent in studying its host-guest complex formation and binding orientation with luminescent dye molecules [6]. Due to the compatible nature of CD, recently, enormous studies were focused on CD and its derivatives to explore the utilisation of this assessible host molecule in different aspects. This chapter confers the new developments of luminescent CDs and the studies of CDs with luminescent guest molecules for top notch applications.

## HOST-GUEST SYSTEMS

Krishnan *et al.* [7] reported the aqueous mediated photophysical studies of resorcinol based acridinedione dyes with β-CD in the presence of urea. The urea and water hydrogen bonding self-assemblies led the formation of microspheres based on different environment, resulting in an effective displacement of dye from the hydrophobic nanocavity of β-CD. Roy *et al.* [8] explored the umbelliferone, a drug with α-CD inclusion complex. The complex (Fig. **1**) has been optimized by molecular docking and increased bioavailability with minimal dosage in the human body.

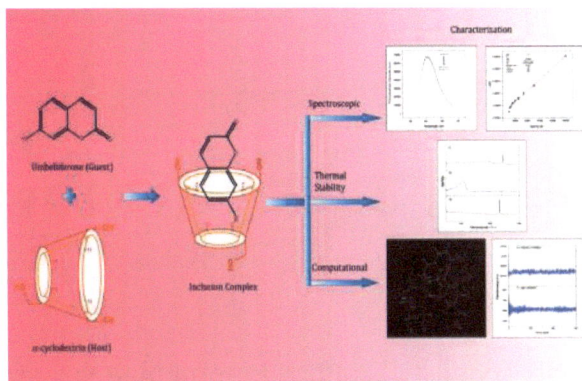

**Fig. (1).** Schematic representation of the studies of umbelliferone and α-CD inclusion complex [8].

Periasamy *et al.* [9] studied the host-guest inclusion complex of β-CD and 4,4' - (1,4-phenylenediisopropylidene) bisaniline (PDB) in solid and solution states by numerous analytical techniques. UV and fluorescence spectral studies confirmed the 1:2 PDB: β-CD complex formation and the molecular docking studies also support this. A detailed spectroscopic investigation of the binding of pyrene with β-CD derivatives and their binary mixtures has been reported by Levine *et al.* [10]. Li *et al.* [11] studied the interaction between CDs and pullulanase. Enzyme activity and kinetic studies exhibited that α-CD, β-CD and γ-CD inhibited pullulanase in a vying manner and fluorescence spectroscopy suggested the formation of CD and pullulanase complexes. Visible-light responsive supramolecular gel (Fig. **2**) has been fabricated by β-CD dimer and tetra-orth--methoxy-substituted azobenzene dimer through the host-guest interaction. The substituted methoxy groups responsive to the shift in wavelengths of trans and cis forms led to the green and blue light regions, respectively [12].

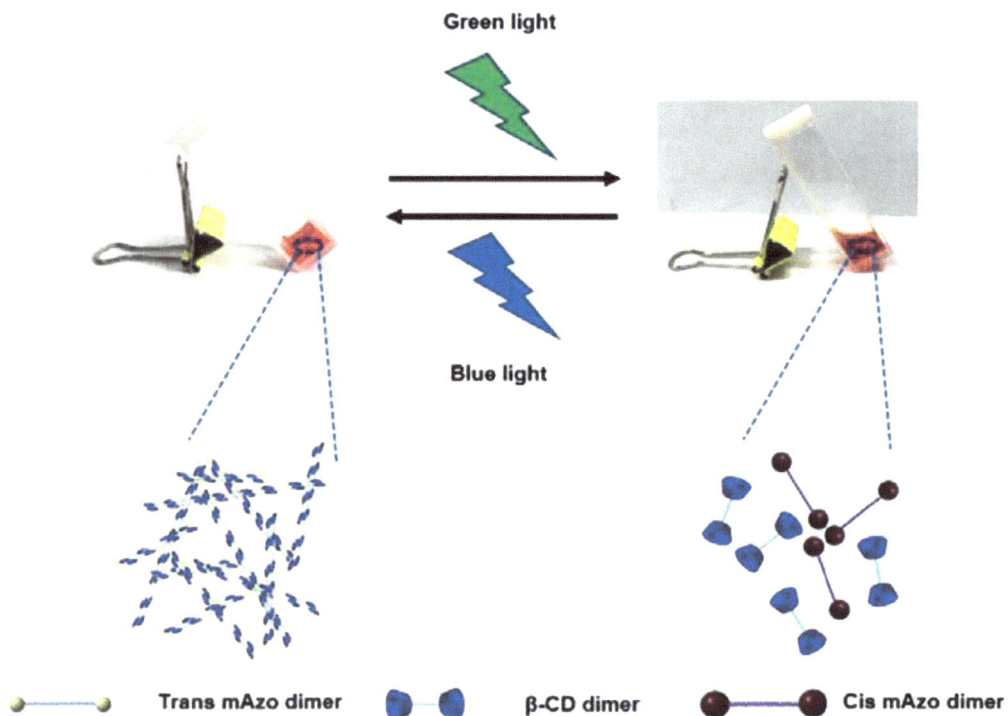

**Fig. (2).** Schematic representation of visible-light responsive β-CD dimer host-guest supramolecular gels [12].

Kim *et al.* [13] reported a synthetic strategy to form high yield CD based intra-nanogap particles (Fig. **3**) with a well-defined ∼1 nm interior gap. They incorporated 10 different fluorescent Raman dyes such as crystal violet, basic fuchsin, bromophenol blue, rhodamine B, methylene blue, Safranin O *etc.*, within the gap using the CD based host−guest chemistry.

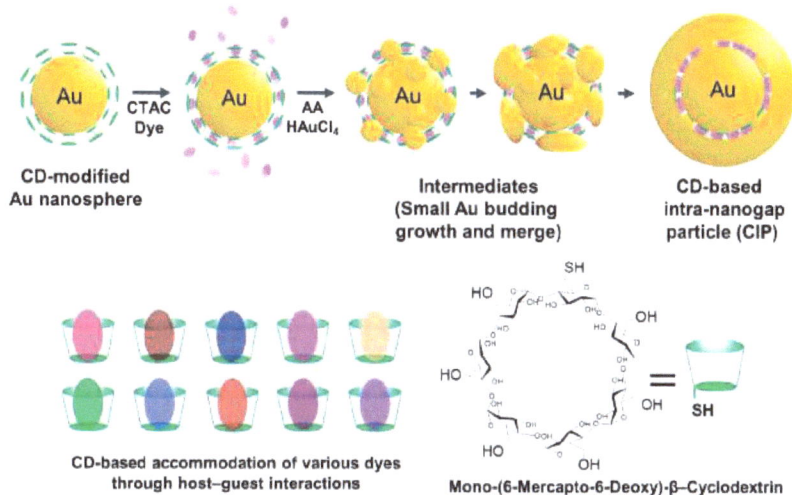

CD-modified
Au nanosphere

Intermediates
(Small Au budding
growth and merge)

CD-based
intra-nanogap
particle (CIP)

CD-based accommodation of various dyes
through host–guest interactions

Mono-(6-Mercapto-6-Deoxy)-β–Cyclodextrin

**Fig. (3).** Schematic representation of the synthesis of cyclodextrin-based intra-nanogap particles which accommodate different Raman dyes inside the intra-nanogap [13].

Thiabaud *et al.* [14] studied the Pt$^{4+}$-ferrocene conjugates (Fc) and their CD host–guest complexes. The complex formation between Fc and β-CD is presumed due to the hydrophobic interactions of the Fc and the inner β-CD cavity. Inclusion complex of [3-(4-methylphenyl)-4,5-dihydro-1,2-oxazole-4,5-diyl]bis(methylene) diacetate and methyl- β-CD have been studied by Ay *et al.* [15] in aqueous media, which exhibited dynamic and static quenching in fluorescence. Singh *et al.* [16] studied the interaction of a cationic 1-pyrene methyl amine with a sulfated β-CD derivative (Fig. **4**). The interaction of a cationic probe with a β-CD derivative has been studied using various photophysical methods.

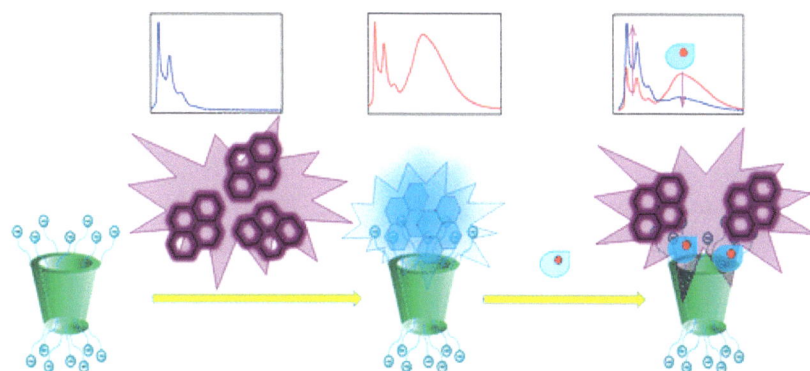

**Fig. (4).** Scheme of the photophysical response of 1-pyrene methyl amine with sulfated β-CD derivative [16].

Hossam *et al.* [17] reported steady state absorption, emission and time resolved fluorescence emission studies of 1-naphthylamine-4-sulfonate and 2-naphthylamine-6-sulfonate compounds with different CD derivatives in neutral aqueous media. The nature of the ionic liquid, 1-decyl-3-methylimidazoliumtera-fluoroborate within β-CD inclusion complex and the binding affinity have been studied in an aqueous medium by various spectroscopic methods [18]. The noncovalent interaction of a benzophenanthridine alkaloid sanguinarine with water soluble sulfobutylether-β-CD has been investigated by absorption and steady-state as well as time-resolved fluorescence techniques which revealed the metal ion recognition [19].

Assaf *et al.* [20] reported host–guest complexation of several cobalt bis (1,2-dicarbollide) anions (COSANs) with CDs in an aqueous medium. With indicator displacement assays combination, a label-free fluorescence-based method was evolved for real-time monitoring of the translocation of COSANs (Fig. **5**) by lipid bilayer membranes.

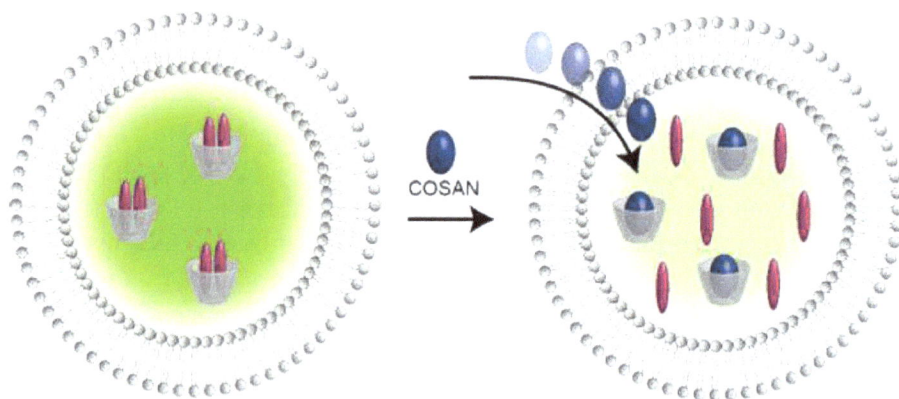

**Fig. (5).** Schematic illustration of COSAN translocation through a vesicular lipid bilayer [20].

Nazarov *et al.* [21] investigated the effect on luminescent properties over the addition of o-carborane or/and adamantane with naphthalene- β-CD complexes.

Host-guest recognition and controlled assembly CD with perylene bisimide based glycocluster indicated the carbohydrate-protein binding interaction. This probe exhibited fluorescence enhancement in HeLa cells when complexing with β-CD [22]. Willems *et al.* [23] reported on-low immobilization of polystyrene particles modified with various degrees of adamantane molecules, and silica surfaces patterned with host-guest interaction of β-CD through microcontact printing (Fig.**6**).

**Fig. (6).** Illustration of the on-flow immobilization of polystyrene microspheres on β-CD [23].

Wei *et al.* [24] demonstrated molecular transmission *via* visible range and rate-controllable photoreactivity as well as host–guest assembly (Fig. 7) based on synergy of aggregation-induced emission.

*Blue and red macrocycles indicate β-CD and γ-CD, respectively. PD = photodimerization.

**Fig. (7).** Suggested mechanism of visible range and rate- controllable photodimerization [24].

Dimirci *et al.* [25] presented polymeric ionic liquid (PIL) brushes via reversible addition–fragmentation chain transfer polymerization for the control of molecular interactions. The anion-exchange of poly(1-vinyl-3-buthylimidazolium bromide) has been used for the control of the complexation. Stereoisomeric β-CD dimers linked with a sulfur atom or an arene spacer have been reported for the creation of a tethered dual CD capsule for precisely manipulating the regio- and enantioselectivities [26].

The thiabendazole encapsulation in hydroxypropyl-β-CD nanofibers has been prepared via polymer-free electrospinning and characterized by Gao *et al.* [27]. A series of amorphous organic small molecules have been prepared with efficient room-temperature phosphorescence (RTP) emission through conveniently modifying phosphor moieties to β-CD (Fig. **8**). The CD derivatives immobilized the phosphors due to the hydrogen bonding which suppressed the nonradiative relaxation and shielded phosphors from quenchers. This aggregation leads to efficient RTP emission with good quantum yields [28].

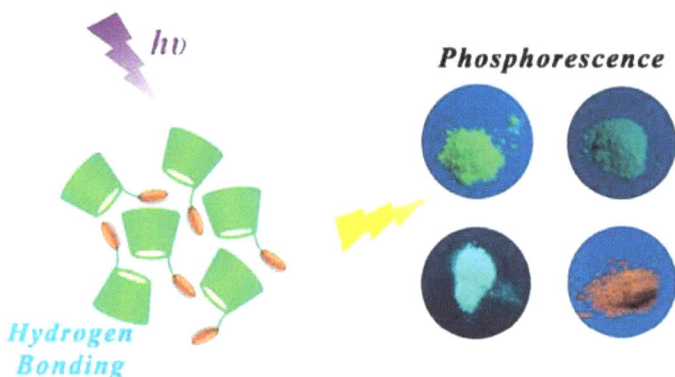

**Fig. (8).** RTP emission of different phosphorescent small molecules [28].

Banerjee *et al.* [29] reported a unique phenomenon of coumarin 6-β-CD inclusion nanostructures on graphene oxide (GO) nanosheets. This induced ground-state electron transfer from the coumarin 6-β-CD composite to GO. The coumarin 6-β-CD composite at first transfers energy to the affixed GO surface and later collides with similar coumarin 6-β-CD@GO adducts leading to dynamic quenching. Qi *et al.* [30] synthesised and studied the adsorption and luminescence characteristics of metallocyclodextrins based on host-guest recognition. Over the solution state, CD host-guest complex develop these gel, membrane assembly and immobilized systems, extending their application into different phases. Similar porphyrin systems are also reported in the literature [31, 32].

## LUMINESCENT CYCLODEXTRINS

Chen *et al.* [33] have reported an anthraquinone-modified β-CD (AQ- β-CD) supramolecular polymer in aqueous solution, constructed through the host–guest interaction. The anthraquinone group of the guest molecule rapidly produced 9,10-anthracenediol, due to the hydrophobic microenvironment of the cyclodextrin cavity and the shielding effect on oxygen molecules. 9,10-anthracenediol emits strong fluorescence by photoreduction (Fig. **9**). This aqueous solution of AQ- β-CD was significantly utilised as reversible emitting counterfeiting ink. AQ- β-CD encrypts the information in the air and decrypts the information by ultraviolet light.

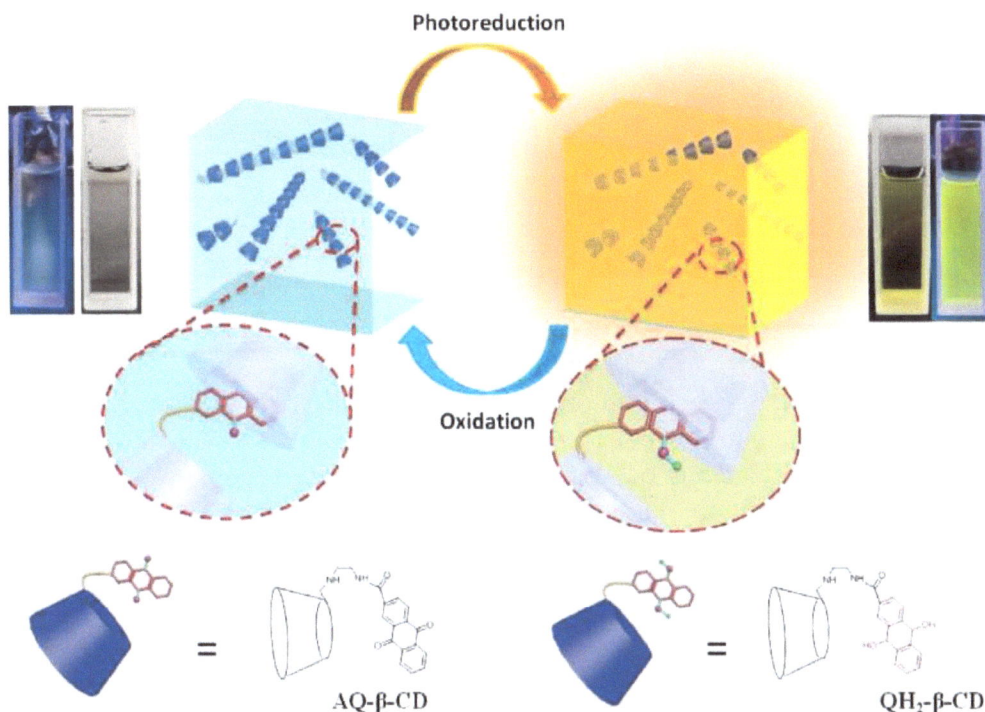

**Fig. (9).** Schematic representation of photoreduction and oxidation of AQ-β-CD supramolecular polymer in aqueous solution [33].

A red-luminescent quaternary supramolecular nanoparticles have been prepared from a dithienylethene derivative, Pluronic F-127, a β-CD-functionalized ruthenium complex and cetrimonium bromide. The results showed that the nanoparticles could be utilised as a photocontrolled cell-imaging agent and as a photoerasable red-luminescent ink [34].

Niu *et al.* [35] reported a multi-color-tunable supramolecular hydrogel from aminoclay, sulfato-β-CD and 4-methyl-styrylpyridinium (Fig. **10**). The formed inclusion complex emitted monomer fluorescence and aminoclay provided a restricted environment for excimer emission. The colour of the supramolecular hydrogel could be adjusted from yellow, white and blue, respectively by adjusting the complex molar ratio.

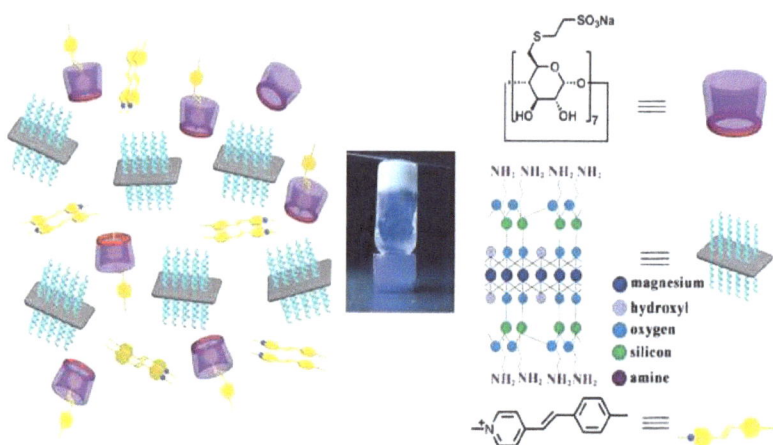

**Fig. (10).** A multi-colour-tunable supramolecular hydrogel [35].

Cheng *et al.* [36] reported fluorescent imprintable hydrogels prepared via organic/inorganic supramolecular co-assembly. Macrocyclic cucurbituril [7] and β-CD rings were capable of the formation host−guest complexes with the surfactant. This led to ternary hybrid hydrogels with luminescent and photoresponsive properties (Fig. **11**). The fluorescence of the hydrogels underwent fast quenching by trans−cis photoisomerization of the cyanostilbene within a short irradiation period in UV light and could be recovered.

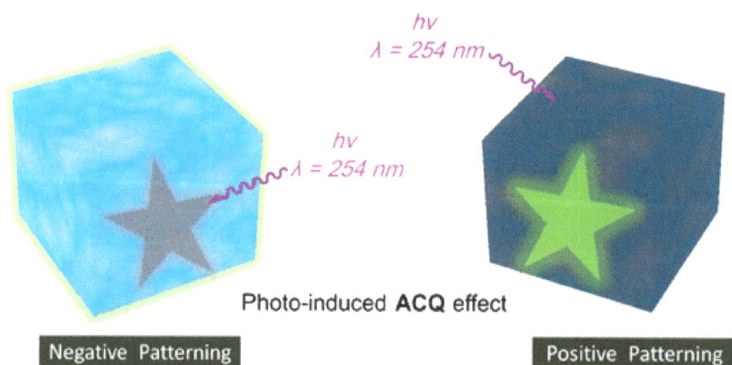

**Fig. (11).** Illustration of fluorescent imprintable hydrogels [36].

Liang *et al.* [37] presented a host-guest complex system built by β-CD, sodium dodecyl sulfate and fluorescent dyes. This exhibited multilevel chirality, consisting molecular chirality of β-CD, a chiral lattice self-assembled nanosheet, induced supramolecular chirality of the complexes, induced chirality of a dye-doped chiral tube and mesoscopic chirality of an assembled helical tube (Fig. **12**). The chiral lattice self-assembly provided intense circularly polarized luminescence which was observed from the achiral dye-doped complexes with a dissymmetry factor up to +0.1.

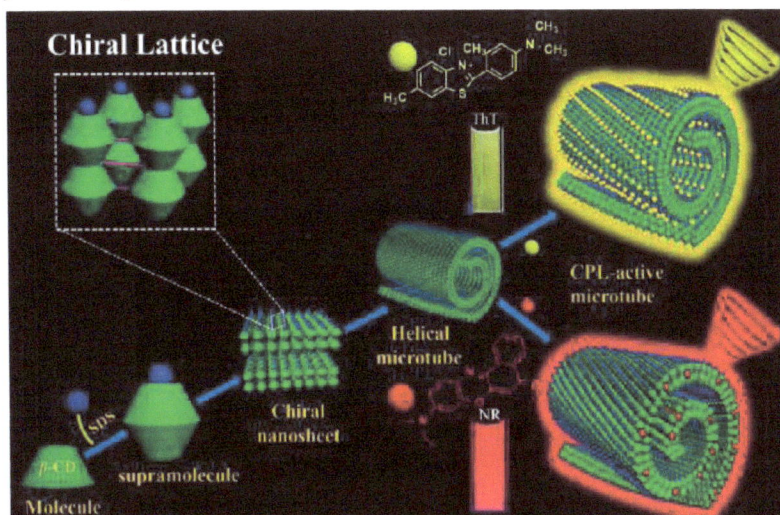

**Fig. (12).** Illustration of the possible self-assembly and hierarchical chirality transfer induced circularly polarized luminescence in the complex assemblies [37].

A novel triazolyl bridged cucurbituril-CD dimer (Fig. **13**) has been synthesized by click reaction. It possessed stable supramolecular inclusion complexes with eminent fluorescence resonance energy transfer, which benefited from the binding of cucurbituril and CD with amantadine- and pyridinium-containing fluorophores instantly [38].

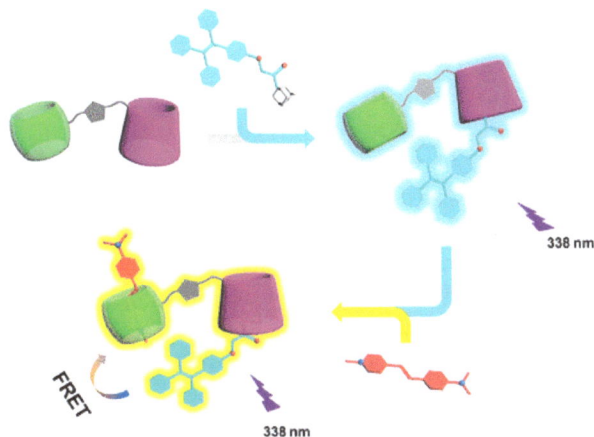

**Fig. (13).** Schematic representation of the intramolecular FRET nature of cucurbit ⌊6⌋uril bridged β-CD dimer [38].

Wang *et al.* [39] prepared composite quantum dots (QDs) using β-CD and its derivatives. It has been utilized to separate and as well as to determine enantiomers in the capillary electrophoresis with laser-induced fluorescence detection integrated system. Wang *et al.* [40] presented a supramolecular strategy for enhancing photochirogenic performance by host or guest modification. They demonstrated the concept with dicationic γ-CD-mediated photocyclodimerization of 2,6- anthracenedicarboxylate (Fig. **14**).

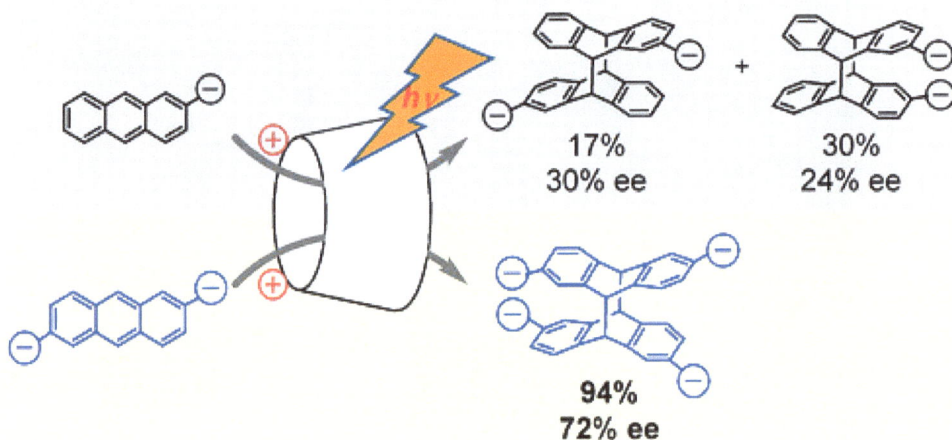

**Fig. (14).** Illustration of a supramolecular strategy to enhance photochirogenic performance [40].

A blue fluorescent supramolecular hydrogel has been prepared from α-CD and benzimidazole derivatives with supramolecular assembly-induced emission properties. The hydrogel empowered the coordination with different lanthanide metal ions. These supramolecular nanoparticles possessed spherical-like morphology, which rendered the xerogel films (Fig. **15**) for lanthanide luminescence. These xerogel films exhibited over luminous transmittance and multicolor photoluminescence. These properties promised the potential use of supramolecular assembly in smart optical film materials [41].

Yu *et al.* [42] investigated the excited state energy transfer process from 4-amid--1,8-naphthalimide to porphyrin in the medium of CD. The results showed the efficient state excited energy transfer of about 99% observed within the artificial nanoassembly. Chen *et al.* [43] designed core−shell structured CD encapsulated with hierarchical dye metal-organic frameworks (MOF) for tunable emission (Fig. **16**). The cavity structure of γ-CD and cage structure of MOF facilitated remarkable synergistic effects on fluorescence enhancement.

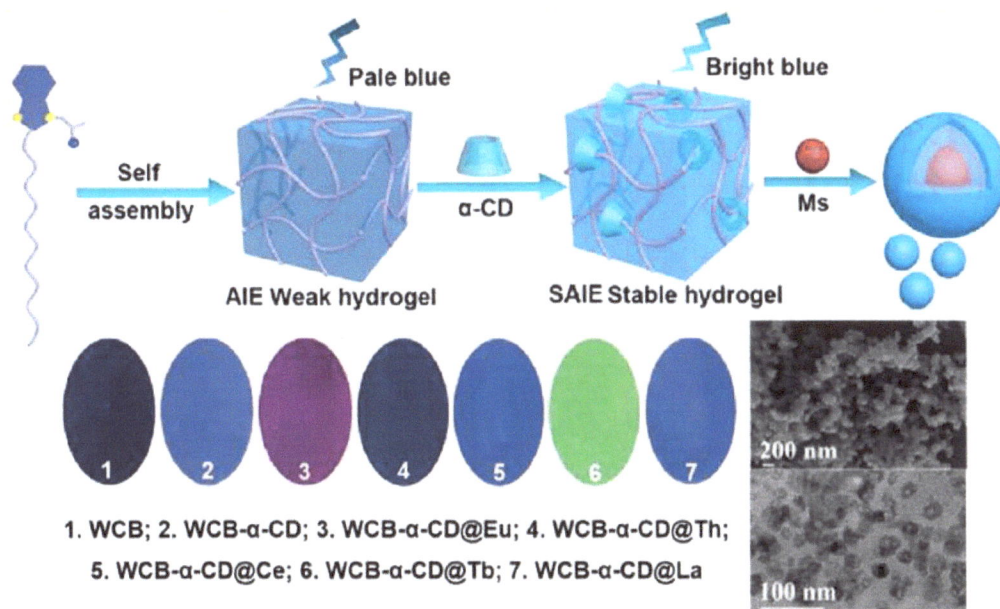

1. WCB; 2. WCB-α-CD; 3. WCB-α-CD@Eu; 4. WCB-α-CD@Th;
5. WCB-α-CD@Ce; 6. WCB-α-CD@Tb; 7. WCB-α-CD@La

**Fig. (15).** Representation of lanthanide-mediated CD based supramolecular assembly for induced emission xerogel films [41].

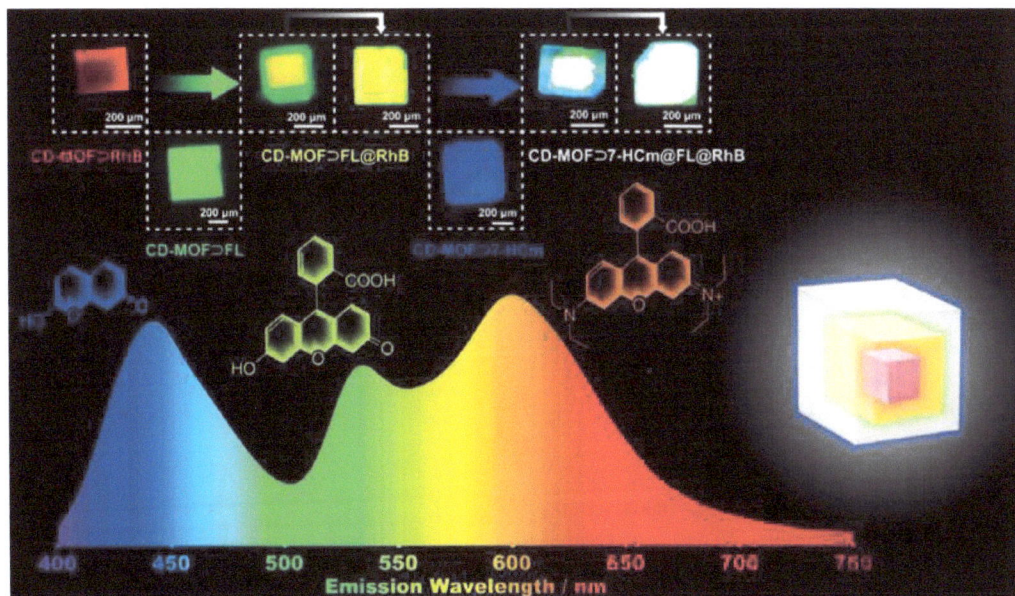

**Fig. (16).** Scheme representing the CD-MOF with dye encapsulation for tunable light emission [43].

Silva *et al.* [44] utilised β-CD as a precursor for holey g-C$_3$N$_4$ graphitic carbon nitride (GCN) nanosheets doped with carbon and studied its application in photocatalytic hydrogen generation. The prepared material exhibited ~5 times higher photocatalytic hydrogen generation under visible light than bulk GCN. Bhunia *et al.* [45] studied the formation of nanospheres and nanocubes of glutathione-coated gold nanoclusters accompanied with CD (Fig. **17**). These suprastructures developed fluorescence on−off composites based of their overall shapes. These observations conveyed that researchers can be able to synthesize appropriate emission range targeted CDs for their requirement.

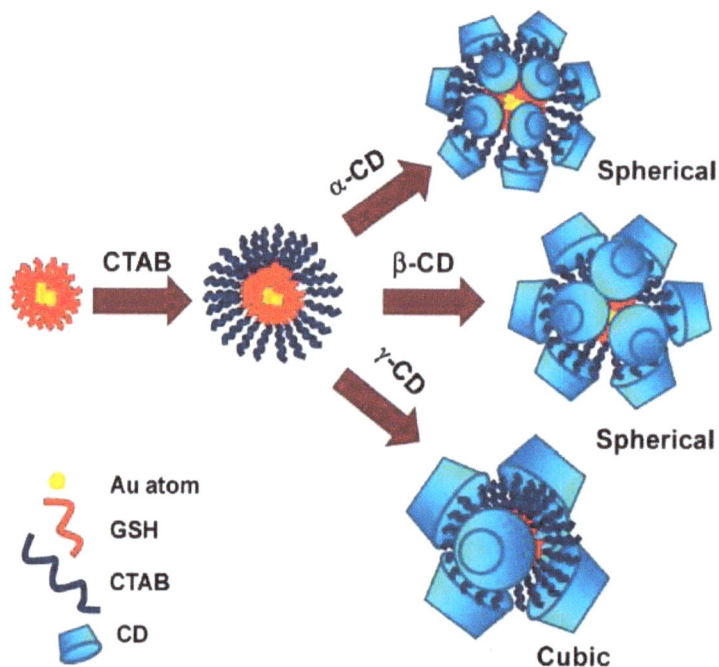

**Fig. (17).** Probable arrangement of the three types of CDs around the glutathione−CTAB-coated Au NCs [45].

## FLUORESCENT SENSORS

Molybdenum disulfide quantum dots modified with 3-aminophenyl boronic acid and functionalized with hydropropyl-β-cyclodextrin represented a novel nanoprobe for the fluorescent detection of parathion-methyl, a typical organophosphorus pesticide inclined and accumulated in water and food products [46]. Kaliyamoorthi *et al.* [47] reported β-CD mediated $Al^{3+}$ ion sensing based on fluorescence quenching with $6.00 \times 10^{-8}$ mol $L^{-1}$ limit of detection. A syn-(methyl,methyl)bimane and β-CD supramolecular complex exhibited sensitive (LOD = 0.60 nM) and selective fluorescence turn-off recognition in the presence of $Co^{2+}$ in aqueous media [48].

Haynes *et al.* [49] used β-CD derivative to promote proximity-induced interactions for the detection of aliphatic alcohols. A common reaction of aliphatic alcohol analytes with brightly coloured organic dye resulted in remarkably analyte-specific colour changes (Fig. **18**) which enabled accurate alcohol identification. In combination with BODIPY and Rhodamine dyes, the colour changes enabled 100% differentiation in colorimetric signals obtained from methanol, ethanol, and isopropanol.

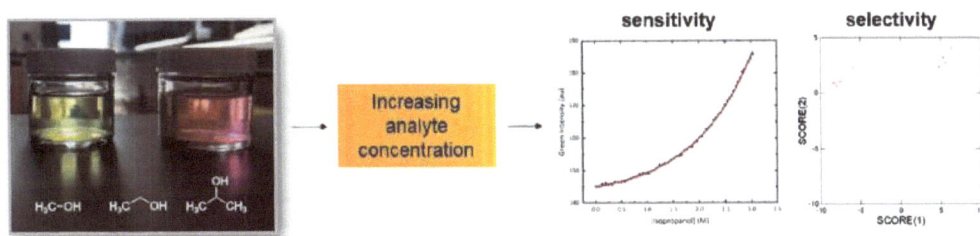

Fig. (**18**). β-CD based colorimetric discrimination of aliphatic alcohols [49].

Prabu *et al.* [50] reported a new "turn-off" fluorescent sensor system for sensitive recognition of $Hg^{2+}$ ion by curcumin/ β-CD inclusion complex. It exhibited selective $Hg^{2+}$ binding and new absorbance and fluorescence peaks with additional prevailing bands. It showed probable colour change from yellow to colourlessness and fluorescent quenching owing to selective binding of $Hg^{2+}$ ion.

Zheng *et al.* [51] demonstrated a method for the detection of berberine in Rhizoma coptidis using β-CD - sensitized fluorescence technique. Berberine is the main extract of medicinal product, that caused an envelope reaction with β-CD to generate fluorescence sensitization. The reported method was simple and environmental friendly with high sensitive and selective determination. Liu *et al.* [52] has developed a fluorescent probe for detection of fungicide 2-aminobenzidazole based on carminic acid and γ-CD.

The influence of β-CD complex formation with antipyrine derivative has been studied on its metal ion sensing behaviour (Fig. **19**). The antipyrine showed a turn-on fluorescence sensing of vanadyl ion, and in CD medium it detected aluminum ion in aqueous solution. The compound show an unfamiliar fluorescence quenching on binding with β-CD. This differential metal ion sensing is presumed to the partial blocking of the chelating moiety by the CD [53].

**Fig. (19).** Scheme shows the differential metal ion sensing by antipyrine with β-CD [53].

Miyagawa *et al.* [54] have demonstrated the molecular recognition nature of the cholic acids with the dansyl-modified CD under hydrostatic pressure. The pre-equilibrium species, naphthyl in and out, acted critical roles in the pressure-dependent chemical sensing. The ultra-trace detection of amphetamine has been efficiently reported by Nazerdeylami *et al.* [55] using 8- hydroxyquinoline-β-CD based fluorescence probe.

Poomalai *et al.* [56] utilised β-CD for the hindrance in excimer formation of anthracene appended piperidone on fluorescence sensing of $Al^{3+}$ ions. A β-CD functionalized N, Zn co-doped carbon dots served for the selective fluorescence detection of fluoroquinolones in milk samples by fluorescence enhancement and induction of red-shift [57]. A hydrophilic hydroxypropyl β-CD cross-linked polymer has been reported and demonstrated as a turn-on fluorescent sensor for selective determination of captopril in biological assays [58].

A fluorescent sensor array (Fig. **20**) has been developed based on three different-colour emitting gold nanoclusters functionalized with three different ligands and a co-capping β-CD ligand for the facile discrimination of three nitrophenol isomers through the linear discriminant analysis of isomer-induced fluorescence quenching. The practicability has been checked and a high accuracy of 98.0% observed on exanimating 51 unknown samples containing a single or a mixture of two isomers [59].

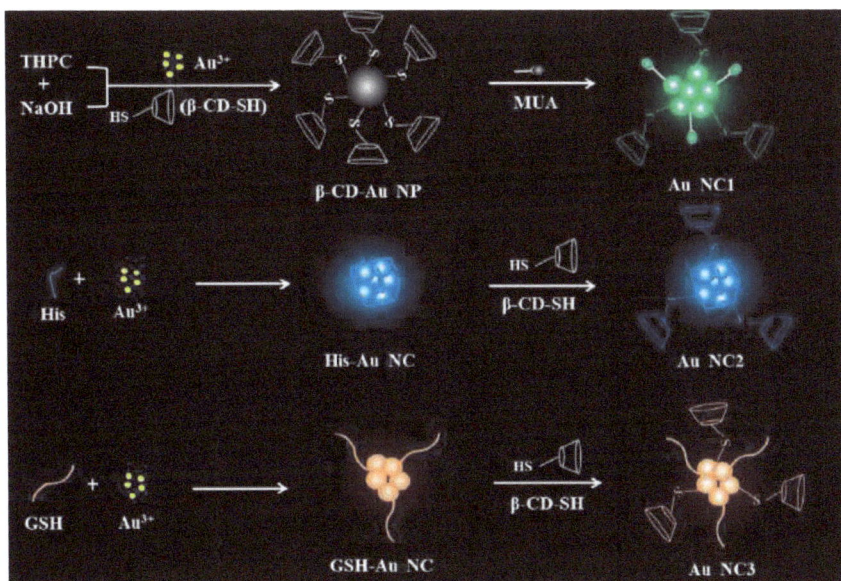

**Fig. (20).** Schematic representation of the synthesis of β-CD mediated dual-ligands cofunctionalized Au NCs [59].

Reported β-CD fluorescent probes facilitates sensing of inorganic toxic metal ions, transition metal ions, simple organic molecules such as common alcohols, nitrophenols to toxic organic pesticides such as parathion-methyl effectively.

## BIOSENSORS

A supramolecular assembly formed between a molecular rotor dye Thioflavin-T and sulfated-β-CD light-up and served as a ratiometric sensor for highly sensitive and selective recognition of glutathione over cystein and homocysteine. Its response towards glutathione in complex biological media of human serum samples demonstrated its potential for practical utility [60]. A β-CD/indocyanine green complex encapsulated poly(N-isopropylacrylamide) nanogel was reported by Liu *et al.* [61] for *ex-vivo*/*in-vivo* deep tissue/HR near infrared ultrasound-switchable fluorescence imaging.

A hepta-dicyanomethylene-4H-pyran appended β-CD (DCM7-β-CD) (Fig. **21**) acted as a delivery enhancing "host" for 1-bromonaphthalene-modified peptides. Interaction between the fluorescent peptides and DCM7-β-CD facilitated the hierarchical formation of distinctive supramolecular architectures. These were termed as supramolecular-peptide-dots [62].

**Fig. (21).** Schematic representation of the sequential second-stage ordered self-assembly between probes and DCM7-β-CD to form supramolecular-peptide-dots [62].

Haynes *et al.* [63] prepared a variety of β-CD derivatives in combo with fluorophore rhodamine 6G to detect distinctly five anabolic steroid analytes. This study exhibited 100% differentiation between structurally similar analytes and micromolar level detection. Haynes *et al.* [64] used CD-promoted fluorescence modulation of a high-quantum yield fluorophore with low concentrations of human growth hormone to detect human growth hormone in low concentrations with a 96% accuracy.

A β- CD and a flavone derivative complex has been designed for a colorimetric and fluorescence turn-on probe to distinguish thiophenol from biothiols in real samples (Fig. **22**). It worked via a combination of excited-state intramolecular proton transfer and photoinduced electron transfer mechanisms [65].

**Fig. (22).** Scheme of β-CD based fluorescence turn-on probe for discriminating thiophenol from biothiols [65].

2-Hydroxypropyl-beta-cyclodextrins supported with L-tryptophan served as a chemosensor for the detection of cholesterol [66]. Due to the excellent fluorescence property and good biocompatibility of silver nanoclusters functionalized with β-CD (CD-AgNCs) used for the sensitive detection of intracellular alkaline phosphatase activity [67].

A switchable chiral discrimination system for amino acid enantiomers has been reported. The biomimetic nanochannel functionalized with N-(1- naphthyl) ethylenediamine and γ-CD acted as a highly chiral choosy gating for L-phenylalanine (Fig. **23**) and regulated well the operation of amino acid transportation [68].

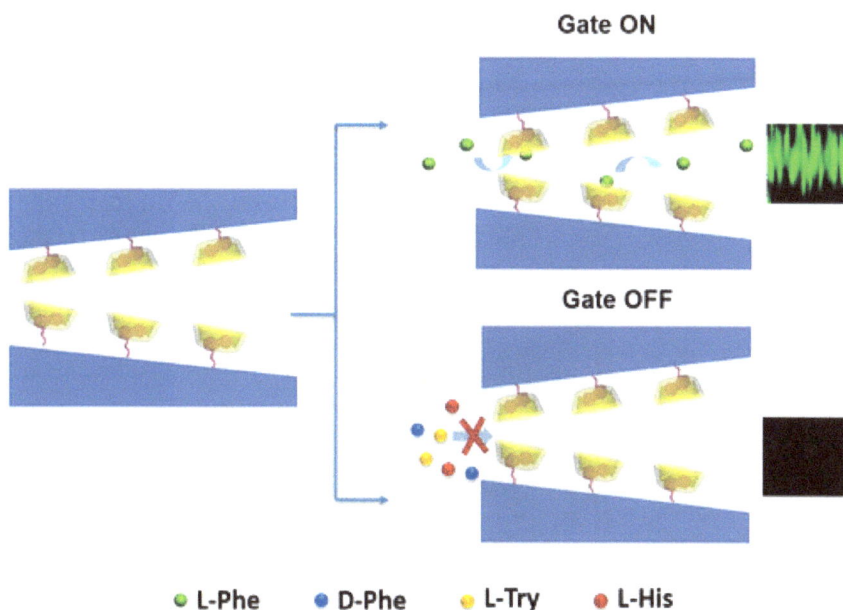

**Fig. (23).** Scheme of selective amino acid transportation based on bioinspired γ-CD pseudorotaxane assembly [68].

β-CD capped ZnO QDs decorated with the vitamin B6 cofactors such as pyridoxal 5′-phosphate and pyridoxal served as novel fluorescent turn-on probes instantaneously detected histamine by fluorescence enhancement [69].

A hypoxia activatable and cytoplasmic protein-powered fluorescence cascade amplifier (HCFA) has been used for imaging hypoxia associated with inflammatory bowel disease *in vivo* (Fig. **24**). The formation of HCFA was based on β-CD polymer combined with an azo dye through a host–guest interaction [70].

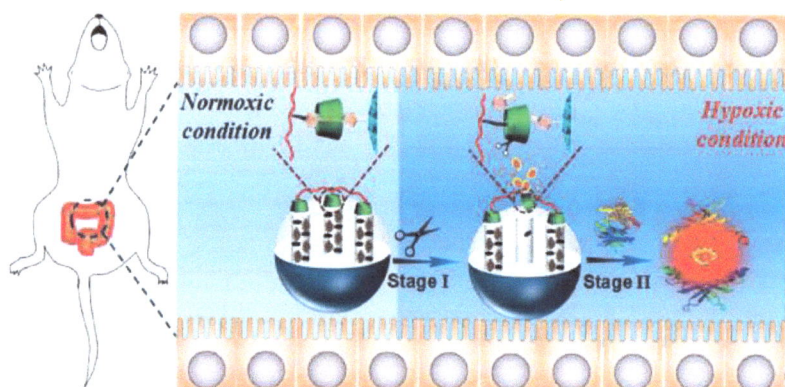

**Fig. (24).** Schematic representation of *in vivo* imaging of hypoxia by β-CD polymer based cytoplasmic protein-powered fluorescence cascade amplifier [70].

Bai *et al.* [71] attached rigid β-CD to the end of a flexible polysiloxane chain to synthesize a novel fluorescent polymer hyperbranched polysiloxane-CD and used it for biological applications. The results exhibited that the fluorescence intensity and quantum yield of the complex was significantly enhanced. It showed that bioimaging and drug delivery results in mouse fibroblast cells (Fig. **25**).

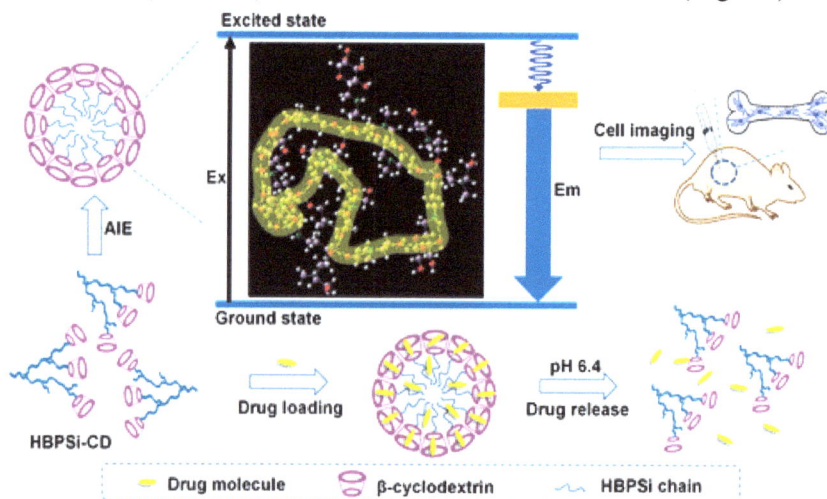

**Fig. (25).** Schematic representation of hyperbranched polysiloxane consisting β-CD for cell imaging and drug delivery [71].

β-CD derivatives aid *ex-vivo* and *in-vivo* sensing of vast biological compounds such as amino acids, steroids, cholesterol and outstretch into diagnostic. Here, the reports dealing β-CD tailoring with bio components are interesting and attractive.

## THERAPEUTIC APPLICATION

Thomsen *et al.* [72] investigated the effectiveness of octakis[6-(--aminoethylthio)-6-deoxy]-γ-CD (γ-Cys), a positively charged, single isomer cyclodextrin derivative, for the delivery improvement of antibiotics to biofilms. This γ-Cys nanocarriers efficiently delivered suitable antibiotics to biofilms and fluorescence microscopy attempted a novel approach for mechanistic investigations. Roozbehi *et al.* [73] demonstrated an enzyme-triggered system based on β-CD to achieve controlled release of hydrophobic drugs in the presence of maltogenic amylase. The triggered release of curcumin in the presence of maltogenic amylase was due to the β-CD degradation by maltogenic amylase, resulting in ring opening and chain scission in β-CD.

A linear alternating supramolecular polymer has been constructed by host-guest inclusion interaction between CD-dimer and bifunctional adamantane-conjugated porphyrin (TPP- Ad2) (Fig. 26). This linear alternating supramolecular polymer exhibited significantly enhanced photodynamic therapy efficacy [74].

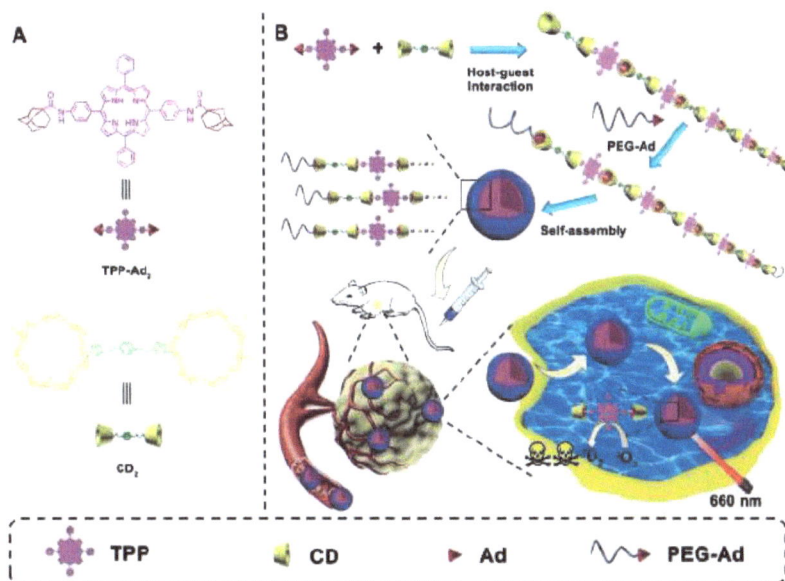

**Fig. (26).** (A) The chemical structures 26 of the (TPP- Ad2) and the CD dimer. (B) Representation of the construction of nanoparticles from linear supramolecular polymers by host- guest interaction, and the enhanced photodynamic therapy process [74].

Gold nanoparticles, amphiphilic cyclodextrin and porphyrins based supramolecular hybrid assemblies have been prepared and studied for their phototherapeutic action by Trapani *et al.* [75]. A feasible and biocompatible self-assembled inclusion complex of indocyanine green and methyl-β-CD (ICG-CD) has been reported for targeted cancer imaging (Fig. **27**). This system has enhanced the fluorescence-guided photothermal cancer therapy. Addressing HT-29 tumours by the ICG-CD complex resulted in a possible reduction in tumour volumes over nine days after the photothermal treatment. After administering a single dose of ICG-CD complex with near IR laser irradiation, no tumour recurrence or body weight loss were observed [76].

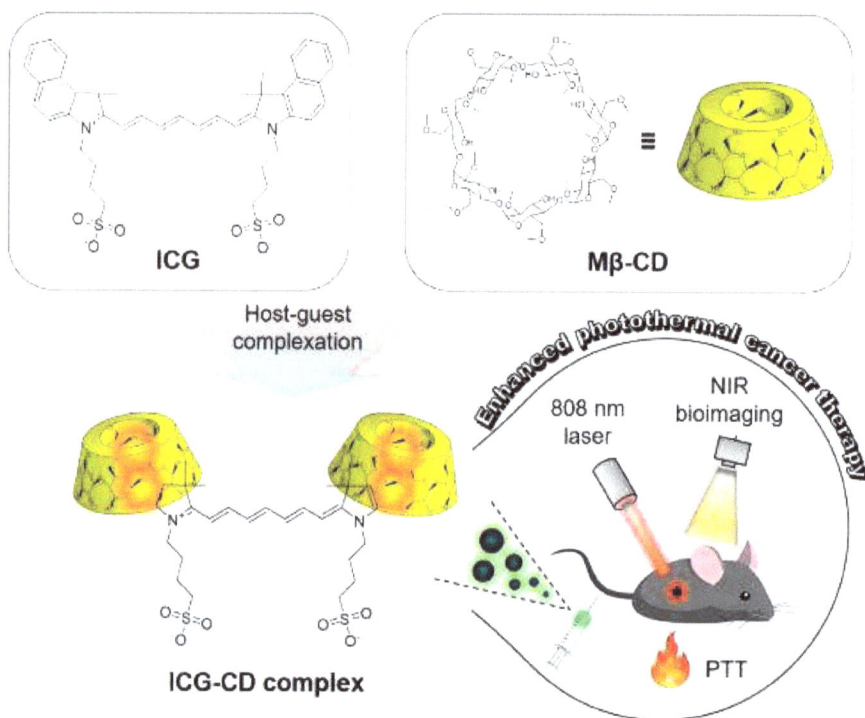

**Fig. (27).** Schematic representation of ICG-CD inclusion complex formation between indocyanine green and methyl-β-cyclodextrin for enhanced photothermal cancer therapy [76].

Zhang *et al.* [77] developed a supramolecular polymer through the host-guest inclusion complexation interactions between a β-CD-graft-poly(2-(dimethylamino) ethyl methacrylate), and a guest polymer, azobenzene modified poly(ε-capr-lactone) (Fig. **28**). They reported it as dual stimuli-responsive supramolecular self-assemblies for potential drug-controlled release carriers.

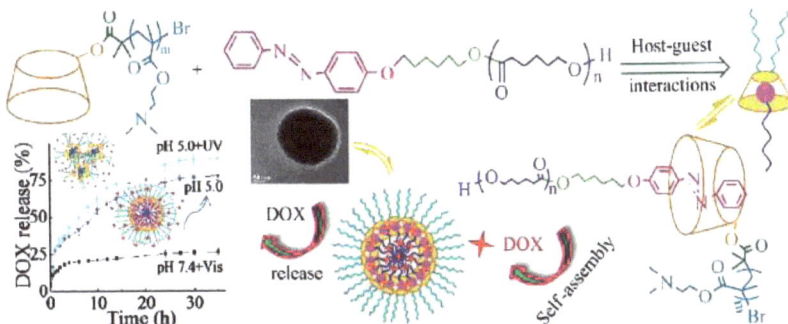

**Fig. (28).** Scheme representing the β-CD self-assemblies for cellular drug-controlled release [77].

Pei *et al.* [78] reported a facile synthetic route of fluorescent hyper-crosslinked β-CD-Carbon QD nanosponges for tumor theranostic usage with exhibited enhanced antitumor efficacy. Welling *et al.* [79] demonstrated the concept to deliver chemical cargo to living cells *in vivo* by CD/adamantane complex targeting the inoculated bacteria in mice (Fig. **29**).

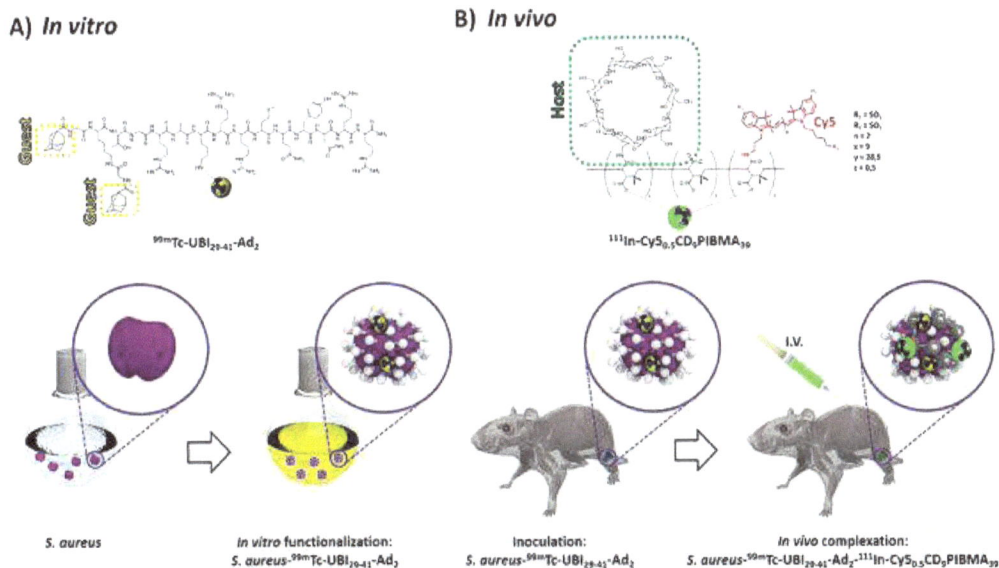

**Fig. (29).** Presumed mechanism of targeting inoculated bacteria in mice using CD/adamantane medium [79].

The interactions between human serum albumin, perfluorooctanoic acid, and β-CD have been investigated to determine the reverse binding of perfluorooctanoic acid to human serum albumin in the presence of β-CD. This phenomenon facilitates potential therapeutic applications for perfluorooctanoic acid in human blood [80]. From these impressions the interest of researchers recently expanded the luminescent β-CDs for therapeutics including controlled drug delivery. Beyond their interest, these complexes were effective and efficient. These could be improved and engaged with imperceptible area of therapeutic in the future.

## CONCLUSION

CD is not just a host material rather it receives importance in many research areas. The diversity of recognition leads to accommodate a variety of guest molecules towards particular applications. Since it is a natural product, also due to the structural features, it is compatible with biosystems, utilized as biosensors and drug delivery vehicles. The controlled drug delivery is a vital task in healing, some of the recent studies show CDs as good drug carriers. In chemosensing, CD attained an important place and more chiral selective sensors were developed which separate isomers. In combination with luminescent nanoparticles, the efficiency of CD has been enhanced and they are now being used in tunable emission driven into light emitting diodes and photoswitches. We presume that this simple oligosaccharide host material will attain more importance in the future as a hybrid luminescent material.

## CONSENT FOR PUBLICATION

Not applicable.

## CONFLICT OF INTEREST

The author declares no conflict of interest, financial or otherwise.

## ACKNOWLEDGEMENTS

This research work was supported by the Department of Science and Technology (DST INSPIRE) [Project number – IFA14/CH-147], India. We also thank RUSA 2.0 (MHRD, India) grant sanctioned vide Letter No. F. 24-51/2014-U, Policy (TNMulti-Gen), Dept. of Edn. Govt. of India, Dt. 09.10.2018 for the additional financial support.

## REFERENCES

[1]    Hu, Q.D.; Tang, G.P.; Chu, P.K. Cyclodextrin-based host-guest supramolecular nanoparticles for delivery: from design to applications. *Acc. Chem. Res.,* **2014,** *47*(7), 2017-2025.
[http://dx.doi.org/10.1021/ar500055s] [PMID: 24873201]

[2]    Wankar, J.; Kotla, N.G.; Gera, S.; Rasala, S.; Pandit, A.; Rochev, Y.A. Recent Advances in Host–Guest Self-Assembled Cyclodextrin Carriers: Implications for Responsive Drug Delivery and Biomedical Engineering. *Adv. Funct. Mater.,* **2020,** *30*(44), 1909049.
[http://dx.doi.org/10.1002/adfm.201909049]

[3]     Ding, L.; He, J.; Huang, L.; Lu, R. Studies on a novel modified β-cyclodextrin inclusion complex. *J. Mol. Struct.,* **2010**, *979*(1-3), 122-127.
[http://dx.doi.org/10.1016/j.molstruc.2010.06.014]

[4]     Li, J.; Loh, X. Cyclodextrin-based supramolecular architectures: Syntheses, structures, and applications for drug and gene delivery. *Adv. Drug Deliv. Rev.,* **2008**, *60*(9), 1000-1017.
[http://dx.doi.org/10.1016/j.addr.2008.02.011] [PMID: 18413280]

[5]     Ikeda, H.; Nakamura, M.; Ise, N.; Oguma, N.; Nakamura, A.; Ikeda, T.; Toda, F.; Ueno, A. Fluorescent Cyclodextrins for Molecule Sensing: Fluorescent Properties, NMR Characterization, and Inclusion Phenomena of *N* -Dansylleucine-Modified Cyclodextrins. *J. Am. Chem. Soc.,* **1996**, *118*(45), 10980-10988.
[http://dx.doi.org/10.1021/ja960183i]

[6]     Baglole, K.N.; Boland, P.G.; Wagner, B.D. Fluorescence enhancement of curcumin upon inclusion into parent and modified cyclodextrins. *J. Photochem. Photobiol. Chem.,* **2005**, *173*(3), 230-237.
[http://dx.doi.org/10.1016/j.jphotochem.2005.04.002]

[7]     Krishnan, A.; Viruthachalam, T.; Rajendran, K. A fluorescence approach on the investigation of urea derivatives interaction with a non-PET based acridinedione dye-beta Cyclodextrin (β-CD) complex in water: Hydrogen-bonding interaction or hydrophobic influences or combined effect? *Spectrochim. Acta A Mol. Biomol. Spectrosc.,* **2021**, *246*, 118990.
[http://dx.doi.org/10.1016/j.saa.2020.118990] [PMID: 33038856]

[8]     Roy, N.; Ghosh, B.; Roy, D.; Bhaumik, B.; Roy, M.N. Exploring the Inclusion Complex of a Drug (Umbelliferone) with α-Cyclodextrin Optimized by Molecular Docking and Increasing Bioavailability with Minimizing the Doses in Human Body. *ACS Omega,* **2020**, *5*(46), 30243-30251.
[http://dx.doi.org/10.1021/acsomega.0c04716] [PMID: 33251458]

[9]     Periasamy, R.; Kothainayaki, S.; Sivakumar, K. Host-guest inclusion complex of β-cyclodextrin and 4,4′-(1,4-phenylenediisopropylidene)bisaniline: Spectral, structural and molecular modeling studies. *J. Mol. Struct.,* **2021**, *1224*, 129050.
[http://dx.doi.org/10.1016/j.molstruc.2020.129050]

[10]    Levine, M.; Smith, B.R. Enhanced Characterization of Pyrene Binding in Mixed Cyclodextrin Systems via Fluorescence Spectroscopy. *J. Fluoresc.,* **2020**, *30*(5), 1015-1023.
[http://dx.doi.org/10.1007/s10895-020-02572-5] [PMID: 32607736]

[11]    Li, X.; Bai, Y.; Ji, H.; Jin, Z. The binding mechanism between cyclodextrins and pullulanase: A molecular docking, isothermal titration calorimetry, circular dichroism and fluorescence study. *Food Chem.,* **2020**, *321*, 126750.
[http://dx.doi.org/10.1016/j.foodchem.2020.126750] [PMID: 32278273]

[12]    Yan, H.; Qiu, Y.; Wang, J.; Jiang, Q.; Wang, H.; Liao, Y.; Xie, X. Wholly Visible-Light-Responsive Host–Guest Supramolecular Gels Based on Methoxy Azobenzene and β-Cyclodextrin Dimers. *Langmuir,* **2020**, *36*(26), 7408-7417.
[http://dx.doi.org/10.1021/acs.langmuir.0c00964] [PMID: 32486643]

[13]    Kim, J.M.; Kim, J.; Ha, M.; Nam, J.M. Cyclodextrin-Based Synthesis and Host–Guest Chemistry of Plasmonic Nanogap Particles with Strong, Quantitative, and Highly Multiplexable Surface-Enhanced Raman Scattering Signals. *J. Phys. Chem. Lett.,* **2020**, *11*(19), 8358-8364.
[http://dx.doi.org/10.1021/acs.jpclett.0c02624] [PMID: 32956585]

[14]    Thiabaud, G.; Harden-Bull, L.; Ghang, Y.J.; Sen, S.; Chi, X.; Bachman, J.L.; Lynch, V.M.; Siddik, Z.H.; Sessler, J.L. Platinum(IV)-Ferrocene Conjugates and Their Cyclodextrin Host–Guest Complexes. *Inorg. Chem.,* **2019**, *58*(12), 7886-7894.
[http://dx.doi.org/10.1021/acs.inorgchem.9b00570] [PMID: 31125214]

[15]    Ay, U.; Sarli, S.E.; Kara, Y.S. Dynamic and static fluorescence quenching on the inclusion complex formed by [3-(4-methylphenyl)-4,5-dihydro-1,2-oxazole-4,5-diyl] bis (methylene) diacetate and methyl-beta-cyclodextrin in aqueous media. *Spectrosc. Lett.,* **2020**, *53*(9), 692-704.

[http://dx.doi.org/10.1080/00387010.2020.1824194]

[16] Singh, G.; Singh, P.K. Stimulus-Responsive Supramolecular Host–Guest Assembly of a Cationic Pyrene Derivative with Sulfated β-Cyclodextrin. *Langmuir,* **2019**, *35*(45), 14628-14638.
[http://dx.doi.org/10.1021/acs.langmuir.9b03083] [PMID: 31609124]

[17] Akl, H.N.; Alazaly, A.M.M.; Salah, D.; Abdel-Samad, H.S.; Abdel-Shafi, A.A. Effects on the photophysical properties of naphthylamine derivatives upon their inclusion in cyclodextrin nanocavities. *J. Mol. Liq.,* **2020**, *311*, 113319.
[http://dx.doi.org/10.1016/j.molliq.2020.113319]

[18] Banjare, M.K.; Behera, K.; Banjare, R.K.; Pandey, S.; Ghosh, K.K. Inclusion complexation of imidazolium-based ionic liquid and β-cyclodextrin: A detailed spectroscopic investigation. *J. Mol. Liq.,* **2020**, *302*, 112530.
[http://dx.doi.org/10.1016/j.molliq.2020.112530]

[19] Kadam, V.; Kakatkar, A.S.; Barooah, N.; Chatterjee, S.; Bhasikuttan, A.C.; Mohanty, J. Supramolecular interaction of sanguinarine dye with sulfobutylether-β-cyclodextrin: modulation of the photophysical properties and antibacterial activity. *RSC Advances,* **2020**, *10*(42), 25370-25378.
[http://dx.doi.org/10.1039/D0RA03823G]

[20] Assaf, K.I.; Begaj, B.; Frank, A.; Nilam, M.; Mougharbel, A.S.; Kortz, U.; Nekvinda, J.; Grüner, B.; Gabel, D.; Nau, W.M. High-Affinity Binding of Metallacarborane Cobalt Bis(dicarbollide) Anions to Cyclodextrins and Application to Membrane Translocation. *J. Org. Chem.,* **2019**, *84*(18), 11790-11798.
[http://dx.doi.org/10.1021/acs.joc.9b01688] [PMID: 31274306]

[21] Nazarov, V.B.; Avakyan, V.G.; Rudyak, V.Y.; Alfimov, M.V.; Vershinnikova, T.G. Luminescent properties and structure of multicomponent naphthalene-β-cyclodextrin complexes. 1. Effect of adding third parties, o-carborane or/and adamantane. *J. Lumin.,* **2011**, *131*(9), 1932-1938.
[http://dx.doi.org/10.1016/j.jlumin.2011.02.004]

[22] Liu, X.; Wang, K.R.; Rong, R.X.; Liu, M.H.; Wang, Q.; Li, X.L. Host-guest controlled assembly and recognition of perylene bisimide-based glycocluster with cyclodextrin. *Dyes Pigments,* **2020**, *174*, 108043.
[http://dx.doi.org/10.1016/j.dyepig.2019.108043]

[23] Willems, S.B.J.; Bunschoten, A.; Wagterveld, R.M.; van Leeuwen, F.W.B.; Velders, A.H. On-Flow Immobilization of Polystyrene Microspheres on β-Cyclodextrin-Patterned Silica Surfaces through Supramolecular Host–Guest Interactions. *ACS Appl. Mater. Interfaces,* **2019**, *11*(39), 36221-36231.
[http://dx.doi.org/10.1021/acsami.9b11069] [PMID: 31487143]

[24] Wei, P.; Li, Z.; Zhang, J.X.; Zhao, Z.; Xing, H.; Tu, Y.; Gong, J.; Cheung, T.S.; Hu, S.; Sung, H.H.Y.; Williams, I.D.; Kwok, R.T.K.; Lam, J.W.Y.; Tang, B.Z. Molecular Transmission: Visible and Rate-Controllable Photoreactivity and Synergy of Aggregation-Induced Emission and Host–Guest Assembly. *Chem. Mater.,* **2019**, *31*(3), 1092-1100.
[http://dx.doi.org/10.1021/acs.chemmater.8b04909]

[25] Demirci, S.; Kinali-Demirci, S.; VanVeller, B. Controlled Supramolecular Complexation of Cyclodextrin-Functionalized Polymeric Ionic Liquid Brushes. *ACS Appl. Polym. Mater.,* **2020**, *2*(2), 751-757.
[http://dx.doi.org/10.1021/acsapm.9b01058]

[26] Ji, J.; Wu, W.; Liang, W.; Cheng, G.; Matsushita, R.; Yan, Z.; Wei, X.; Rao, M.; Yuan, D.Q.; Fukuhara, G.; Mori, T.; Inoue, Y.; Yang, C. An Ultimate Stereocontrol in Supramolecular Photochirogenesis: Photocyclodimerization of 2-Anthracenecarboxylate Mediated by Sulfur-Linked β-Cyclodextrin Dimers. *J. Am. Chem. Soc.,* **2019**, *141*(23), 9225-9238.
[http://dx.doi.org/10.1021/jacs.9b01993] [PMID: 31117644]

[27] Gao, S.; Liu, Y.; Jiang, J.; Li, X.; Zhao, L.; Fu, Y.; Ye, F. Encapsulation of thiabendazole in hydroxypropyl- *β* -cyclodextrin nanofibers via polymer-free electrospinning and its characterization.

*Pest Manag. Sci.,* **2020**, *76*(9), 3264-3272.
[http://dx.doi.org/10.1002/ps.5885] [PMID: 32378331]

[28] Li, D.; Lu, F.; Wang, J. *et al.* Amorphous Metal-Free Room-Temperature Phosphorescent Small Molecules with Multicolor Photoluminescence via a Host–Guest and Dual-Emission Strategy. *J. Am. Chem. Soc.,* **2018**, *140*(5), 1916-1923.
[http://dx.doi.org/10.1021/jacs.7b12800] [PMID: 29300466]

[29] Banerjee, R.; Sinha, R.; Purkayastha, P. β-Cyclodextrin Encapsulated Coumarin 6 on Graphene Oxide Nanosheets: Impact on Ground-State Electron Transfer and Excited-State Energy Transfer. *ACS Omega,* **2019**, *4*(14), 16153-16158.
[http://dx.doi.org/10.1021/acsomega.9b02335] [PMID: 31592159]

[30] Qi, Y.; Wang, X.; Chen, H. *et al.* A family of metallocyclodextrins: synthesis, absorption and luminescence characteristic studies based on host–guest recognition. *Supramol. Chem.,* **2015**, *27*(1-2), 44-51.
[http://dx.doi.org/10.1080/10610278.2014.904867]

[31] Hu, X.; Zhu, Z.; Dong, H. *et al.* Inorganic and Metal–Organic Nanocomposites for Cascade-Responsive Imaging and Photochemical Synergistic Effects. *Inorg. Chem.,* **2020**, *59*(7), 4617-4625.
[http://dx.doi.org/10.1021/acs.inorgchem.9b03719] [PMID: 32207928]

[32] Zhu, Z.; Li, H.; Xiang, Y. *et al.* Pyridinium porphyrins and AuNPs mediated bionetworks as SPR signal amplification tags for the ultrasensitive assay of brain natriuretic peptide. *Mikrochim. Acta,* **2020**, *187*(6), 327.
[http://dx.doi.org/10.1007/s00604-020-04289-5] [PMID: 32405667]

[33] Chen, L.; Chen, Y.; Fu, H.G.; Liu, Y. Reversible Emitting Anti-Counterfeiting Ink Prepared by Anthraquinone-Modified *β* -Cyclodextrin Supramolecular Polymer. *Adv. Sci. (Weinh.),* **2020**, *7*(14), 2000803.
[http://dx.doi.org/10.1002/advs.202000803] [PMID: 32714771]

[34] Fu, H.G.; Chen, Y.; Dai, X.Y.; Liu, Y. Quaternary Supramolecular Nanoparticles as a Photoerasable Luminescent Ink and Photocontrolled Cell-Imaging Agent. *Adv. Opt. Mater.,* **2020**, *8*(15), 2000220.
[http://dx.doi.org/10.1002/adom.202000220]

[35] Niu, J.; Chen, Y.; Liu, Y. Supramolecular hydrogel with tunable multi-color and white-light fluorescence from sulfato-β-cyclodextrin and aminoclay. *Soft Matter,* **2019**, *15*(17), 3493-3496.
[http://dx.doi.org/10.1039/C9SM00450E] [PMID: 30932126]

[36] Cheng, Q.; Cao, Z.; Hao, A.; Zhao, Y.; Xing, P. Fluorescent Imprintable Hydrogels via Organic/Inorganic Supramolecular Coassembly. *ACS Appl. Mater. Interfaces,* **2020**, *12*(13), 15491-15499.
[http://dx.doi.org/10.1021/acsami.0c04418] [PMID: 32156108]

[37] Liang, J.; Guo, P.; Qin, X. *et al.* Hierarchically Chiral Lattice Self-Assembly Induced Circularly Polarized Luminescence. *ACS Nano,* **2020**, *14*(3), 3190-3198.
[http://dx.doi.org/10.1021/acsnano.9b08408] [PMID: 32129981]

[38] Shen, F.F.; Zhang, Y.M.; Dai, X.Y.; Zhang, H.Y.; Liu, Y. Alkyl-Substituted Cucurbit[6]uril Bridged β-Cyclodextrin Dimer Mediated Intramolecular FRET Behavior. *J. Org. Chem.,* **2020**, *85*(9), 6131-6136.
[http://dx.doi.org/10.1021/acs.joc.9b03513] [PMID: 32264676]

[39] Wang, T.; Cheng, Y.; Zhang, Y. *et al.* β-cyclodextrin modified quantum dots as pseudo-stationary phase for direct enantioseparation based on capillary electrophoresis with laser-induced fluorescence detection. *Talanta,* **2020**, *210*, 120629.
[http://dx.doi.org/10.1016/j.talanta.2019.120629] [PMID: 31987180]

[40] Wang, Q.; Liang, W.; Wei, X. *et al.* A Supramolecular Strategy for Enhancing Photochirogenic Performance through Host/Guest Modification: Dicationic γ-Cyclodextrin-Mediated Photocyclodimerization of 2,6-Anthracenedicarboxylate. *Org. Lett.,* **2020**, *22*(24), 9757-9761.

[http://dx.doi.org/10.1021/acs.orglett.0c03848] [PMID: 33284623]

[41] Yao, H.; Kan, X.T.; Zhou, Q. *et al.* Lanthanide-Mediated Cyclodextrin-Based Supramolecular Assembly-Induced Emission Xerogel Films: A Transparent Multicolor Photoluminescent Material. *ACS Sustain. Chem.& Eng.,* **2020**, *8*(34), 13048-13055.
[http://dx.doi.org/10.1021/acssuschemeng.0c04490]

[42] Yu, J.; Xiao, S.; Chen, F.; Li, L.; Xu, G. Cyclodextrin mediated efficient energy transfer between 4-amido-1,8-naphthalimide and porphyrin. *Dyes Pigments,* **2020**, *180*, 108518.
[http://dx.doi.org/10.1016/j.dyepig.2020.108518]

[43] Chen, Y.; Yu, B.; Cui, Y.; Xu, S.; Gong, J. Core–Shell Structured Cyclodextrin Metal–Organic Frameworks with Hierarchical Dye Encapsulation for Tunable Light Emission. *Chem. Mater.,* **2019**, *31*(4), 1289-1295.
[http://dx.doi.org/10.1021/acs.chemmater.8b04126]

[44] Da Silva, E.S.; Moura, N.M.M.; Coutinho, A. *et al.* β-Cyclodextrin as a Precursor to Holey C-Doped g-C$_3$N$_4$ Nanosheets for Photocatalytic Hydrogen Generation. *ChemSusChem,* **2018**, *11*(16), 2681-2694.
[http://dx.doi.org/10.1002/cssc.201801003] [PMID: 29975819]

[45] Bhunia, S.; Kumar, S.; Purkayastha, P. Gold Nanocluster-Grafted Cyclodextrin Suprastructures: Formation of Nanospheres to Nanocubes with Intriguing Photophysics. *ACS Omega,* **2018**, *3*(2), 1492-1497.
[http://dx.doi.org/10.1021/acsomega.7b01914] [PMID: 31458475]

[46] Yi, Y.; Zeng, W.; Zhu, G. β-Cyclodextrin functionalized molybdenum disulfide quantum dots as nanoprobe for sensitive fluorescent detection of parathion-methyl. *Talanta,* **2021**, *222*, 121703.
[http://dx.doi.org/10.1016/j.talanta.2020.121703] [PMID: 33167292]

[47] Kaliyamoorthi, K.; Maniraj, S.; Govindaraj, T.S.; Ramasamy, S.; Paulraj, M.S.; Enoch, I.V.M.V.; Melchior, A. Unusual Fluorescence Quenching-Based Al$^{3+}$ Sensing by an Imidazolylpiperazine Derivative. β-Cyclodextrin Encapsulation-Assisted Augmented Sensing. *J. Fluoresc.,* **2020**, *30*(3), 445-453.
[http://dx.doi.org/10.1007/s10895-020-02511-4] [PMID: 32125570]

[48] Pramanik, A.; Amer, S.; Grynszpan, F.; Levine, M. Highly sensitive detection of cobalt through fluorescence changes in β-cyclodextrin-bimane complexes. *Chem. Commun. (Camb.),* **2020**, *56*(81), 12126-12129.
[http://dx.doi.org/10.1039/D0CC05812B] [PMID: 32914795]

[49] Haynes, A.; Halpert, P.; Levine, M. Colorimetric Detection of Aliphatic Alcohols in β-Cyclodextrin Solutions. *ACS Omega,* **2019**, *4*(19), 18361-18369.
[http://dx.doi.org/10.1021/acsomega.9b02612] [PMID: 31720538]

[50] Prabu, S.; Mohamad, S. Curcumin/beta-cyclodextrin inclusion complex as a new "turn-off" fluorescent sensor system for sensitive recognition of mercury ion. *J. Mol. Struct.,* **2020**, *1204*, 127528.
[http://dx.doi.org/10.1016/j.molstruc.2019.127528]

[51] Zheng, Y.H.; Li, W.H.; Chen, P.; Zhou, Y.; Lu, W.; Ma, Z.C. Determination of berberine in Rhizoma coptidis using a β-cyclodextrin-sensitized fluorescence method. *RSC Advances,* **2020**, *10*(66), 40136-40141.
[http://dx.doi.org/10.1039/D0RA07573F]

[52] Liu, L.; Yi, G.; Yang, L. *et al.* Designing and preparing supramolecular fluorescent probe based on carminic acid and γ-cyclodextrins and studying their application for detection of 2-aminobenzidazole. *Carbohydr. Polym.,* **2020**, *241*, 116367.
[http://dx.doi.org/10.1016/j.carbpol.2020.116367] [PMID: 32507167]

[53] Selvan, G.T.; Poomalai, S.; Ramasamy, S. *et al.* Differential Metal Ion Sensing by an Antipyrine Derivative in Aqueous and β-Cyclodextrin Media: Selectivity Tuning by β-Cyclodextrin. *Anal. Chem.,*

**2018**, *90*(22), 13607-13615.
[http://dx.doi.org/10.1021/acs.analchem.8b03810] [PMID: 30412380]

[54]   Miyagawa, A.; Yoneda, H.; Mizuno, H.; Numata, M.; Okada, T.; Fukuhara, G. Hydrostatic-Pressur-
-Controlled Molecular Recognition: A Steroid Sensing Case Using Modified Cyclodextrin.
*ChemPhotoChem,* **2021**, *5*(2), 118-122.
[http://dx.doi.org/10.1002/cptc.202000204]

[55]   Nazerdeylami, S.; Ghasemi, J.B.; Amiri, A.; Mohammadi Ziarani, G.; Badiei, A. A highly sensitive
fluorescence measurement of amphetamine using 8-hydroxyquinoline-β-cyclodextrin grafted on
graphene oxide. *Diamond Related Materials,* **2020**, *109*, 108032.
[http://dx.doi.org/10.1016/j.diamond.2020.108032]

[56]   Poomalai, S.; Soundrapandian, S.; Ramasamy, S.; Selvakumar Paulraj, M.; Enoch, I.V.M.V.
Fluorescence sensing of Al3+ by an anthracene appended piperidone. Hindrance in excimer formation
offered by β-cyclodextrin complexation. *Chem. Phys. Lett.,* **2020**, *751*, 137551.
[http://dx.doi.org/10.1016/j.cplett.2020.137551]

[57]   Zhu, Y.; Lu, Y.; Shi, L.; Yang, Y. β-Cyclodextrin functionalized N,Zn codoped carbon dots for
specific fluorescence detection of fluoroquinolones in milk samples. *Microchem. J.,* **2020**, *153*,
104517.
[http://dx.doi.org/10.1016/j.microc.2019.104517]

[58]   Shi, Y.; Peng, J.; Meng, X.; Huang, T.; Zhang, J.; He, H. Turn-on fluorescent detection of captopril in
urine samples based on hydrophilic hydroxypropyl β-cyclodextrin polymer. *Anal. Bioanal. Chem.,*
**2018**, *410*(28), 7373-7384.
[http://dx.doi.org/10.1007/s00216-018-1343-9] [PMID: 30191274]

[59]   Yang, H.; Lu, F.; Sun, Y.; Yuan, Z.; Lu, C. Fluorescent Gold Nanocluster-Based Sensor Array for
Nitrophenol Isomer Discrimination via an Integration of Host–Guest Interaction and Inner Filter
Effect. *Anal. Chem.,* **2018**, *90*(21), 12846-12853.
[http://dx.doi.org/10.1021/acs.analchem.8b03394] [PMID: 30296826]

[60]   Singh, V.R.; Singh, P.K. A novel supramolecule-based fluorescence turn-on and ratiometric sensor for
highly selective detection of glutathione over cystein and homocystein. *Mikrochim. Acta,* **2020**,
*187*(11), 631.
[http://dx.doi.org/10.1007/s00604-020-04602-2] [PMID: 33125575]

[61]   Liu, R.; Yao, T.; Liu, Y.; Yu, S.; Ren, L.; Hong, Y.; Nguyen, K.T.; Yuan, B. Temperature-sensitive
polymeric nanogels encapsulating with β-cyclodextrin and ICG complex for high-resolution deep-
tissue ultrasound-switchable fluorescence imaging. *Nano Res.,* **2020**, *13*(4), 1100-1110.
[http://dx.doi.org/10.1007/s12274-020-2752-6]

[62]   Jiao, J.B.; Wang, G.Z.; Hu, X.L. *et al.* Cyclodextrin-Based Peptide Self-Assemblies (Spds) That
Enhance Peptide-Based Fluorescence Imaging and Antimicrobial Efficacy. *J. Am. Chem. Soc.,* **2020**,
*142*(4), 1925-1932.
[http://dx.doi.org/10.1021/jacs.9b11207] [PMID: 31884796]

[63]   Haynes, A.Z.; Levine, M. Detection of anabolic steroids *via* cyclodextrin-promoted fluorescence
modulation. *RSC Advances,* **2020**, *10*(42), 25108-25115.
[http://dx.doi.org/10.1039/D0RA03485A]

[64]   Haynes, A.Z.; Levine, M. Detection of Human Growth Hormone (hGH) via Cyclodextrin-Promoted
Fluorescence Modulation. *Anal. Lett.,* **2021**, *54*(11), 1871-1880.
[http://dx.doi.org/10.1080/00032719.2020.1828445]

[65]   Xu, T.; Zhao, S.; Wu, X.; Zeng, L.; Lan, M. β-Cyclodextrin-Promoted Colorimetric and Fluorescence
Turn-on Probe for Discriminating Highly Toxic Thiophenol from Biothiols. *ACS Sustain. Chem.&
Eng.,* **2020**, *8*(16), 6413-6421.
[http://dx.doi.org/10.1021/acssuschemeng.0c00766]

[66]   Holkar, A.; Ghodke, S.; Bangde, P.; Dandekar, P.; Jain, R. Fluorescence-Based Detection of

Cholesterol Using Inclusion Complex of Hydroxypropyl-β-Cyclodextrin and l-Tryptophan as the Fluorescence Probe. *J. Pharm. Innov.,* **2020**.

[67]  Wang, M.; Wang, S.; Li, L.; Wang, G.; Su, X. β-Cyclodextrin modified silver nanoclusters for highly sensitive fluorescence sensing and bioimaging of intracellular alkaline phosphatase. *Talanta,* **2020,** *207,* 120315.
[http://dx.doi.org/10.1016/j.talanta.2019.120315] [PMID: 31594591]

[68]  Zhang, X.; Zhang, F.; Zhu, F.; Zhang, X.; Tian, D.; Johnson, R.P.; Li, H. Bioinspired γ-Cyclodextrin Pseudorotaxane Assembly Nanochannel for Selective Amino Acid Transport. *ACS Appl. Bio Mater.,* **2019,** *2*(8), 3607-3612.
[http://dx.doi.org/10.1021/acsabm.9b00473] [PMID: 35030747]

[69]  Yadav, A.; Upadhyay, Y.; Bera, R.K.; Sahoo, S.K. Vitamin $B_6$ cofactors guided highly selective fluorescent turn-on sensing of histamine using beta-cyclodextrin stabilized ZnO quantum dots. *Food Chem.,* **2020,** *320,* 126611.
[http://dx.doi.org/10.1016/j.foodchem.2020.126611] [PMID: 32199201]

[70]  Zhou, Y.; Yang, S.; Guo, J.; Dong, H.; Yin, K.; Huang, W.T.; Yang, R. *in vivo* Imaging of Hypoxia Associated with Inflammatory Bowel Disease by a Cytoplasmic Protein-Powered Fluorescence Cascade Amplifier. *Anal. Chem.,* **2020,** *92*(8), 5787-5794.
[http://dx.doi.org/10.1021/acs.analchem.9b05278] [PMID: 32192346]

[71]  Bai, L.; Yan, H.; Bai, T.; Feng, Y.; Zhao, Y.; Ji, Y.; Feng, W.; Lu, T.; Nie, Y. High Fluorescent Hyperbranched Polysiloxane Containing β-Cyclodextrin for Cell Imaging and Drug Delivery. *Biomacromolecules,* **2019,** *20*(11), 4230-4240.
[http://dx.doi.org/10.1021/acs.biomac.9b01217] [PMID: 31633916]

[72]  Thomsen, H.; Agnes, M.; Uwangue, O. *et al.* Increased antibiotic efficacy and noninvasive monitoring of Staphylococcus epidermidis biofilms using per-cysteamine-substituted γ-cyclodextrin – A delivery effect validated by fluorescence microscopy. *Int. J. Pharm.,* **2020,** *587,* 119646.
[http://dx.doi.org/10.1016/j.ijpharm.2020.119646] [PMID: 32679261]

[73]  Roozbehi, S.; Dadashzadeh, S.; Sajedi, R.H. An enzyme-mediated controlled release system for curcumin based on cyclodextrin/cyclodextrin degrading enzyme. *Enzyme Microb. Technol.,* **2021,** *144,* 109727.
[http://dx.doi.org/10.1016/j.enzmictec.2020.109727] [PMID: 33541570]

[74]  Tian, J.; Xia, L.; Wu, J.; Huang, B.; Cao, H.; Zhang, W. Linear Alternating Supramolecular Photosensitizer for Enhanced Photodynamic Therapy. *ACS Appl. Mater. Interfaces,* **2020,** *12*(29), 32352-32359.
[http://dx.doi.org/10.1021/acsami.0c07333] [PMID: 32584539]

[75]  Trapani, M.; Romeo, A.; Parisi, T.; Sciortino, M.T.; Patanè, S.; Villari, V.; Mazzaglia, A. Supramolecular hybrid assemblies based on gold nanoparticles, amphiphilic cyclodextrin and porphyrins with combined phototherapeutic action. *RSC Advances,* **2013,** *3*(16), 5607-5614.
[http://dx.doi.org/10.1039/c3ra40204e]

[76]  Jo, G.; Lee, B.Y.; Kim, E.J.; Park, M.H.; Hyun, H. Indocyanine Green and Methyl-β-Cyclodextrin Complex for Enhanced Photothermal Cancer Therapy. *Biomedicines,* **2020,** *8*(11), 476.
[http://dx.doi.org/10.3390/biomedicines8110476] [PMID: 33167365]

[77]  Zhang, J.; Zhou, Z.H.; Li, L. *et al.* Dual Stimuli-Responsive Supramolecular Self-Assemblies Based on the Host–Guest Interaction between β-Cyclodextrin and Azobenzene for Cellular Drug Release. *Mol. Pharm.,* **2020,** *17*(4), 1100-1113.
[http://dx.doi.org/10.1021/acs.molpharmaceut.9b01142] [PMID: 32125862]

[78]  Pei, M.; Pai, J.Y.; Du, P.; Liu, P. Facile Synthesis of Fluorescent Hyper-Cross-Linked β-Cyclodextri--Carbon Quantum Dot Hybrid Nanosponges for Tumor Theranostic Application with Enhanced Antitumor Efficacy. *Mol. Pharm.,* **2018,** *15*(9), 4084-4091.
[http://dx.doi.org/10.1021/acs.molpharmaceut.8b00508] [PMID: 30040427]

[79]    Welling, M.M.; Duszenko, N.; van Willigen, D.M.; Smits, W.K.; Buckle, T.; Roestenberg, M.; van Leeuwen, F.W.B. Cyclodextrin/Adamantane-Mediated Targeting of Inoculated Bacteria in Mice. *Bioconjug. Chem.,* **2021**, *32*(3), 607-614.
[http://dx.doi.org/10.1021/acs.bioconjchem.1c00061] [PMID: 33621052]

[80]    Weiss-Errico, M.J.; Miksovska, J.; O'Shea, K.E. β-Cyclodextrin Reverses Binding of Perfluorooctanoic Acid to Human Serum Albumin. *Chem. Res. Toxicol.,* **2018**, *31*(4), 277-284.
[http://dx.doi.org/10.1021/acs.chemrestox.8b00002] [PMID: 29589912]

# CHAPTER 2

# Calixarene Based Luminescent Systems

**Bosco Christin Maria Arputham Ashwin**[1]**, Venkatesan Sethuraman**[2] **and Paulpandian Muthu Mareeswaran**[2,*]

[1] *Department of Chemistry, Pioneer Kumaraswamy College, Nagercoil629 003, Tamilnadu, India.*

[2] *Department of Industrial Chemistry, Alagappa University, Karaikudi630 003, Tamilnadu, India*

**Abstract:** This chapter focuses on the recent developments in calixarenes chemistry using luminescence aspect. This aspect comprises host guest system of luminescent guests with calixarenes as well as luminescent organic moieties tethered calixarene systems. The utilization of organic moieties tethered calixarene systems towards sensor applications for anions, cations and biomolecules is explored. The calixarene systems are used as a platform as well as convenient receptors for the luminescent sensor systems. The ease of modification in the rims of calixarenes renders numerous synthetic possibilities to recognise a particular target molecule and also to fabricate solid state luminescent sensor systems.

**Keywords:** Aggregation, Amino acids, Anion, Bioimaging, Biomolecule, Calixarene, Cation, Chemosensor, Device, Drug, Dye, Emission, Fluorescence, Host-guest, Interaction, Lanthanoids, Luminescence, Macrocycle, Nanoparticles, Recognition.

## INTRODUCTION

Calixarenes, the vase structured phenol-based macromolecules, have great attention among the most investigated scaffolds in supramolecular chemistry due to their easy modification as per the utilization [1, 2]. Calixarene is made up of phenol moieties connected by ethylene bridge [3]. They have more advantages especially (a) facile lower and upper rims functionalization, (b) a well-defined nonpolar cavity, (c) substituent and guest dependent definite conformations and (d) distinct target binding sites [4, 5]. The well-defined as well as flexible cavities are efficient to receive guest molecules [6,7]. The hydrophobic cavity can interact with guest molecules and ions by means of cation-$\pi$ interactions [8, 9]. The upper and lower rims can be synthetically tuned to target the specific guest molecules by

---

[*] **Corresponding author Paulpandian Muthu Mareeswaran**: Department of Industrial Chemistry, Alagappa University, Karaikudi 630 003, Tamilnadu, India; Tel: +919790963437; E-mails: muthumareeswaran@gmail.com and mareeswaran@alagappauniversity.ac.in

**Paulpandian Muthu Mareeswaran, Palaniswamy Suresh and Seenivasan Rajagopal (Eds.)**

means of desired interactions for the specific binding [10,11]. Calixarene itself is a non-luminescent molecule and the luminescent properties can be manipulated either by the inclusion of luminescent guest molecules or by tethering luminescent moieties on the rims of calixarenes (Fig. **1**). The substituents enhanced the luminescent property of the calixarenes and have a variety of applications with specific targets than the guest induced luminescent systems. The recent research studies of the guest induced luminescent properties of calixarenes and predominantly luminescent moieties tethered calixarenes systems in detail are presented in this chapter.

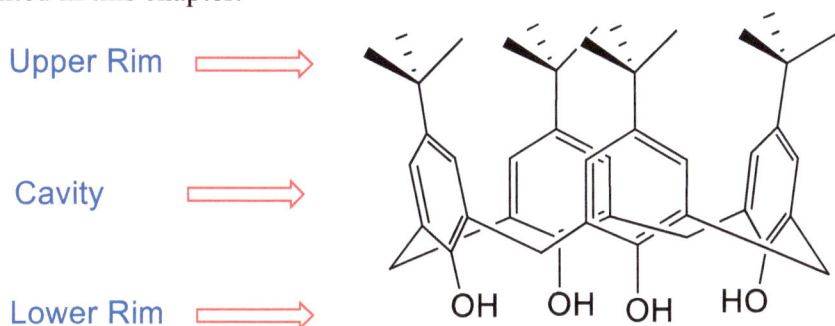

**Fig. (1).** Structure of t-butyl calix [4]arene and their parts are represented. Most of the fluorophore guests were accommodated within the cavity.

## LUMINESCENT HOST-GUEST SYSTEMS

Studying the nature of non-bonding interactions and revealing the binding structure of the macrocyclic host molecule with guest molecules is an interesting and difficult one [12]. Utilizing the luminescent nature of the guest molecule of the host-guest complex with calixarene is a well preferred technique for understanding these aspects by means of the luminescence changes observed [13, 14]. In calixarene derivatives the aqueous soluble *p*-sulfonatocalix [4]arene has much reports with vast luminescent guest molecules [15] such as amino acids, proteins [16,17], curcumin [18], safranin T [14], viologen derivatives [19], 1,8-diaminonaphthalene [20], coumarin 460 [21], 7-methoxycoumarin [22], gallic acid [23], n-(4-hydroxyphenyl)-imidazole [24], vitamin E [25], thioflavin-T [26], triphenylpyrylium cation [27], diphenylamine [28], acenaphthene-1,2-dione [29] and more. Compared with advanced techniques developed, the studies based on luminescence change have great attraction due to quick and facile interpretations experimentally. Most of these findings support the theoretically stimulated results [22].

## Fluorescent Calixarene

Oueslati *et al.* [30] studied the luminescent nature of the elongated nanoporous micro crystals of distal calix [4]arene dimethylester derivative. The hydrogen bonds and van der Waals interactions within these molecules provided a stable linear nanoscale tubular polymeric structure. The crystals showed luminescence at room temperature due to the monomer fluorescence, dimer fluorescence and monomer phosphorescence with different excited state lifetimes. Zhu *et al.* [31] investigated the fluorescence properties of *p*-sulfonatocalix[n]arene (SC[n], n=4, 6 and 8)-cetyltrimethylammonium bromide (CTAB) supra-amphiphiles. Both SC[n] and CTAB were non-fluorescent but the formed SC[n]-CTAB complexes emitted fluorescence. When the molar ratio of CTAB to SC[n] reached the stoichiometry, the strongest fluorescence intensity was observed. An approach to inducing luminescence behaviour by incorporating dye molecules in calixarene has been reported [32]. Making use of this method as a solid-state sensor device consisting of calix [4]arene crafted on ruthenium dye doped silica nanoparticles has been reported [33]. It exhibited excellent recognition of glyphosate in an aqueous medium based on FRET [34].

## ION SENSING APPLICATIONS

The ion sensing applications using calixarene based luminescent systems is a widely studied aspect of sensor applications of cations. Kim *et al.* [35] synthesised a calixarene based calix [4]azacrown, a fluorescent molecule containing anthracenyl unit. It displayed chelation-enhanced fluorescence with $Cs^+$, $Rb^+$ and $K^+$ metal ions. The interesting "molecular taekwondo" processes between metal ion pairs were observed by the change in fluorescence. The luminescence nature of a water soluble calix [4]arene derivative, 5,11,17,23-tetr--sulfonate-25,26,27,28-tetra-carboxymethoxycalix [4]arene with a lanthanoid ion ($Tb^{3+}$) complexation has been studied in gelation medium. The calix [4]arene derivative formed an efficient energy transfer complex with $Tb^{3+}$ ion and the binding mechanism in gelatine was studied in detail by fluorescence [36].

In 2007 Kim *et al.* [37] reported a review of calixarene derived fluorescent probes. It exposed the luminescent calixarene sensors reported before in detail and the insight mechanisms were discussed. The unique topology of calixarenes offers a wide range of scaffolds, facilitating them to encapsulate plentiful different toxic cations and biologically relevant anions. Chang *et al.* [38] synthesised a triazole-modified calix [4]crown (**1**) in the 1,3-alternate conformation and studied sensing behaviour of metal ions. This chemosensor established that a metal ion exchange can trigger an on−off switchable fluorescent chemosensor (Fig. **2**) for analyte ion.

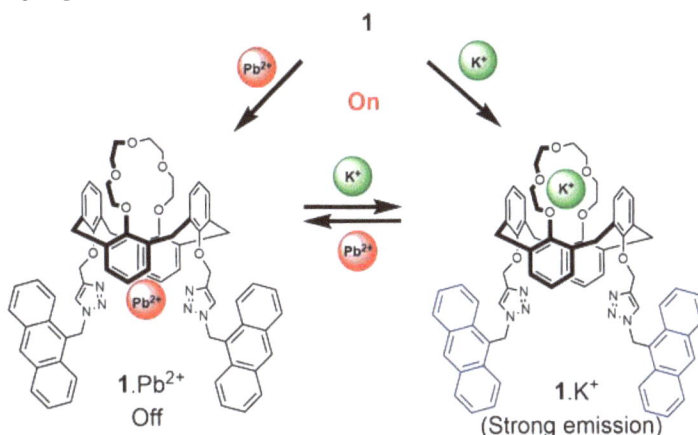

**Fig. (2).** Schematic representation of an on−off switchable fluorescent chemosensor of Pb$^{2+}$ upon addition of different amounts of K$^+$ ion [38].

Dhir *et al.* [39] reported a fluorescence based sensor of dansyl moiety substituted calix [4]arene with partial cone confirmation for selective recognition of Hg$^{2+}$ and Cu$^{2+}$ ions in two different modes. Ocak *et al.* [40] reported fluorophores having two dansyl groups in lower-rim and upper-rim allyl substituted di-ionized calix [4]arenes for metal ion complexation studies. The same group recorded upper-rim benzyl-substituted, di-ionized calix [4]arenes for metal ion complexation later [41]. Amide linked lower rim 1,3-dibenzimidazole derivative of calix [4]arene has shown sensitive and selective recognition to Hg$^{2+}$ ion in aqueous acetonitrile solution on fluorescence spectroscopy [42]. Fluorescent sensor for Cd$^{2+}$ and Zn$^{2+}$ ions by a pyrenyl appended triazole based calix [4]arene molecule (Fig. **3**) has been reported [43].

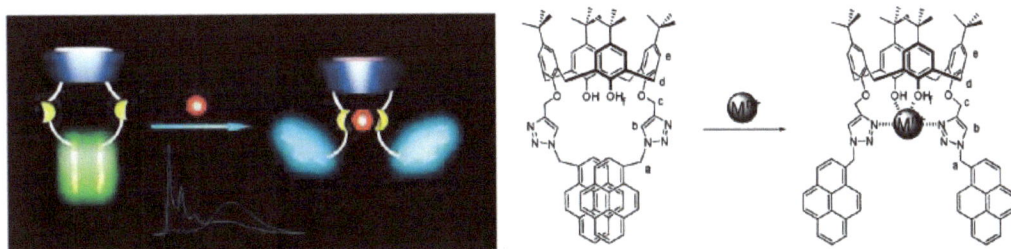

**Fig. (3).** Scheme and proposed structure for pyrenyl appended triazole based calix [4]arene with metal ions [43].

A 1,3-alternate calix [4]arene with bis-triazoles and bis-enaminones located on each end was synthesised (Fig. **4**), which acted as a Ag$^+$ homobinuclear ditopic fluorescent chemosensor [44]. Zhu *et al.* [45] synthesized two pyrene-armed calix [4]arenes with triazole connection *via* "click" chemistry. This molecule exhibited an integrated logic gate based fluorescent sensor for $Zn^{2+}$ and $Cu^{2+}$ ions. Pyrenyl appended triazole-based thiacalix [4]arenes were reported for selective fluorescent enhancement by $Ag^+$ ion over other metal ions in neutral solution [46].

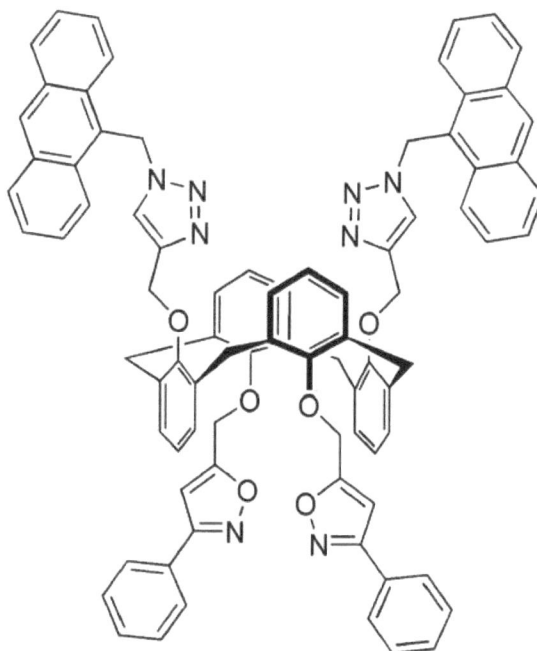

**Fig. (4).** Proposed structure of 1,3-alternate calix [4]arene for the sensing of $Ag^{1+}$ ion [44].

A calix [4] arene with lower-rim proximal triazolylpyrenes (Fig. **5**), was reported for a ratiometric fluorescent chemosensor for $Ag^+$ ion with higher sensitivity [47]. 1,2,3-Triazole ring based on 1,3-alternate-thiacalix [4]arene has been synthesised and their fluorescence behaviour over $Ag^+$ ion was studied. The electronic properties of triazole ring like large dipole moment and the presence of electropositive group are valuable in designing tools for self-assembly and cation-binding sites of receptors [48]. 1,3,4-Oxadiazoles based calix [4]arene conjugates were reported as fluorescent chemosensor for the specific recognition of $Cu^{2+}$ ions [49, 50].

Similar thiacalix [4] arene based fluorescent chemosensor was developed for sensitive low ion recognition of $Cu^{2+}$ ion in ethanol [51]. A thiacalix [4]arene-cinnamaldehyde derivative underwent a red shift in fluorescence with $Ag^+$ ions (Fig. **6**). This is due to the binding of $Ag^+$ ions with imino nitrogens, which induces intramolecular charge transfer (ICT) process from nitrogen atoms of the dimethylamino moiety to the imino moiety [52].

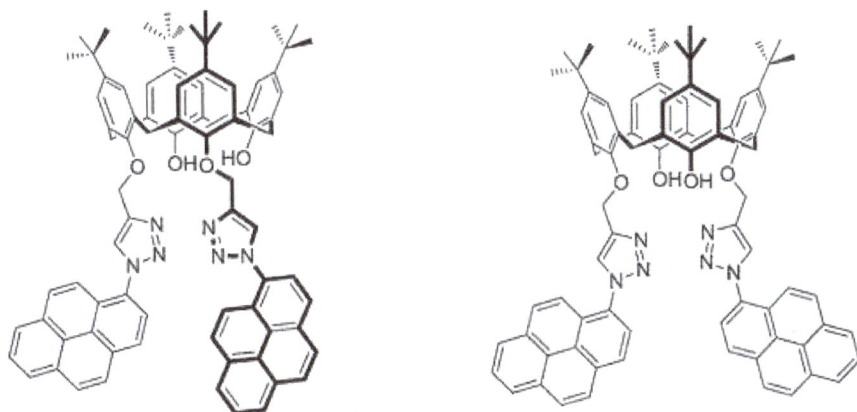

**Fig. (5).** Structures of the lower-rim proximal triazolylpyrenes containing calix [4]arenes for Ag+ ion sensor [47].

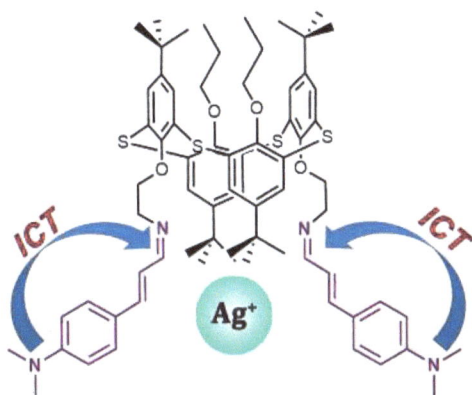

**Fig. (6).** Scheme of ICT-induced detection of Ag+ by thiacalix [4]arene-cinnamaldehyde derivative [52].

Pathak *et al.* [53] reported a triazole linked thiophene calix [4]arene for selective sensing of $Zn^{2+}$ (Fig. 7) and as a biomimetic model, which supported the metal detoxification and oxidative stress involving metallothionein (MT). Pyridyl-based triazole-linked calix [4]arene conjugates were synthesized by the same group which exhibited sensitive switch-on sensing for $Zn^{2+}$ in HEPES (N-(--Hydroxyethyl)piperazine-N'-(2-ethanesulfonic acid)) buffer medium [54]. Ni *et al.* [55] reported a pyrene-linked triazole-modified homooxacalix [3]arene for highly selective sensor for $Pb^{2+}$ ion over other metal ions by enhancement of the pyrene monomer emission.

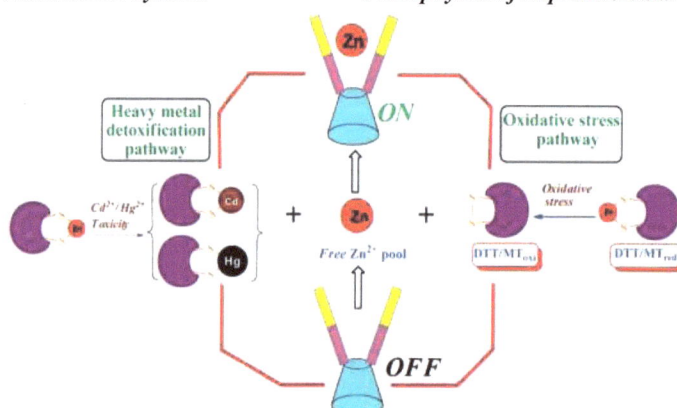

**Fig. (7).** Scheme of $Zn^{2+}$ ion recognition and biomimetic model of triazole linked thiophene conjugate of calix [4]arene. DTT represents dithiothreotol [53].

Two rhodamine B lactams substituted at 1,3-alternate to a thiacalix [4]arene was reported for highly sensitive $Fe^{3+}$ and $Cr^{3+}$ ion-induced fluorescent sensor. The $Fe^{3+}$ and $Cr^{3+}$ ion induced reversible spiro ring-opening gave fluorescence and colorimetric enhancement [56]. Zhan *et al.* [57] synthesised an anthraquinone-modified calix [4]arene by "click" chemistry. It showed a logic gate sensitive recognition of $Ca^{2+}$ and $F^-$ ions based on photoinduced electron transfer from the 1,2,3-triazole ring. Zhan *et al.* [58] reported a pyridyl-appended calix [4]arene which exhibited selective binding with $Fe^{3+}$ ions. Furthermore, this calix [4]arene was employed in synthesizing silver nanoparticles (NPs) with a distinct colorimetric response to $Fe^{3+}$ ions (Fig. **8**).

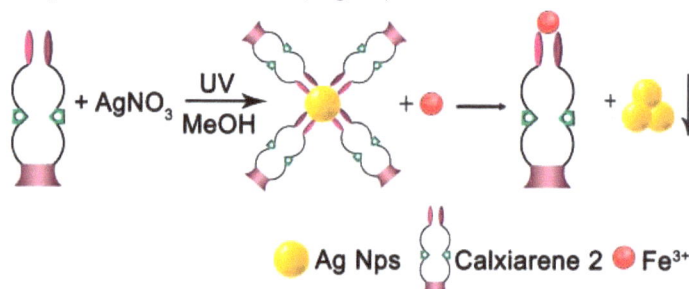

**Fig. (8).** Synthesis of pyridyl-appended calix [4] arene – Ag NPs and a scheme of the $Fe^{3+}$ ion induced aggregation [58].

Patra *et al.* [59] synthesised a series of fluorescent calix [4]arenes based on coumarin as fluorogenic unit and studied their metal ion-binding properties for chemosensor application. A coumarin combined calix [4]arene was presented for its $Cu^{2+}$ ion fluorescent sensing [60]. Cho *et al.* [61] reported a pyrene-appended calix [2]triazole [2]arene as a bimodal fluorescent sensor for $Zn^{2+}$ (or $Cd^{2+}$) and $Fe^{2+}$. Similar pyrenyl-appended calix [2]triazole [2]arene displayed excellent selectivity for Cu$^{2+}$ ion. This system was reported as the initial Cu$^{2+}$ ion-induced fluorescent turn-on system for imaging Wilson's disease copper trafficking (Fig. **9**) over time in a model [62].

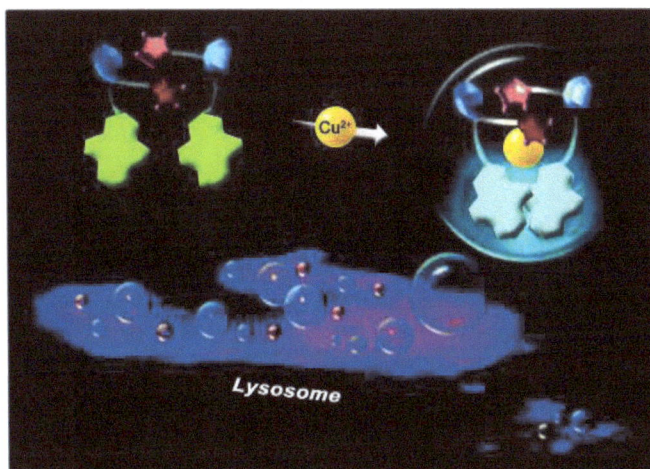

**Fig. (9).** Illustration of Cu$^{2+}$ fluorescent turn-on system utilized for copper trafficking pathway in Wilson's disease [62].

A phenanthrene-based calix [4] arene having cone conformation has been synthesized and reported as a fluorescent sensor for Cu$^{2+}$ and F$^-$ ions [63]. Song *et al.* [64] presented a review comprising of the calixarene based chemosensors based on click-derived triazoles. A series of metal ion receptors, $C_3$-symmetric triazole derivatives based on hexahomotrioxacalix [3]arene were synthesised by click chemistry (Fig. **10**) and their binding properties were studied by Jin *et al.* [65].

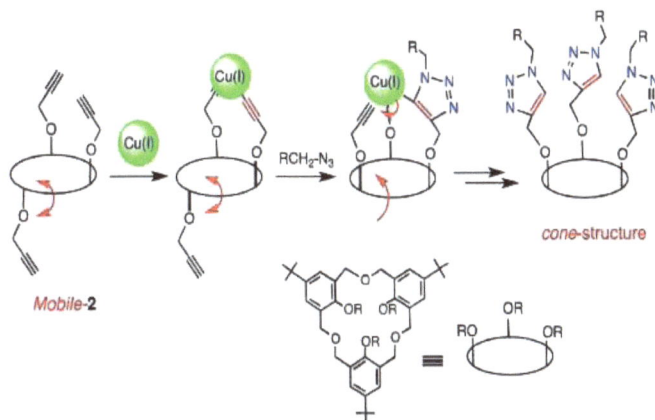

**Fig. (10).** The Cu(I) catalysed "click" reaction mechanisms for hexahomotrioxacalix [3]arene derivatives from mobile- to cone-structure [65].

Anthryl-1,2,4-oxadiazole substituted calix [4]arenes exhibited highly selective fluorescent chemodosimeters for $Fe^{3+}$ ion with high sensitivity [66]. A $Fe^{2+}$ or $Fe^{3+}$ ion selective fluorescent chemosensor obtained from sugar-thiacrown-ether appended calix [4]arene, coupled with pyrene units (Fig. **11**) was reported by Ling *et al.* [67]. A pyrene-appended ditopic homooxacalix [3]arene was reported for ratiometric fluorescent chemosensor for heavy and transition metal ions [68].

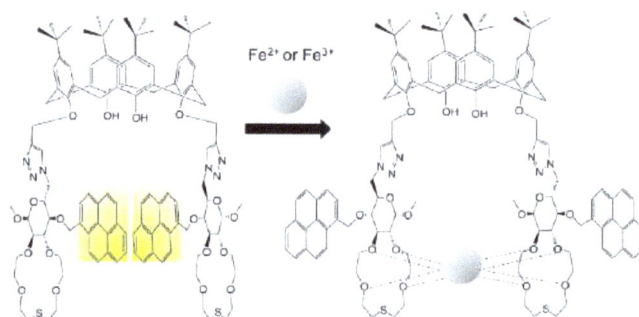

**Fig. (11).** The proposed binding structure of sugar-thiacrown-ether appended calix [4]arene with $Fe^{2+}/Fe^{3+}$ [67].

Jiang *et al.* [69] reported a fluorescent sensor of a hexahomotrioxacalix [3]arene scaffold having triazole rings (Fig. **12**) as the binding sites on the lower rim in a cone conformation. The detection of $Zn^{2+}$ and $Cd^{2+}$ in the presence of excess other metal ions at low ion concentration, a fluorescence enhancement was observed due to the cavity shape of calixarene changing to a more-upright form from a 'flattened-cone' and suppression of photoinduced electron transfer (PET). This molecule showed high sensitivity and selectivity for $Cu^{2+}$ and $Fe^{3+}$ ions by fluorescence quenching due to the PET or heavy atom effect.

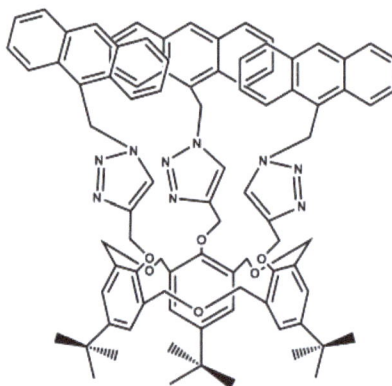

**Fig. (12).** Structure of the hexahomotrioxacalix [3]arene scaffold for the recognition of $Zn^{2+}$ ion by inhibition of PET [69].

Ullmann *et al.* [70] reported a macrocyclic ligand containing two *o,o'*-bis(iminomethyl)phenol and two calix [4]arene head units. This fluorescent calix [4]arene exhibited increase in fluorescence intensity by a factor of 10 in the presence of $Zn^{2+}$ with a nanomolar detection limit. A triazole appended benzaldehydic lower rim calix [4]arene derivative has been reported for the sensing of $Fe^{3+}$ ion by means of fluorescence quenching [71]. Anandababu *et al.* [72] reported a *p*-tert-butylcalix [4]arene thiospirolactam rhodamine b based ligand (Fig. **13**) for effective off-on sensor for $Hg^{2+}$ ion in aqueous medium and bioimaging into HeLa cells.

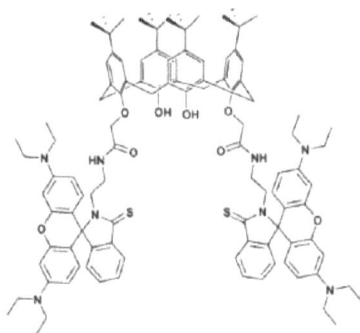

**Fig. (13).** Proposed structure of rhodamine b modified *p*-tert-butylcalix [4]arene, a probe for $Hg^{2+}$ ion [72].

Four naphthyl "capped" 1,2,3-triazole-linked calix [4]arenes have been synthesized by Rahman *et al.* [73] and reported as fluorescent chemosensors towards $Fe^{3+}$ and $Hg^{2+}$ ions. Two bis(naphthyl)methane "capped" 1,2,3-triazole linked calix [4]arenes have been synthesised, which exhibited $Fe^{3+}$ ion recognition by fluorescence quenching [74]. Similar triazole-bridged pyrene-appended calix [4]arene (Fig. **14**) was reported as $Zn^{2+}$ and $Cd^{2+}$ metal ion fluorescent chemosensor [75].

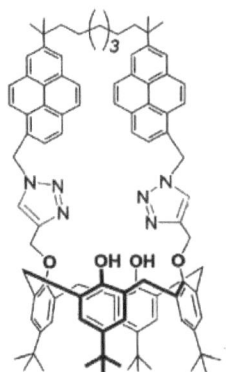

**Fig. (14).** Triazole-bridged pyrene-appended calix [4]arene for $Zn^{2+}$ and $Cd^{2+}$ ions detection [75].

Two crab-like chemosensors based on calix [4]arene bearing (thio)barbituric acid groups have been presented for their highly selective recognition of $Hg^{2+}$ and $Cu^{2+}$ ions by fluorescence [76]. Zadmard *et al.* [77] reported the synthesis and recognition properties of highly functionalized calix [4]arenes by means of multicomponent reactions. Akbarzadeh *et al.* [78] synthesized a sensitive fluorescent sensor for $Cu^{2+}$ ion by combining two minoxidil molecules at the lower-rim of calix [4]arene annulus through the formation of two urea linkages. Calix [4]arene bearing BODIPY and triazolenaphthylene groups has been reported as a good fluorescent sensor for $Hg^{2+}$ due to the coordination between $Hg^{2+}$ and the triazole rings [79]. Modi *et al.* [80] presented the synthesis of thiacalixphenyl [4]arene tetra-N-(3-propyl) phthalimide (Fig. **15**) for the selective nanomolar fluorescent recognition of $Hg^{2+}$ ion. Bhatti *et al.* [81] reported a water soluble *p*-sulphonatocalix [4]arene appended with rhodamine for selective chemosensor of $Hg^{2+}$ ion.

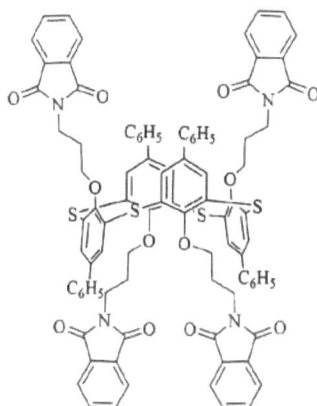

**Fig. (15).** Structure of propyl phthalimide modified thiacalixphenyl [4]arene for the sensing of $Hg^{2+}$ ions [80].

Yilmaz *et al.* [82] reported three facile calixarene derivatives having four rhodamine units at the upper rim as an integrated sensing system for selective fluorometric detection of $Hg^{2+}$ ions in living cells. A pyrazole and triazole functionalized 1,3-alternate calix [4] arenes (Fig. **16**) exhibited efficient fluor-escent sensing of $Hg^{2+}$ and $Ag^{+}$ ions [83]. Bis(spiropyran)-conjugated calix [4]arene molecules through a 1,2,3-triazole linker were synthesised and studied for their cooperativity bindings of a divalent metal cation series [84]. Sister calix [4]arene derivatives containing two 7-nitro-benzofurazan units in 1,3-(distal) derivatives were studied for ion sensing and bioimaging applications [85].

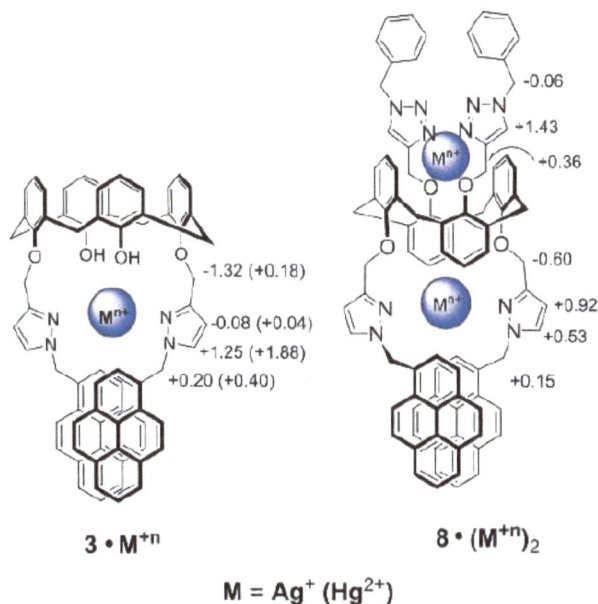

| | -0.06 |
| | +1.43 |
| | +0.36 |
| | -0.60 |
| | +0.92 |
| | +0.53 |
| | +0.15 |

-1.32 (+0.18)
-0.08 (+0.04)
+1.25 (+1.88)
+0.20 (+0.40)

**3 • M$^{+n}$**                                                    **8 • (M$^{+n}$)$_2$**

**M = Ag$^+$ (Hg$^{2+}$)**

**Fig. (16).** Two possible binding modes of pyrazole and triazole functionalized 1,3-alternate calix [4]arene with Ag$^+$ and Hg$^{2+}$ ions with the chemical shift differences of Ag$^+$ [83].

A pyrene-allied calix [4]arene displayed chelation enhanced fluorescence (CHEF)-PET based highly selective and sensitive fluorescence recognition of Zn$^{2+}$, Hg$^{2+}$ and I$^-$ ions. A low cost and disposable paper-based sensing strip was also developed [86]. Another tritopic calix [4]arene based CHEF-PET fluorescence probe was reported for La$^{3+}$, Cu$^{2+}$, and Br$^-$ ions [87]. A calix [4]arene based CHEF-PET probe for Mn$^{2+}$, Cr$^{3+}$ and F$^-$ ions have been reported by Sutariya *et al.* [88]. A calix [4]arene scaffold having two fluorogenic aminoquinoline moieties has been synthesized and studied as a selective and sensitive sensor for Sr$^{2+}$ and CN$^-$ ions over a broad range of ions. This probe unveiled trace level detection and paper-based single analytical strip for quick detection of Sr$^{2+}$ and CN$^-$ ion [89].

A micelle formed by the self-assembly of fluorescent amino thiadiazole calix [4]arene derivative was reported as a selective chemosensor for Pb$^{2+}$ ion. The fluorescence quenching of pyrene was attributed to the close proximity of Pb$^{2+}$ ions in the complex to the pyrene by fluorescent internal filter effect (IFE) [90]. A triazole-derivative of spiro-indoline-linked calix [4]arene has been reported for its ratiometric Cu$^{2+}$ binding, bioimaging, mitochondrial targeting and anticancer activity [91]. Calix [4]arene-oxacyclophane architectures were reported for the effective fluorescent probes for Cu$^{2+}$ ion detection [92]. Jin *et al.* [93] reported the quantum chemical characterization of the binding behaviour of bis(bora)calix

[4]arene, a known fluorescent sensor with F$^-$ ion (Fig. **17**). The theoretical calculations indicated that the fluorescence quenching nature when F$^-$ binds with bis(bora)calix [4]arene was due to the cooperative *endo-* and *exo-* bindings.

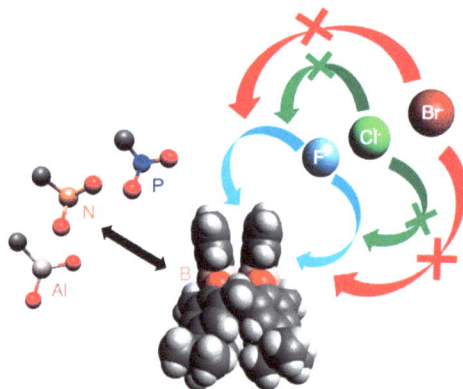

**Fig. (17).** Scheme representing the theoretical studies of the bis(bora)calix [4]arene for F$^-$ ion sensing [93].

Chen *et al.* [94] observed the spectral Cu$^{2+}$ ion sensing properties of *p*-ter--butylthiacalix [4]arene (TCA) using theoretical Time-Dependent Density Functional Theory (TD-DFT) (Fig. **18**). The addition of Cu$^{2+}$ to TCA established a series of low-lying excited states involving copper d orbitals. It resulted in the overlap between TCA complexes absorption and emission of perylene. These indicated that the fluorescence quenching of perylene was due to the fluorescence resonance energy transfer (FRET).

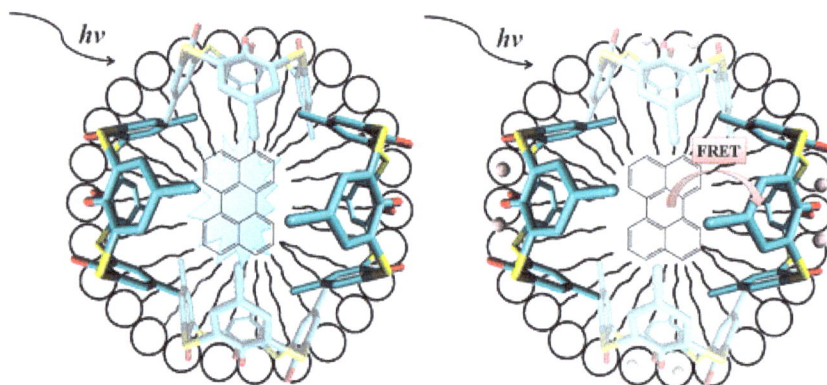

**Fig. (18).** Illustration of the Cu$^{2+}$ fluorescent recognition by TCA using TD-DFT [94].

## Anion Sensors

Excimer emission based fluorescent sensor for F$^-$ ion has been designed using two pyrene moieties as fluorophore and cone and 1,3-alternate conformation as ionophore incorporating calix [4]arene derivatives [95]. Triphenylamine unit containing calix [4]arene exhibited significant fluorescent quenching towards F⁻ ion [96]. A triazole-modified calix [4]arene derivative exhibited distinct sensing modes toward different anions [97]. Chawla *et al.* [98] reported selective recognition of CN⁻ ion by fluorescent calix [4]arenes (Fig. **19**). The *in situ* generated calix [4]arene-Cu²⁺ complex distinctly recognizes CN⁻ ion by distinct "off–on" behaviour over another anions. A bis-naphthalimidocalix [4]arene-Cu²⁺ supramolecular complex exhibited an eminent fluorescent recognition system for CN⁻ ions in an aqueous medium [99].

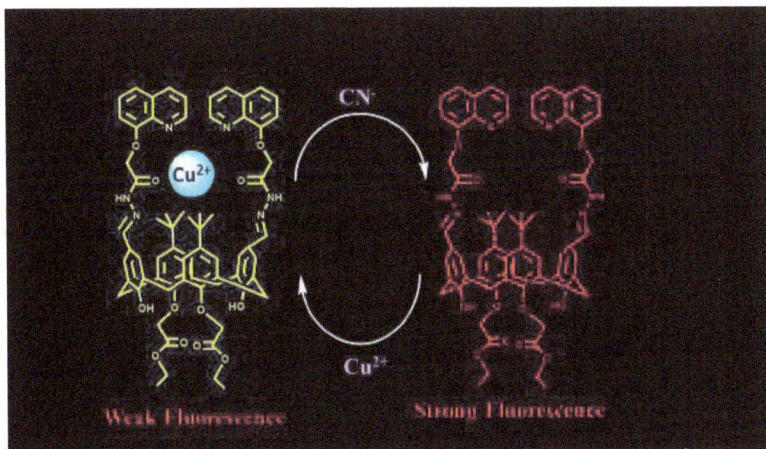

**Fig. (19).** Illustration of the selective recognition of both Cu²⁺ and CN⁻ ions through "on–off–on" fluorescence response by fluorescent calixarene [98].

## Solid State Sensor

A bimodal triazolyl-linked anththracenyl and 3-propylthio-acetate functionalized calix [4]arene formed an effective self-assembled monolayer on Au in a multi-arrayed microcantilever (Fig. **20**) instrument and showed a sensitive recognition of low concentrations of Hg²⁺ ion [100].

Li *et al.* [101] reported levodopa (L-DOPA) molecular sensor by thiolated calix [6]arene grafted on C-dots/Ir/Au. Calix [6]arene as a double recognition element in combination with a molecularly imprinted polymer (MIP) showed successful L-DOPA determination.

## MOLECULAR RECOGNITION

Zadmard *et al.* [102] reported the synthesis of a series of five calix [4]arene-hexaamide derivatives (Fig. **21**) and they exhibited considerable binding affinity towards β-lactoglobulin protein confirmed by fluorescence studies.

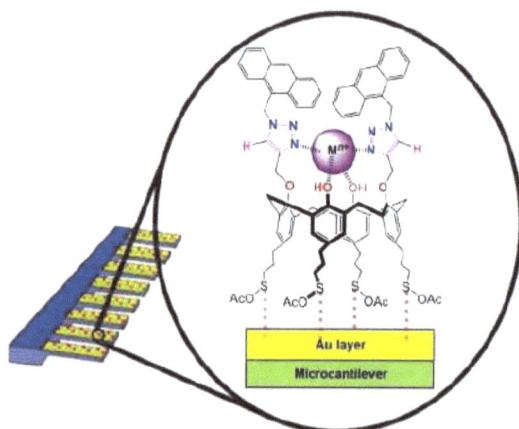

**Fig. (20).** Schematic representation of a microcantilever array consisting of 8 cantilevers. *Inset:* the chemical structure of calix [4]arene derivative along with the site of the binding of the target cation ($Hg^{2+}$) [100].

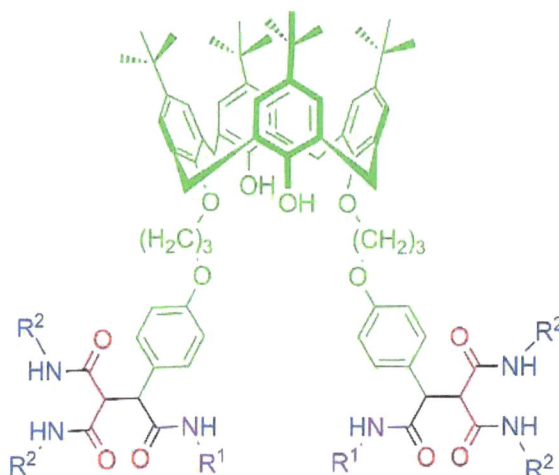

**Fig. (21).** Structure of calix [4]arene-hexaamide derivatives used as β-lactoglobulin fluorescent sensor [102].

A fluorescent "turn off-on" mechanism based scrupulous recognition of biologically important amino acids is attributed by thiacalix [4]arene tetrahydrazide reduced silver nanoparticles [103]. A bimodal fluorescent guanidinium calix [4]arene conjugate showed DNA binding ability and transfection of red fluorescent protein encoded plasmid in cancer cells (Fig. **22**). It demonstrated superior transfection ability without helper adjuvant in the MCF-7 cells. This advantage was due to lipophilic linker, conformational rigidity, and fluorescent moiety. It was reported as a promising cargo for the genetic material delivery into the biological cells [104].

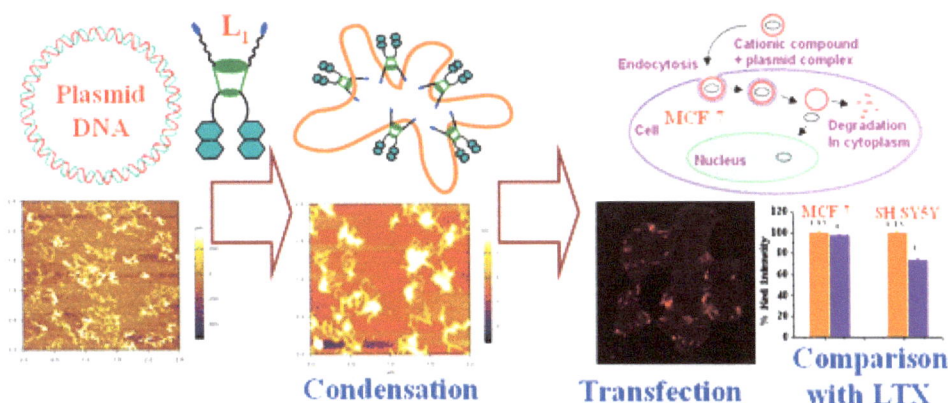

**Fig. (22).** Scheme of bi-functional calix [4]arene system suitable to transfect DNA [104].

Kaur *et al.* [105] developed a nano-aggregate-$Fe^{3+}$ complex based on dipodal 2-(--aminophenyl)benzimidazole scaffold located at the lower rim of calix [4]arene for amplified fluorescence detection of biomolecule adenosine diphosphate (ADP) in an aqueous medium. The fluorescence quenching with the addition of $Fe^{3+}$ ions could be attributed to the conglomeration of the nano-aggregates. They were separated or further assembled while ADP was added to the ensemble. This sensor exhibited excellent application in monitoring hydrolysis of ADP to AMP by apyrase, which led to its potential use for ADP-pertinent biological activities.

Ozyilmaz *et al.* [106] synthesized iron magnetic nanoparticles capped with fluorescent calixarene derivatives and encapsulated with a lipase. The encapsulated lipase on nanoparticles exhibited higher thermal and operational stability over the encapsulated lipase without calix [4]arene derivative. The encapsulated lipase nanoparticle with calixarene moieties showed an increase in activity, reusability, stability and enhanced kinetic resolution stereoselectivity. A fluorescent 1,3-di-naphthalimide conjugate of calix [4]arene exhibited trinitrophenol sensing property [107].

## LUMINESCENT LANTHANOID CALIXARENE COMPLEXES

In the last decade, numerous studies reported on the complexation of luminescent lanthanoids with calixarenes. Lanthanide compounds have many applications in materials and life science, exploiting their light-emitting and magnetic properties [108]. Their effective lanthanoid emission necessarily required sensitisation by an "antenna" ligand to overcome the low absorption cross section of lanthanide cation. The lower rim carbonyl substituted calix [4]arene was a well-established ionophoric motif. It provided the calixarene phenyl rings to use as the antenna by their $\pi\pi^*$ electronic transitions (Fig. **23**) [109]. Viana *et al.* [110] synthesized and investigated the photophysical properties of $Tb^{3+}$, $Eu^{3+}$ and $Gd^{3+}$ hybrids based on calix [4] arene-tetracarboxylate.

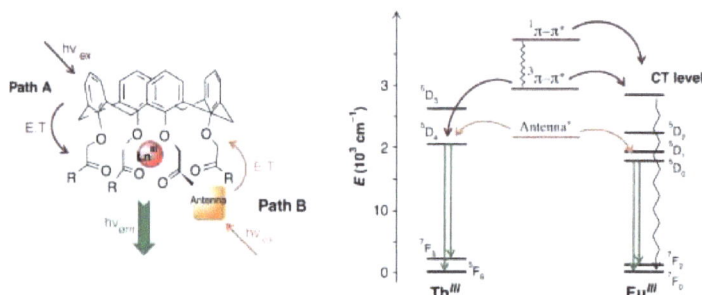

**Fig. (23).** A simple illustration of lanthanide emission sensitisation process in a typical lanthanoid calixarene complex [109].

The complex formation ability of bis-1,3-diketones towards $Ni^{2+}$, $Al^{3+}$, $Cu^{2+}$ and lanthanide ions ($Tb^{3+}$, $Nd^{3+}$, $Eu^{3+}$) has been investigated (Fig. **24**) by fluorescence spectroscopy [111]. A terbium-*p*-tert-butylthiacalix [4]arene coordination chain showed selective sensing of $Fe^{3+}$ and $Cr_2O_7^{2-}$ in an aqueous medium and nitroaromatic reagents in a DMF solution [112].

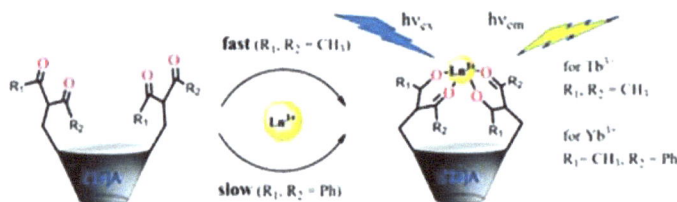

**Fig. (24).** Schematic representation of lanthanide ion binding bis-1,3-diketone calix [4]arenes [111].

Tetra(carboxymethyl)thiacalix [4]arene mixed with Tb/Dy coordination ladders exhibited a new pathway towards luminescent molecular nanomagnets [113].

## CONCLUSION

The facile modification of lower and upper rims of calixarene molecules has been utilized for sensing cations, anions and biologically important molecules. It could be extended for bio sensing and gas sensing applications. There is a huge scope in fabricating luminescent architecture by means of calixarene scaffolds. From the reports available, the researchers mainly focused on the calix [4]arene derivatives and there is a void for developing luminescent calixarene with a different number of aromatic units other than four. Extending these luminescent calixarenes into an aqueous medium is an effective one. In the future, luminescence property of these calixarenes could be utilized for the applications in organic light emitting diode, organic framework and energy devices.

## CONSENT OF PUBLICATION

The author grants the publisher the sole and exclusive license of the full copyright of this material.

## CONFLICT OF INTEREST

The author declares no conflict of interest, financial or otherwise.

## ACKNOWLEDGEMENTS

This research work was supported by the Department of Science and Technology (DST INSPIRE) [Project number – IFA14/CH-147], India. We also thank RUSA 2.0 (MHRD, India) grant sanctioned vide Letter No. F. 24-51/2014-U, Policy (TNMulti-Gen), Dept. of Edn. Govt. of India, Dt. 09.10.2018 for the additional financial support.

## REFERENCES

[1]   Böhmer, V. Calixarenes, Macrocycles with(Almost) Unlimited Possibilities. *Angew. Chem. Int. Ed. Engl.,* **1995,** *34*(7), 713-745.
       [http://dx.doi.org/10.1002/anie.199507131]

[2]   Neri, P.; Sessler, J.L.; Wang, M.X. *Calixarenes and Beyond*; Springer International Publishing: Cham, **2016**.
       [http://dx.doi.org/10.1007/978-3-319-31867-7]

[3]   Ashwin, B.C.M.A.; Shanmugavelan, P.; Muthu Mareeswaran, P. Electrochemical aspects of cyclodextrin, calixarene and cucurbituril inclusion complexes. *J. Incl. Phenom. Macrocycl. Chem.,* **2020,** *98*(3-4), 149-170.
       [http://dx.doi.org/10.1007/s10847-020-01028-4]

[4]   Gutsche, C.D.; Dhawan, B.; No, K.H.; Muthukrishnan, R. Calixarenes. 4. The synthesis, characterization, and properties of the calixarenes from p-tert-butylphenol. *J. Am. Chem. Soc.,* **1981,** *103*(13), 3782-3792.
       [http://dx.doi.org/10.1021/ja00403a028]

[5]   Español, E.; Villamil, M. Calixarenes: Generalities and Their Role in Improving the Solubility, Biocompatibility, Stability, Bioavailability, Detection, and Transport of Biomolecules. *Biomolecules,* **2019,** *9*(3), 90.
       [http://dx.doi.org/10.3390/biom9030090] [PMID: 30841659]

[6]   Rebek, J., Jr Host–guest chemistry of calixarene capsules. *Chem. Commun. (Camb.),* **2000,** (8), 637-643.
       [http://dx.doi.org/10.1039/a910339m]

[7]   Dong, H.; Zou, F.; Hu, X.; Zhu, H.; Koh, K.; Chen, H. Analyte induced AuNPs aggregation enhanced surface plasmon resonance for sensitive detection of paraquat. *Biosens. Bioelectron.,* **2018,** *117*, 605-612.
       [http://dx.doi.org/10.1016/j.bios.2018.06.057] [PMID: 30005380]

[8]    Mutihac, L.; Lee, J.H.; Kim, J.S.; Vicens, J. Recognition of amino acids by functionalized calixarenes. *Chem. Soc. Rev.,* **2011**, *40*(5), 2777-2796.
[http://dx.doi.org/10.1039/c0cs00005a] [PMID: 21321724]

[9]    Shinkai, S.; Araki, K.; Manabe, O. Does the calixarene cavity recognise the size of guest molecules? On the 'hole-size selectivity' in water-soluble calixarenes. *J. Chem. Soc. Chem. Commun.,* **1988**, (3), 187-189.
[http://dx.doi.org/10.1039/C39880000187]

[10]   Nimse, S.B.; Kim, T. Biological applications of functionalized calixarenes. *Chem. Soc. Rev.,* **2013**, *42*(1), 366-386.
[http://dx.doi.org/10.1039/C2CS35233H] [PMID: 23032718]

[11]   Wang, X.; Du, D.; Dong, H.; Song, S.; Koh, K.; Chen, H. para-Sulfonatocalix[4]arene stabilized gold nanoparticles multilayers interfaced to electrodes through host-guest interaction for sensitive ErbB2 detection. *Biosens. Bioelectron.,* **2018**, *99*, 375-381.
[http://dx.doi.org/10.1016/j.bios.2017.08.011] [PMID: 28802750]

[12]   Konovalov, A.I.; Antipin, I.S. Supramolecular systems based on calixarenes. *Mendeleev Commun.,* **2008**, *18*(5), 229-237.
[http://dx.doi.org/10.1016/j.mencom.2008.09.001]

[13]   Grady, T.; Harris, S.J.; Smyth, M.R.; Diamond, D.; Hailey, P. Determination of the enantiomeric composition of chiral amines based on the quenching of the fluorescence of a chiral calixarene. *Anal. Chem.,* **1996**, *68*(21), 3775-3782.
[http://dx.doi.org/10.1021/ac960383c] [PMID: 21619250]

[14]   Ashwin, B.C.M.A.; Vinothini, A.; Stalin, T.; Muthu Mareeswaran, P. Synthesis of a Safranin T - *p* - Sulfonatocalix[4]arene Complex by Means of Supramolecular Complexation. *ChemistrySelect,* **2017**, *2*(3), 931-936.
[http://dx.doi.org/10.1002/slct.201601939]

[15]   Guo, D.S.; Liu, Y. Supramolecular chemistry of p-sulfonatocalix[n]arenes and its biological applications. *Acc. Chem. Res.,* **2014**, *47*(7), 1925-1934.
[http://dx.doi.org/10.1021/ar500009g] [PMID: 24666259]

[16]   Muthu Mareeswaran, P.; Prakash, M.; Subramanian, V.; Rajagopal, S. Recognition of aromatic amino acids and proteins with *p* -sulfonatocalix[4]arene - A luminescence and theoretical approach. *J. Phys. Org. Chem.,* **2012**, *25*(12), 1217-1227.
[http://dx.doi.org/10.1002/poc.2996]

[17]   Song, S.; Shang, X.; Zhao, J. *et al.* Sensitive and selective determination of caspase-3 based on calixarene functionalized reduction of graphene oxide assisted signal amplification. *Sens. Actuators B Chem.,* **2018**, *267*, 357-365.
[http://dx.doi.org/10.1016/j.snb.2018.03.185]

[18]   Muthu Mareeswaran, P.; Babu, E.; Sathish, V.; Kim, B.; Woo, S.I.; Rajagopal, S. p-Sulfonatocalix[4]arene as a carrier for curcumin. *New J. Chem.,* **2014**, *38*(3), 1336-1345.
[http://dx.doi.org/10.1039/c3nj00935a]

[19]   Madasamy, K.; Gopi, S.; Kumaran, M.S.; Radhakrishnan, S.; Velayutham, D.; Mareeswaran, P.M.; Kathiresan, M. A Supramolecular Investigation on the Interactions between Ethyl terminated Bis-viologen Derivatives with Sulfonato Calix[4]arenes. *ChemistrySelect,* **2017**, *2*(3), 1175-1182.
[http://dx.doi.org/10.1002/slct.201601818]

[20]   Saravanan, C.; Senthilkumaran, M.; Ashwin, B.C.M.A.; Suresh, P.; Muthu Mareeswaran, P. Spectral and electrochemical investigation of 1,8-diaminonaphthalene upon encapsulation of p-sulfonatocalix[4]arene. *J. Incl. Phenom. Macrocycl. Chem.,* **2017**, *88*(3-4), 239-246.
[http://dx.doi.org/10.1007/s10847-017-0729-1]

[21]   Ashwin, B.C.M.A.; Chitumalla, R.K.; Herculin Arun Baby, A.; Jang, J.; Muthu Mareeswaran, P.

Spectral, electrochemical and computational investigations on the host–guest interaction of Coumarin-460 with p-sulfonatocalix[4]arene. *J. Incl. Phenom. Macrocycl. Chem.,* **2018**, *90*(1-2), 51-60.
[http://dx.doi.org/10.1007/s10847-017-0762-0]

[22]  Ashwin, B.C.M.A.; Herculin Arun Baby, A.; Prakash, M.; Hochlaf, M.; Muthu Mareeswaran, P. A combined experimental and theoretical study on *p*- sulfonatocalix[4]arene encapsulated 7-methoxy coumarin. *J. Phys. Org. Chem.,* **2018**, *31*(4), e3788.
[http://dx.doi.org/10.1002/poc.3788]

[23]  Saravanan, C.; Chitumalla, R.K.; Ashwin, B.C.M.A.; Senthilkumaran, M.; Suresh, P.; Jang, J.; Muthu Mareeswaran, P. Effectual binding of gallic acid with p-sulfonatocalix[4]arene: An experimental and theoretical interpretation. *J. Lumin.,* **2018**, *196*, 392-398.
[http://dx.doi.org/10.1016/j.jlumin.2017.12.063]

[24]  Senthilkumaran, M.; Maruthanayagam, K.; Vigneshkumar, G.; Chitumalla, R.K.; Jang, J.; Muthu Mareeswaran, P. Spectral, Electrochemical and Computational Investigations of Binding of n-(--Hydroxyphenyl)-imidazole with p-Sulfonatocalix[4]arene. *J. Fluoresc.,* **2017**, *27*(6), 2159-2168.
[http://dx.doi.org/10.1007/s10895-017-2155-6] [PMID: 28887595]

[25]  Ashwin, B.C.M.A.; Saravanan, C.; Senthilkumaran, M.; Sumathi, R.; Suresh, P.; Muthu Mareeswaran, P. Spectral and electrochemical investigation of *p* -sulfonatocalix[4]arene-stabilized vitamin E aggregation. *Supramol. Chem.,* **2018**, *30*(1), 32-41.
[http://dx.doi.org/10.1080/10610278.2017.1351612]

[26]  Saravanan, C.; Ashwin, B.C.M.A.; Senthilkumaran, M.; Mareeswaran, P.M. Supramolecular Complexation of Biologically Important Thioflavin-T with p-Sulfonatocalix[4]arene. *ChemistrySelect,* **2018**, *3*(9), 2528-2535.
[http://dx.doi.org/10.1002/slct.201702937]

[27]  Senthilkumaran, M.; Chitumalla, R.K.; Vigneshkumar, G.; Rajkumar, E.; Muthu Mareeswaran, P.; Jang, J. Investigation of the upper rim binding of triphenylpyrylium cation with p-sulfonatocalix[4]arene. *J. Incl. Phenom. Macrocycl. Chem.,* **2018**, *91*(3-4), 161-169.
[http://dx.doi.org/10.1007/s10847-018-0809-x]

[28]  Saravanan, C.; Kumar, R.; Yuvakumar, R.; Shanmugavelan, P. Experimental and Theoretical Investigations on the Host - Guest Interaction of Diphenylamine with p - Sulfonatocalix [ 4 ] Arene. *Indian J. Chem.,* **2020**, *59*, 929-938.

[29]  Senthilkumaran, M.; Saravanan, C.; Ashwin, B.C.M.A.; Shanmugavelan, P.; Muthu Mareeswaran, P.; Prakash, M. Inclusion induced water solubility and binding investigation of acenaphthene-1,2-dione with p-sulfonatocalix[4]arene. *J. Incl. Phenom. Macrocycl. Chem.,* **2020**, *98*(1-2), 105-115.
[http://dx.doi.org/10.1007/s10847-020-01017-7]

[30]  Oueslati, I.; Coleman, A.W.; de Castro, B.; Berberan-Santos, M.N. A fluorescent and phosphorescent nanoporous solid: crystalline calix[4]arene. *J. Fluoresc.,* **2008**, *18*(6), 1123-1129.
[http://dx.doi.org/10.1007/s10895-008-0362-x] [PMID: 18437542]

[31]  Zhu, J.; Chen, L.; Guo, X. Distinctive spectroscopic properties and adsorption behaviors of p-sulfonatocalixarene-cetyltrimethylammonium bromide supra-amphiphilic systems. *Colloids Surf. A Physicochem. Eng. Asp.,* **2020**, *601*, 125029.
[http://dx.doi.org/10.1016/j.colsurfa.2020.125029]

[32]  Muthu Mareeswaran, P.; Babu, E.; Rajagopal, S. Optical recognition of anions by ruthenium(II)-bipyridine-calix[4]arene system. *J. Fluoresc.,* **2013**, *23*(5), 997-1006.
[http://dx.doi.org/10.1007/s10895-013-1226-6] [PMID: 23715935]

[33]  Ashwin, B.C.M.A.; Mareeswaran, P.M. Grafting of Calixarene on Ruthenium(II)bipyridine Doped Luminescent Silica Nanoparticle. *Mater. Today Proc.,* **2019**, *14*, 379-385.
[http://dx.doi.org/10.1016/j.matpr.2019.04.160]

[34]  Ashwin, B.C.M.A.; Saravanan, C.; Stalin, T.; Muthu Mareeswaran, P.; Rajagopal, S. FRET-based Solid-state Luminescent Glyphosate Sensor Using Calixarene-grafted Ruthenium(II)bipyridine Doped

Silica Nanoparticles. *ChemPhysChem,* **2018**, *19*(20), 2768-2775.
[http://dx.doi.org/10.1002/cphc.201800447] [PMID: 29989285]

[35]   Kim, J.S.; Noh, K.H.; Lee, S.H.; Kim, S.K.; Kim, S.K.; Yoon, J. Molecular taekwondo. 2. A new calix[4]azacrown bearing two different binding sites as a new fluorescent ionophore. *J. Org. Chem.,* **2003**, *68*(2), 597-600.
[http://dx.doi.org/10.1021/jo020538i] [PMID: 12530890]

[36]   Zhang, N.; Tang, S.H.; Liu, Y. Luminescence behavior of a water soluble calix[4]arene derivative complex with terbium ion(III) in gelation solution. *Spectrochim. Acta A Mol. Biomol. Spectrosc.,* **2003**, *59*(5), 1107-1112.
[http://dx.doi.org/10.1016/S1386-1425(02)00288-3] [PMID: 12633728]

[37]   Kim, J.S.; Quang, D.T. Calixarene-derived fluorescent probes. *Chem. Rev.,* **2007**, *107*(9), 3780-3799.
[http://dx.doi.org/10.1021/cr068046j] [PMID: 17711335]

[38]   Chang, K.C.; Su, I.H.; Senthilvelan, A.; Chung, W.S. Triazole-modified calix[4]crown as a novel fluorescent on-off switchable chemosensor. *Org. Lett.,* **2007**, *9*(17), 3363-3366.
[http://dx.doi.org/10.1021/ol071337+] [PMID: 17650010]

[39]   Dhir, A.; Bhalla, V.; Kumar, M. Ratiometric sensing of Hg2+ based on the calix[4]arene of partial cone conformation possessing a dansyl moiety. *Org. Lett.,* **2008**, *10*(21), 4891-4894.
[http://dx.doi.org/10.1021/ol801984y] [PMID: 18831557]

[40]   Ocak, Ü.; Ocak, M.; Surowiec, K.; Liu, X.; Bartsch, R.A. Metal ion complexation in acetonitrile by upper-rim allyl-substituted, di-ionized calix[4]arenes bearing two dansyl fluorophores. *Tetrahedron,* **2009**, *65*(34), 7038-7047.
[http://dx.doi.org/10.1016/j.tet.2009.06.038]

[41]   Ocak, Ű.; Ocak, M.; Shen, X.; Gorman, A.H.; Surowiec, K.; Bartsch, R.A. Metal ion complexation in acetonitrile by upper-rim benzyl-substituted, di-ionized calix[4]arenes bearing two dansyl fluorophores. *ARKIVOC,* **2010**, *2010*(7), 81-97.
[http://dx.doi.org/10.3998/ark.5550190.0011.707]

[42]   Joseph, R.; Ramanujam, B.; Acharya, A.; Khutia, A.; Rao, C.P. Experimental and computational studies of selective recognition of Hg2+ by amide linked lower rim 1,3-dibenzimidazole derivative of calix[4]arene: species characterization in solution and that in the isolated complex, including the delineation of the nanostructures. *J. Org. Chem.,* **2008**, *73*(15), 5745-5758.
[http://dx.doi.org/10.1021/jo800073g] [PMID: 18590337]

[43]   Park, S.Y.; Yoon, J.H.; Hong, C.S. *et al.* A pyrenyl-appended triazole-based calix[4]arene as a fluorescent sensor for Cd2+ and Zn2+. *J. Org. Chem.,* **2008**, *73*(21), 8212-8218.
[http://dx.doi.org/10.1021/jo8012918] [PMID: 18817447]

[44]   Ho, I.T.; Haung, K.C.; Chung, W.S. 1,3-Alternate calix[4]arene as a homobinuclear ditopic fluorescent chemosensor for Ag+ ions. *Chem. Asian J.,* **2011**, *6*(10), 2738-2746.
[http://dx.doi.org/10.1002/asia.201100023] [PMID: 21506282]

[45]   Zhu, L.N.; Gong, S.L.; Gong, S.L. *et al.* Novel Pyrene-armed Calix[4]arenes through Triazole Connection: Ratiometric Fluorescent Chemosensor for Zn $^{2+}$ and Promising Structure for Integrated Logic Gates. *Chin. J. Chem.,* **2008**, *26*(8), 1424-1430.
[http://dx.doi.org/10.1002/cjoc.200890259]

[46]   Ni, X.L.; Zeng, X.; Redshaw, C.; Yamato, T. Synthesis and evaluation of a novel pyrenyl-appended triazole-based thiacalix[4]arene as a fluorescent sensor for Ag+ ion. *Tetrahedron,* **2011**, *67*(18), 3248-3253.
[http://dx.doi.org/10.1016/j.tet.2011.03.008]

[47]   Wang, N.J.; Sun, C.M.; Chung, W.S. A highly selective fluorescent chemosensor for Ag+ based on calix[4]arene with lower-rim proximal triazolylpyrenes. *Sens. Actuators B Chem.,* **2012**, *171-172*, 984-993.
[http://dx.doi.org/10.1016/j.snb.2012.06.014]

[48]   Ni, X.L.; Zeng, X.; Hughes, D.L.; Redshaw, C.; Yamato, T. Synthesis, crystal structure and complexation behaviour of a thiacalix[4]arene bearing 1,2,3-triazole groups. *Supramol. Chem.*, **2011**, *23*(10), 689-695.
       [http://dx.doi.org/10.1080/10610278.2011.601603]

[49]   Han, J.; Wang, F.L.; Liu, Y.X.; Zhang, F.Y.; Meng, J-B.; He, Z.J. Calix[4]arene-Based 1,3,4-Oxadiazoles: Novel Fluorescent Chemosensors for Specific Recognition of Cu2+. *ChemPlusChem,* **2012**, *77*(3), 196-200.
       [http://dx.doi.org/10.1002/cplu.201200004]

[50]   Xie, D.H.; Wang, X.J.; Sun, C.; Han, J. Calix[4]arene based 1,3,4-oxadiazole as a fluorescent chemosensor for copper(II) ion detection. *Tetrahedron Lett.*, **2016**, *57*(51), 5834-5836.
       [http://dx.doi.org/10.1016/j.tetlet.2016.11.051]

[51]   Zhao, J.L.; Tomiyasu, H.; Wu, C.; Cong, H.; Zeng, X.; Rahman, S.; Georghiou, P.E.; Hughes, D.L.; Redshaw, C.; Yamato, T. Synthesis, crystal structure and complexation behaviour study of an efficient Cu2+ ratiometric fluorescent chemosensor based on thiacalix[4]arene. *Tetrahedron*, **2015**, *71*(45), 8521-8527.
       [http://dx.doi.org/10.1016/j.tet.2015.09.038]

[52]   Kumar, M.; Kumar, N.; Bhalla, V. Thiacalix[4]arene-cinnamaldehyde derivative: ICT-induced preferential nanomolar detection of Ag+ among different transition metal ions. *Org. Biomol. Chem.,* **2012**, *10*(9), 1769-1774.
       [http://dx.doi.org/10.1039/c1ob06294h] [PMID: 22234392]

[53]   Pathak, R.K.; Hinge, V.K.; Mondal, M.; Rao, C.P. Triazole-linked-thiophene conjugate of calix[4]arene: its selective recognition of Zn2+ and as biomimetic model in supporting the events of the metal detoxification and oxidative stress involving metallothionein. *J. Org. Chem.,* **2011**, *76*(24), 10039-10049.
       [http://dx.doi.org/10.1021/jo201865x] [PMID: 22053726]

[54]   Kumar Pathak, R.; Kumar Hinge, V.; Rai, A.; Panda, D.; Pulla Rao, C. Imino-phenolic-pyridyl conjugates of calix[4]arene (L1 and L2) as primary fluorescence switch-on sensors for Zn2+ in solution and in HeLa cells and the recognition of pyrophosphate and ATP by [ZnL2]. *Inorg. Chem.,* **2012**, *51*(9), 4994-5005.
       [http://dx.doi.org/10.1021/ic202426v] [PMID: 22519733]

[55]   Ni, X.L.; Wang, S.; Zeng, X.; Tao, Z.; Yamato, T. Pyrene-linked triazole-modified homooxacalix[3]arene: a unique C3 symmetry ratiometric fluorescent chemosensor for Pb2+. *Org. Lett.,* **2011**, *13*(4), 552-555.
       [http://dx.doi.org/10.1021/ol102914t] [PMID: 21194233]

[56]   Zheng, X.; Zhang, W.; Mu, L.; Zeng, X.; Xue, S.; Tao, Z.; Yamatob, T. A novel rhodamine-based thiacalix[4]arene fluorescent sensor for Fe3+ and Cr3+. *J. Incl. Phenom. Macrocycl. Chem.,* **2010**, *68*(1-2), 139-146.
       [http://dx.doi.org/10.1007/s10847-010-9759-7]

[57]   Zhan, J.; Fang, F.; Tian, D.; Li, H. Anthraquinone-modified calix[4]arene: click synthesis, selective calcium ion fluorescent chemosensor and INHIBIT logic gate. *Supramol. Chem.,* **2012**, *24*(4), 272-278.
       [http://dx.doi.org/10.1080/10610278.2012.656124]

[58]   Zhan, J.; Wen, L.; Miao, F.; Tian, D.; Zhu, X.; Li, H. Synthesis of a pyridyl-appended calix[4]arene and its application to the modification of silver nanoparticles as an Fe [3+] colorimetric sensor. *New J. Chem.,* **2012**, *36*(3), 656-661.
       [http://dx.doi.org/10.1039/C2NJ20776A]

[59]   Patra, S.; Lo, R.; Chakraborty, A.; Gunupuru, R.; Maity, D.; Ganguly, B.; Paul, P. Calix[4]arene based fluorescent chemosensor bearing coumarin as fluorogenic unit: Synthesis, characterization, ion-binding property and molecular modeling. *Polyhedron,* **2013**, *50*(1), 592-601.

[http://dx.doi.org/10.1016/j.poly.2012.12.002]

[60] Chawla, H.M.; Goel, P.; Shukla, R. Calix[4]arene based molecular probe for sensing copper ions. *Tetrahedron Lett.,* **2014**, *55*(14), 2173-2176. [http://dx.doi.org/10.1016/j.tetlet.2014.02.021]

[61] Cho, J.; Pradhan, T.; Kim, J.S.; Kim, S. Bimodal calix[2]triazole[2]arene fluorescent ionophore. *Org. Lett.,* **2013**, *15*(16), 4058-4061. [http://dx.doi.org/10.1021/ol401469z] [PMID: 23915313]

[62] Cho, J.; Pradhan, T.; Lee, Y.M.; Kim, J.S.; Kim, S. A calix[2]triazole[2]arene-based fluorescent chemosensor for probing the copper trafficking pathway in Wilson's disease. *Dalton Trans.,* **2014**, *43*(43), 16178-16182. [http://dx.doi.org/10.1039/C4DT02208D] [PMID: 25260151]

[63] Sahin, O.; Akceylan, E. A phenanthrene-based calix[4]arene as a fluorescent sensor for Cu2+ and F−. *Tetrahedron,* **2014**, *70*(39), 6944-6950. [http://dx.doi.org/10.1016/j.tet.2014.07.100]

[64] Song, M.; Sun, Z.; Han, C.; Tian, D.; Li, H.; Kim, J.S. Calixarene-based chemosensors by means of click chemistry. *Chem. Asian J.,* **2014**, *9*(9), 2344-2357. [http://dx.doi.org/10.1002/asia.201400024] [PMID: 24898975]

[65] Jin, C.C.; Kinoshita, T.; Cong, H.; Ni, X-L.; Zeng, X.; Hughes, D.L.; Redshaw, C.; Yamato, T. Synthesis and inclusion properties of C3-symmetric triazole derivatives based on hexahomotrioxacalix[3]arene. *New J. Chem.,* **2012**, *36*(12), 2580-2586. [http://dx.doi.org/10.1039/c2nj40599g]

[66] Chen, Y.J.; Yang, S.C.; Tsai, C.C.; Chang, K.C.; Chuang, W.H.; Chu, W.L.; Kovalev, V.; Chung, W.S. Anthryl-1,2,4-oxadiazole-substituted calix[4]arenes as highly selective fluorescent chemodosimeters for Fe(3+). *Chem. Asian J.,* **2015**, *10*(4), 1025-1034. [http://dx.doi.org/10.1002/asia.201403265] [PMID: 25620418]

[67] Ling, I.; Hashim, R.; Sabah, K.J. Sugar thiacrown-ether appended calix[4]arene as a selective chemosensor for Fe $^{2+}$ and Fe $^{3+}$ ions. *RSC Advances,* **2015**, *5*(107), 88038-88044. [http://dx.doi.org/10.1039/C5RA15448K]

[68] Wu, Y.; Ni, X.L.; Mou, L.; Jin, C.C.; Redshaw, C.; Yamato, T. Synthesis of a ditopic homooxacalix[3]arene for fluorescence enhanced detection of heavy and transition metal ions. *Supramol. Chem.,* **2015**, *27*(7-8), 501-507. [http://dx.doi.org/10.1080/10610278.2014.1002841]

[69] Jiang, X.K.; Ikejiri, Y.; Jin, C.C. *et al.* Synthesis and evaluation of a novel fluorescent sensor based on hexahomotrioxacalix[3]arene for Zn2+ and Cd2+. *Tetrahedron,* **2016**, *72*(32), 4854-4858. [http://dx.doi.org/10.1016/j.tet.2016.06.055]

[70] Ullmann, S.; Schnorr, R.; Handke, M.; Laube, C.; Abel, B.; Matysik, J.; Findeisen, M.; Rüger, R.; Heine, T.; Kersting, B. Zn $^{2+}$ -Ion Sensing by Fluorescent Schiff Base Calix[4]arene Macrocycles. *Chemistry,* **2017**, *23*(16), 3824-3827. [http://dx.doi.org/10.1002/chem.201700253] [PMID: 28195665]

[71] Nag, R.; Vashishtha, M.; Rao, C.P. Switching the Ion Selectivity from Fe $^{3+}$ to Al $^{3+}$ by a Triazole-Appended Calix[4]arene-Based Amphiphile †. *ChemistrySelect,* **2018**, *3*(4), 1248-1256. [http://dx.doi.org/10.1002/slct.201702999]

[72] Anandababu, A.; Anandan, S.; Ashokkumar, M. A Simple Discriminating p-tert-Butylcalix[4]arene Thiospirolactam Rhodamine B Based Colorimetric and Fluorescence Sensor for Mercury Ion and Live Cell Imaging Applications. *ChemistrySelect,* **2018**, *3*(16), 4413-4420. [http://dx.doi.org/10.1002/slct.201800044]

[73] Rahman, S.; Assiri, Y.; Alodhayb, A.N.; Beaulieu, L.Y.; Oraby, A.K.; Georghiou, P.E. Naphthyl "capped" triazole-linked calix[4]arene hosts as fluorescent chemosensors towards Fe $^{3+}$ and Hg $^{2+}$ : an

experimental and DFT computational study. *New J. Chem.,* **2016**, *40*(1), 434-440.
[http://dx.doi.org/10.1039/C5NJ01362C]

[74]   Aljabri, M.D.; Rahman, S.; Georghiou, P.E. Synthesis, Complexation and DFT Computational Studies of Bis(naphthyl)methane-"Capped" Triazole-Linked Calix[4]arenes as Fe $^{3+}$ Fluorescent Chemosensors. *ChemistrySelect,* **2017**, *2*(3), 1214-1218.
[http://dx.doi.org/10.1002/slct.201601923]

[75]   Georghiou, P.E.; Rahman, S.; Alrawashdeh, A. *et al.* Synthesis, supramolecular complexation and DFTstudies of a bis(pyrene)-appended 'capped' triazole-linked calix[4]arene as Zn $^{2+}$ and Cd $^{2+}$ fluorescent chemosensors. *Supramol. Chem.,* **2020**, *32*(5), 325-333.
[http://dx.doi.org/10.1080/10610278.2020.1739686]

[76]   Zadmard, R.; Akbari-Moghaddam, P.; Darvishi, S. Calix[4]arene-based crab-like molecular sensors for highly selective detection of mercury and copper ions. *Supramol. Chem.,* **2017**, *29*(1), 17-23.
[http://dx.doi.org/10.1080/10610278.2016.1161195]

[77]   Zadmard, R.; Akbarzadeh, A.; Jalali, M.R. Highly functionalized calix[4]arenes *via* multicomponent reactions: synthesis and recognition properties. *RSC Advances,* **2019**, *9*(34), 19596-19605.
[http://dx.doi.org/10.1039/C9RA03354H]

[78]   Akbarzadeh, A.; Zadmard, R.; Jalali, M.R. Synthesis of novel 6-piperidin-1-ylpyrimidine-2,4-diamine 3-oxide substituted calix[4]arene as a highly selective and sensitive fluorescent sensor for Cu2+ in aqueous samples. *Tetrahedron Lett.,* **2020**, *61*(13), 151658.
[http://dx.doi.org/10.1016/j.tetlet.2020.151658]

[79]   Sulak, M.; Kursunlu, A. N.; Girgin, B.; Karakuş, Ö. Ö.; Güler, E. **2017**.

[80]   Modi, K.; Panchal, U.; Mehta, V.; Panchal, M.; Kongor, A.; Jain, V.K. Propyl phthalimide-modified thiacalixphenyl[4]arene as a "turn on" chemosensor for Hg(II) ions. *J. Lumin.,* **2016**, *179*, 378-383.
[http://dx.doi.org/10.1016/j.jlumin.2016.07.019]

[81]   Bhatti, A.A.; Oguz, M.; Yilmaz, M. New water soluble p-sulphonatocalix[4]arene chemosensor appended with rhodamine for selective detection of Hg2+ ion. *J. Mol. Struct.,* **2020**, *1203*, 127436.
[http://dx.doi.org/10.1016/j.molstruc.2019.127436]

[82]   Yilmaz, B.; Keskinates, M.; Bayrakci, M. Novel integrated sensing system of calixarene and rhodamine molecules for selective colorimetric and fluorometric detection of Hg $^{2+}$ ions in living cells. *Spectrochim. Acta A Mol. Biomol. Spectrosc.,* **2021**, *245*, 118904.
[http://dx.doi.org/10.1016/j.saa.2020.118904] [PMID: 32932034]

[83]   Chen, Y.J.; Chen, M.Y.; Lee, K.T.; Shen, L.C.; Hung, H.C.; Niu, H.C.; Chung, W.S. 1,3-Alternate Calix[4]arene Functionalized With Pyrazole and Triazole Ligands as a Highly Selective Fluorescent Sensor for Hg $^{2+}$ and Ag $^{+}$ Ions. *Front Chem.,* **2020**, *8*, 593261.
[http://dx.doi.org/10.3389/fchem.2020.593261] [PMID: 33282834]

[84]   Lee, M.; Cho, D.; Kim, I.; Lee, J.; Lee, J.Y.; Satheeshkumar, C.; Song, C. Cooperative Binding of Metal Cations to a Spiropyran-Conjugated Calix[4]arene. *ChemistrySelect,* **2017**, *2*(12), 3527-3533.
[http://dx.doi.org/10.1002/slct.201700222]

[85]   Alizada, M.; Gul, A.; Oguz, M.; Kursunlu, A.N.; Yilmaz, M. Ion sensing of sister sensors based-on calix[4]arene in aqueous medium and their bioimaging applications. *Dyes Pigments,* **2021**, *184*, 108741.
[http://dx.doi.org/10.1016/j.dyepig.2020.108741]

[86]   Sutariya, P.G.; Soni, H.; Gandhi, S.A.; Pandya, A. Luminescent behavior of pyrene-allied calix[4]arene for the highly pH-selective recognition and determination of Zn $^{2+}$, Hg $^{2+}$ and I $^{-}$*via* the CHEF-PET mechanism: computational experiment and paper-based device. *New J. Chem.,* **2019**, *43*(25), 9855-9864.
[http://dx.doi.org/10.1039/C9NJ01388A]

[87]   Sutariya, P.G.; Soni, H.; Gandhi, S.A.; Pandya, A. Novel tritopic calix[4]arene CHEF-PET

fluorescence paper based probe for La3+, Cu2+, and Br−: Its computational investigation and application to real samples. *J. Lumin.,* **2019**, *212*, 171-179.
[http://dx.doi.org/10.1016/j.jlumin.2019.04.029]

[88]   Sutariya, P.G.; Soni, H.; Gandhi, S.A.; Pandya, A. Novel luminescent paper based calix[4]arene chelation enhanced fluorescence- photoinduced electron transfer probe for Mn2+, Cr3+ and F-. *J. Lumin.,* **2019**, *208*, 6-17.
[http://dx.doi.org/10.1016/j.jlumin.2018.12.009]

[89]   Sutariya, P.G.; Soni, H.; Gandhi, S.A.; Pandya, A. Turn on fluorescence strip based sensor for recognition of Sr$^{2+}$ and CN$^-$ via lowerrim substituted calix[4]arene and its computational investigation. *Spectrochim. Acta A Mol. Biomol. Spectrosc.,* **2020**, *238*, 118456.
[http://dx.doi.org/10.1016/j.saa.2020.118456] [PMID: 32417642]

[90]   Guo, Z.; Li, Y.; Ding, C.; Hu, X.; Wen, Y.; Wang, B. A novel Pb2+-selective micellar self-assembled fluorescent chemosensor based on amino thiadiazole calix[4]arene derivative. *Inorg. Chem. Commun.,* **2019**, *103*, 43-46.
[http://dx.doi.org/10.1016/j.inoche.2019.03.009]

[91]   Nag, R.; Polepalli, S.; Althaf Hussain, M.; Rao, C.P. Ratiometric Cu $^{2-}$ Binding, Cell Imaging, Mitochondrial Targeting, and Anticancer Activity with Nanomolar IC $_{50}$ by Spiro-Indoline-Conjugated Calix[4]arene. *ACS Omega,* **2019**, *4*(8), 13231-13240.
[http://dx.doi.org/10.1021/acsomega.9b01402] [PMID: 31460450]

[92]   Costa, A.I.; Barata, P.D.; Fialho, C.B.; Prata, J.V. Highly Sensitive and Selective Fluorescent Probes for Cu(II) Detection Based on Calix[4]arene-Oxacyclophane Architectures. *Molecules,* **2020**, *25*(10), 2456.
[http://dx.doi.org/10.3390/molecules25102456] [PMID: 32466180]

[93]   Jin, J.; Park, J.Y.; Lee, Y.S. Optical Nature and Binding Energetics of Fluorescent Fluoride Sensor Bis(bora)calix[4]arene and Design Strategies of Its Homologues. *J. Phys. Chem. C,* **2016**, *120*(42), 24324-24334.
[http://dx.doi.org/10.1021/acs.jpcc.6b06729]

[94]   Chen, S.; Hu, X.; Li, Y. TD-DFT Study on Thiacalix[4]arene, the Receptor of a Fluorescent Chemosensor for Cu $^{2+}$. *J. Phys. Chem. A,* **2017**, *121*(37), 6942-6948.
[http://dx.doi.org/10.1021/acs.jpca.7b05733] [PMID: 28841318]

[95]   Maity, D.; Chakraborty, A.; Gunupuru, R.; Paul, P. Calix[4]arene based molecular sensors with pyrene as fluorogenic unit: Effect of solvent in ion selectivity and colorimetric detection of fluoride. *Inorg. Chim. Acta,* **2011**, *372*(1), 126-135.
[http://dx.doi.org/10.1016/j.ica.2011.01.053]

[96]   Erdemir, S.; Malkondu, S.; Kocyigit, O.; Alıcı, O. A novel colorimetric and fluorescent sensor based on calix[4]arene possessing triphenylamine units. *Spectrochim. Acta A Mol. Biomol. Spectrosc.,* **2013**, *114*, 190-196.
[http://dx.doi.org/10.1016/j.saa.2013.05.069] [PMID: 23770508]

[97]   Hung, H.C.; Chang, Y.Y.; Luo, L.; Hung, C.H.; Diau, E.W.G.; Chung, W.S. Different sensing modes of fluoride and acetate based on a calix[4]arene with 25,27-bistriazolylmethylpyrenylacetamides. *Photochem. Photobiol. Sci.,* **2014**, *13*(2), 370-379.
[http://dx.doi.org/10.1039/c3pp50175b] [PMID: 24385051]

[98]   Chawla, H.M.; Shukla, R.; Goel, P. Sensitive recognition of cyanide through supramolecularly complexed new calix[4]arenes. *New J. Chem.,* **2014**, *38*(11), 5264-5267.
[http://dx.doi.org/10.1039/C4NJ00065J]

[99]   Shahid, M.; Chawla, H.M.; Bhatia, P. Novel calix[4]arene based metallo-supramolecular complex for recognition of cyanide ions in aqueous medium. *Supramol. Chem.,* **2017**, *29*(4), 290-295.
[http://dx.doi.org/10.1080/10610278.2016.1222074]

[100]  Alodhayb, A.N.; Braim, M.; Beaulieu, L.Y.; Valluru, G.; Rahman, S.; Oraby, A.K.; Georghiou, P.E.

Metal ion binding properties of a bimodal triazolyl-functionalized calix[4]arene on a multi-array microcantilever system. Synthesis, fluorescence and DFT computation studies. *RSC Advances,* **2016**, *6*(6), 4387-4396.
[http://dx.doi.org/10.1039/C5RA12685A]

[101] Li, S.; Pang, C.; Ma, X.; Zhao, M.; Li, H.; Wang, M.; Li, J.; Luo, J. Chiral drug fluorometry based on a calix[6]arene/molecularly imprinted polymer double recognition element grafted on nano--dots/Ir/Au. *Mikrochim. Acta,* **2020**, *187*(7), 394.
[http://dx.doi.org/10.1007/s00604-020-04356-x] [PMID: 32556561]

[102] Zadmard, R.; Akbari-Moghaddam, P.; Darvishi, S.; Mirza-Aghayan, M. An Efficient Multi-Component Synthesis of Highly Functionalized Calix[4]arenes with Pronounced Binding Affinity toward β-Lactoglobulin. *Eur. J. Org. Chem.,* **2016**, *2016*(22), 3894-3899.
[http://dx.doi.org/10.1002/ejoc.201600543]

[103] Darjee, S.M.; Bhatt, K.D.; Panchal, U.S.; Jain, V.K. Scrupulous recognition of biologically important acids by fluorescent "turn off-on" mechanism of thiacalix reduced silver nanoparticles. *Chin. Chem. Lett.,* **2017**, *28*(2), 312-318.
[http://dx.doi.org/10.1016/j.cclet.2016.07.026]

[104] Samanta, K.; Ranade, D.S.; Upadhyay, A.; Kulkarni, P.P.; Rao, C.P. A Bimodal, Cationic, and Water-Soluble Calix[4]arene Conjugate: Design, Synthesis, Characterization, and Transfection of Red Fluorescent Protein Encoded Plasmid in Cancer Cells. *ACS Appl. Mater. Interfaces,* **2017**, *9*(6), 5109-5117.
[http://dx.doi.org/10.1021/acsami.6b14656] [PMID: 28103012]

[105] Kaur, H.; Singh, N.; Kaur, N.; Jang, D.O. Nano-aggregate-Fe3+ complex based on benzimidazole-modified calix[4]arene for amplified fluorescence detection of ADP in aqueous media. *Sens. Actuators B Chem.,* **2019**, *284*, 193-201.
[http://dx.doi.org/10.1016/j.snb.2018.12.116]

[106] Ozyilmaz, E.; Cetinguney, S.; Yilmaz, M. Encapsulation of lipase using magnetic fluorescent calix[4]arene derivatives; improvement of enzyme activity and stability. *Int. J. Biol. Macromol.,* **2019**, *133*, 1042-1050.
[http://dx.doi.org/10.1016/j.ijbiomac.2019.04.182] [PMID: 31042560]

[107] Narula, A.; Hussain, M.A.; Upadhyay, A.; Rao, C.P. 1,3-Di-naphthalimide Conjugate of Calix[4]arene as a Sensitive and Selective Sensor for Trinitrophenol and This Turns Reversible when Hybridized with Carrageenan as Beads. *ACS Omega,* **2020**, *5*(40), 25747-25756.
[http://dx.doi.org/10.1021/acsomega.0c03060] [PMID: 33073100]

[108] Bradberry, S.J.; Savyasachi, A.J.; Martinez-Calvo, M.; Gunnlaugsson, T. Development of responsive visibly and NIR luminescent and supramolecular coordination self-assemblies using lanthanide ion directed synthesis. *Coord. Chem. Rev.,* **2014**, *273-274*, 226-241.
[http://dx.doi.org/10.1016/j.ccr.2014.03.023]

[109] Massi, M.; Ogden, M. Luminescent Lanthanoid Calixarene Complexes and Materials. *Materials (Basel),* **2017**, *10*(12), 1369.
[http://dx.doi.org/10.3390/ma10121369] [PMID: 29182546]

[110] Viana, R.S.; Oliveira, C.A.F.; Chojnacki, J.; Barros, B.S.; Alves-Jr, S.; Kulesza, J. Structural and spectroscopic investigation of new luminescent hybrid materials based on calix[4]arene-tetracarboxylate and Ln 3+ ions (Ln = Gd, Tb or Eu). *J. Solid State Chem.,* **2017**, *251*, 26-32.
[http://dx.doi.org/10.1016/j.jssc.2017.04.002]

[111] Podyachev, S.N.; Sudakova, S.N.; Gimazetdinova, G.S.; Shamsutdinova, N.A.; Syakaev, V.V.; Barsukova, T.A.; Iki, N.; Lapaev, D.V.; Mustafina, A.R. Synthesis, metal binding and spectral properties of novel bis-1,3-diketone calix[4]arenes. *New J. Chem.,* **2017**, *41*(4), 1526-1537.
[http://dx.doi.org/10.1039/C6NJ03381D]

[112] Han, H.; Zhang, G.; Li, K.; Liao, W. A Tb-calixarene coordination chain for luminescent sensing of

Fe3+, Cr2O72− and 2,4-DNT. *Polyhedron,* **2019**, *163*, 84-90.
[http://dx.doi.org/10.1016/j.poly.2019.01.067]

[113]  Ovsyannikov, A.S.; Khariushin, I.V.; Solovieva, S.E.; Antipin, I.S.; Komiya, H.; Marets, N.; Tanaka, H.; Ohmagari, H.; Hasegawa, M.; Zakrzewski, J.J.; Chorazy, S.; Kyritsakas, N.; Hosseini, M.W.; Ferlay, S. Mixed Tb/Dy coordination ladders based on tetra(carboxymethyl)thiacalix[4]arene: a new avenue towards luminescent molecular nanomagnets. *RSC Advances,* **2020**, *10*(20), 11755-11765.
[http://dx.doi.org/10.1039/D0RA01263G]

# CHAPTER 3

# Resorcinarene Crowns as Versatile Host Molecules and Their Potential Applications

**Selvaraj Devi**[1], **Somasundaram Anbu Anjugam Vandarkuzhali**[2,*] and **Vairaperumal Tharmaraj**[3,*]

[1] *P. G. Department of Chemistry, Cauvery College for Women, Tiruchirappalli-620018, Tamilnadu, India*

[2] *Department of Chemistry, Saveetha School of Engineering, Saveetha Institute of Medical and Technical Sciences, Saveetha University, Chennai-600005, Tamilnadu, India*

[3] *Environmental Science and Technology Research Group, Department of Chemical Engineering, SRM Institute of Science and Technology, Kattankulathur-603203, Tamilnadu, India*

**Abstract:** Resorcinarene crowns are significant building blocks for supramolecular chemistry. Resorcinarenes are part of the calixarenes family and are macrocyclic, bowl-shaped molecules. Derived from four resorcinol subunits, they have hydroxy groups at the wide rim of the bowl. These cavities were utilized for their potential recognition of racemic guests and catalysis applications. In this chapter, we focused on the overview of synthesis, conformational properties of resorcinarenes crown and their potential applications such as separation technique using high performance liquid chromatography (HPLC), gas chromatography (GC) and ion chromatrography (IC). In addition, they are also used as chemo sensors, antibacterial and antioxidant agents, contrast agents, nanoparticles synthesis and catalytic systems. Finally, we concluded the chapter with the significance of resorcinarenes crown.

**Keywords:** Antibacterial and antioxidant activity, Catalysts, Resorcinarene crown, Sensors, Separation technique.

## INTRODUCTION

Resorcinarenes are a structurally versatile group of macrocyclic supramolecular host molecules that are closely related to the calixarenes. In 1940, Niederl and Vogel [1] proposed the structure of resorcinarenes with the molecular ratio of the aldehyde and the resorcinol as 4:4.

---

[*] **Corresponding author Vairaperumal Tharmaraj and Somasundaram Anbu Anjugam Vandarkuzhali**: Environmental Science and Technology Research Group, Department of Chemical Engineering, SRM Institute of Science and Technology, Kattankulathur-603203, Tamil Nadu, India; Tel: +919789249631; and Department of Chemistry, Saveetha School of Engineering, Saveetha Institute of Medical and Technical Sciences, Saveetha University, Chennai- 600005, Tamilnadu, India; E-mails: tharmachem@gmail.com and anbu.anju@gmail.com

**Paulpandian Muthu Mareeswaran, Palaniswamy Suresh and Seenivasan Rajagopal (Eds.)**

Finally, the structure of resorcinarene was proved in 1968 by single crystal X-ray analysis [2, 3] as shown in structure 1. The IUPAC name of the resorcinarene is 2,8,14,20-tetraalkylpentacyclo-[19.3.1.13,7.19,13.115,19]-octacosa-1-(25),3,5,7 (28),9,11,13(27),15,17,19(26),21, 23-dodecaene-4,6,10,12,16,18,22,24-octol.

Resorcinarenes have five different isomeric forms such as crown ($C_{4v}$), boat ($C_{2v}$), chair ($C_{2h}$), diamond ($C_s$) and saddle ($D_{2d}$). Two major conformations such as chair ($C_2v$) and crown ($C_4v$) are preferred as shown in structure 2. The two resorcinol rings are almost vertical and the remaining two are aligned as horizontal, indicating the boat form of resorcinarene. The crown conformation is highly ordered with the formation of hydrogen bond network (Fig. (**1**)).

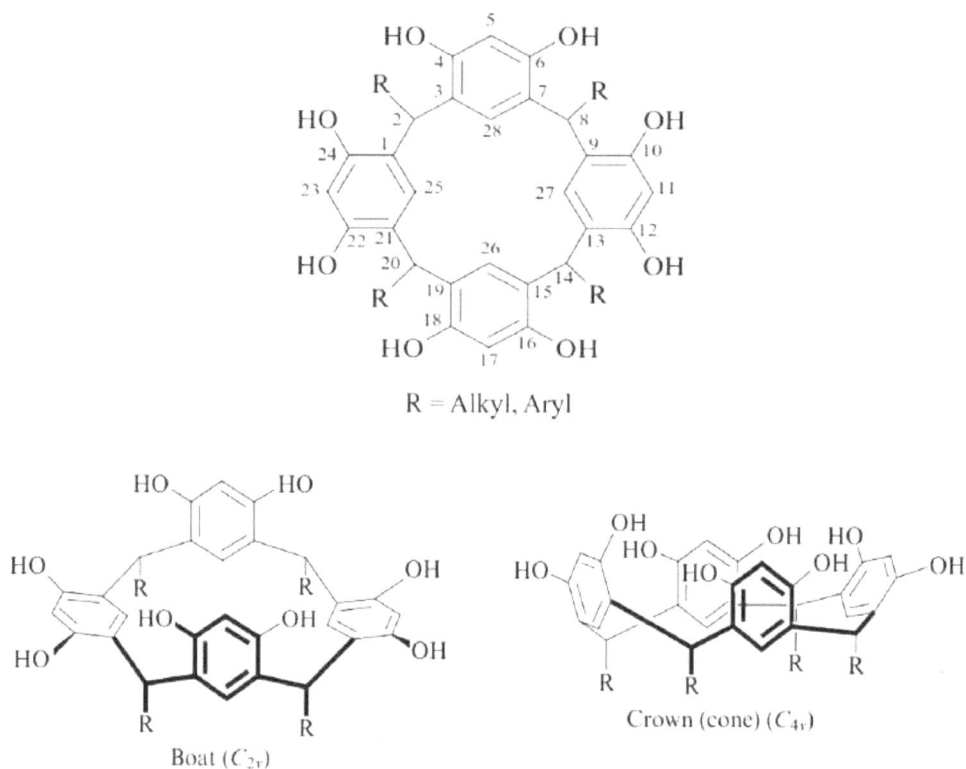

R = Alkyl, Aryl

Boat ($C_{2v}$)

Crown (cone) ($C_{4v}$)

**Fig. (1).** Conformations of resorcinarenes.

The upper rim of resorcinarene cavity has a more hydrophilic nature because of the –OH group derived from the resorcinol. In addition, the upper rim is readily available for further functionalization with various molecules to improve the binding affinity and selectivity. Therefore, the major forms of Chair ($C_2v$) and Crown ($C_4v$) resorcinarenes have provided good binding properties but comparatively crown form has more significance in functionalization, binding properties and structural features due to the hydroxyl groups of the upper rim having cavity structure. It is more suitable for modification with guest binding molecules [4].

Resorcinarenes crown, a macrocyclic supramolecular hosts molecule are capable of binding with various guest molecules or ions [5 - 7]. Therefore, resorcinarenes crown have been used in various fields of applications such as drug/ gene delivery [8, 9], catalyst [10], liquid crystals [11] and detection of volatile organic chemicals [12].

In addition, the host-guest properties of resorcinarenes crown allowed it to apply for environmental applications due to the changes in the structure of resorcinarenes from vesicles to micelles according to pH [13 - 15]. Resorcinarene crowns form complex with alkali and alkaline earth metal cations like $K^+$, $Cs^+$, $Rb^+$ and $Ag^+$, *etc.* It shows many interesting structural properties and forms multilayers of capsules and nanomaterials [16 - 18]. This kind of nanomaterials based on resorcinarene crowns is used for antibacterial activity [19].

This chapter mainly focuses on the synthesis, structural conformation and complexation properties of selective functionalized resorcinarene crown. The potential applications of resorcinarene crowns in the construction of supramolecular assemblies, sensors and catalytic applications, *etc.* have been discussed. Finally, this chapter provides a conclusion with future perspectives on this field.

## Synthesis of Resorcinarenes

Resorcinarenes are a different form of calixarenes, constructed from resorcinol and various aliphatic or aromatic aldehydes by acid catalyzed condensation reaction as shown in Fig. (2).

In addition, all the possible functionalization has been designed and synthesized [20 - 26]. The upper and lower rims are readily available for functionalization with suitable functional groups either directly during the acid-catalyzed condensation reaction or after synthesis of basic resorcinarene to modify the lower rim.

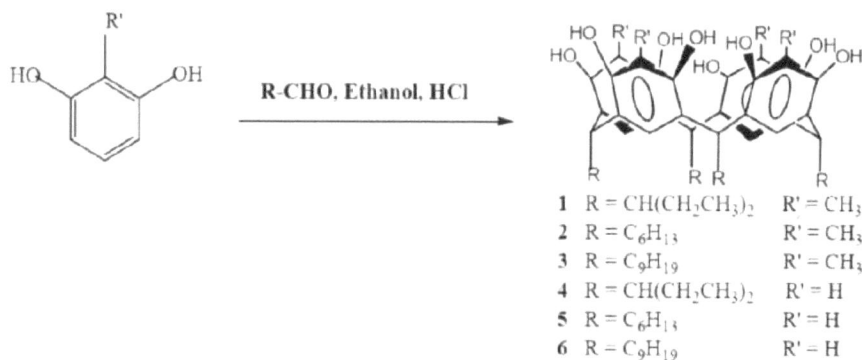

|   | | |
|---|---|---|
| 1 | R = CH(CH$_2$CH$_3$)$_2$ | R' = CH$_3$ |
| 2 | R = C$_6$H$_{13}$ | R' = CH$_3$ |
| 3 | R = C$_9$H$_{19}$ | R' = CH$_3$ |
| 4 | R = CH(CH$_2$CH$_3$)$_2$ | R' = H |
| 5 | R = C$_6$H$_{13}$ | R' = H |
| 6 | R = C$_9$H$_{19}$ | R' = H |

**Fig. (2).** Condensation reaction between resorcinol and aldehyde [20 - 26].

In recent years, researchers have developed various types of resorcinarenes through many new synthetic routes. C-alkoxycarbonylmethylcalix [4] resorcinarenes have been developed by using 2,4-dimethoxycinnamates in the presence of BF$_3$ as a Lewis acid catalyst as shown in Fig. (3) [27, 28].

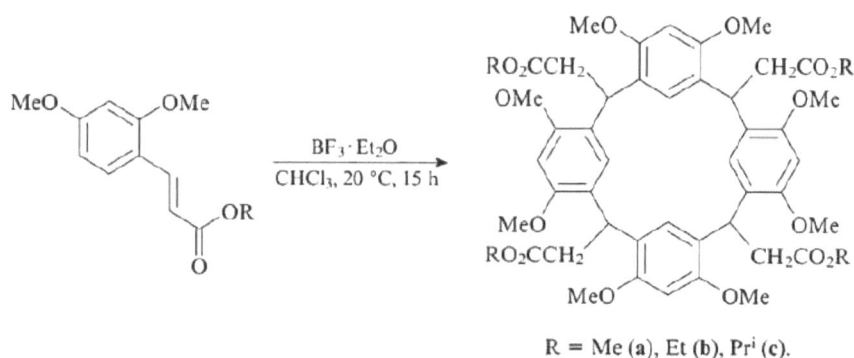

R = Me (**a**), Et (**b**), Pr$^i$ (**c**).

**Fig. (3).** BF$_3$ Lewis acid catalyst condensation based resorcinarene [27, 28].

Iwanek and Syzdol have achieved the formation of resorcinarene crown selectively with about 85% yield using SnCl$_4$ Lewis acids [29, 30]. Cyclocondensation reaction between 1, 3-dimethoxybenzene with isovaleraldehyde formed selective resorcinarene crown as shown in Fig. (4). Konishi *et al.* [31], also reported the synthesis of resorcin [4]arene octaalkyl ethers using Sc(OTf)$_3$ as a catalyst in the cyclocondensation of 1,3-dialkoxybenzenes with 1,3,5-trioxane.

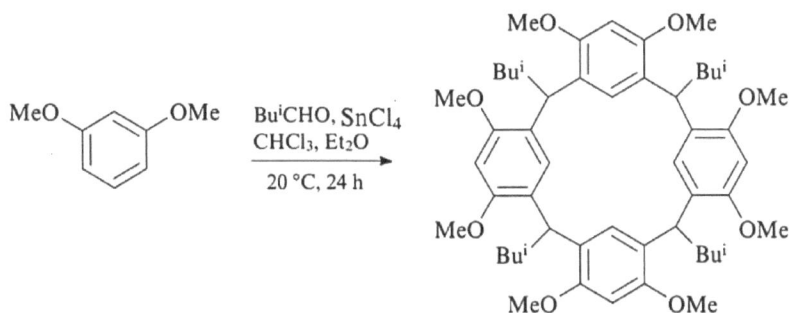

**Fig. (4).** SnCl$_4$ Lewis acid catalyzed cyclocondensation reaction for resorcinarene synthesis [29, 30].

Watson *et al.*, [32] have developed octa-O-alkyl resorcin [4]- arenes using a Brønsted acid catalyst acetic acid H$_2$SO$_4$ or HCl (Fig. (**5**)).

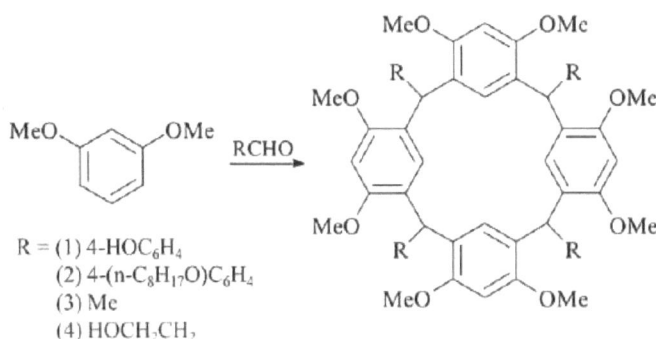

R = (1) 4-HOC$_6$H$_4$
(2) 4-(n-C$_8$H$_{17}$O)C$_6$H$_4$
(3) Me
(4) HOCH$_2$CH$_2$

**Fig. (5).** SnCl$_4$ Lewis acid catalyst in condensation reaction for resorcinarenes synthesis [32].

Recently, Parac-Vogt *et al.*, [33 - 36] have developed environmentally friendly lanthanide(III) tosylates and lanthanide(III) nitrobenzenesulfonates based catalyst for synthesizing calix [4]resorcinarenes as shown in Fig. (**6**). Also, ytterbium(III) and bismuth(III) trifluoromethanesulfonates (triflates) have been used as efficient catalysts for the synthesis of calix [4]resorcinarenes [37, 38]. One-step synthesis of resorcinarene has been reported and used for cation exchange resin-induced by m-phenylenediyldioxydiacetate with p-tolualdehyde [39].

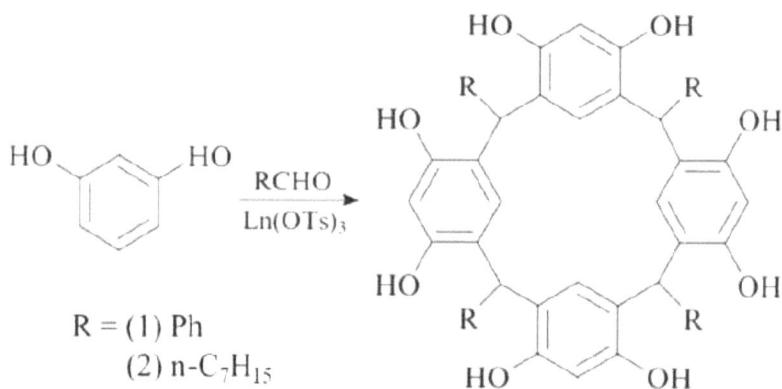

R = (1) Ph

      (2) n-C$_7$H$_{15}$

**Fig. (6).** Ln(OTs)$_3$ catalyzed condensation reaction for resorcinarene synthesis [33 - 36].

## Selective Functionalization of Upper Rim Resorcinarene Crowns

The effective methods were developed for the selective functionalization of upper rim of resorcinarene hydroxyl groups with crown ether groups to extend the multifunctional cavities. Commonly, these kinds of functionalization were done by nucleophilic substitution reaction using deprotonating base like sodium hydride (NaH or HNa) followed by the addition of tosylate as bridging unit. Finally, bis-crown derivatives and a tetra-crown derivative was obtained. The whole process is shown in Fig. (7) [40, 41].

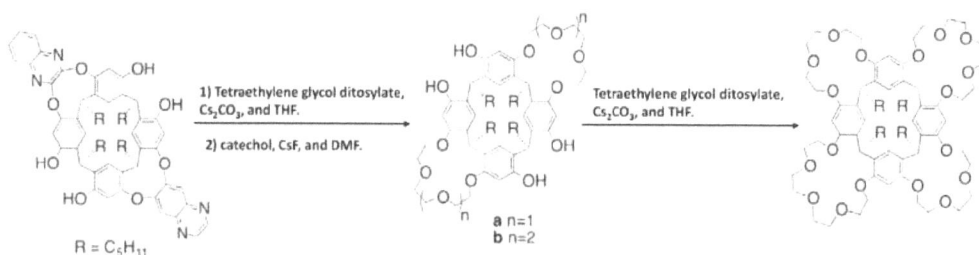

**Fig. (7).** Upper rim selectively modified 'cloverleaf' resorcinarene crown ethers from diquinoxaline cavitand [40, 41].

Nissinen *et al.* [42], have developed the upper rim modifications of mono-crown and bis-crown derivatives as shown in Fig. (**8**). Resorcinarene was allowed to react with Cs$_2$CO$_3$ base for 15 minutes to obtain only bis-crown product while increasing the reaction time to 60 minutes, both mono- and bis-crown products are obtained. The complete deprotonation of the parent resorcinarene at the early stage reaction acts as a phase transfer catalyst (PTC) to obtain bis crown products. Interestingly, mono-crown species have two different binding sites with crown pocket at one end and another side has two hydroxyl groups in the opposite side of resorcinarene cavity, offering the selective potential receptor for biological molecules.

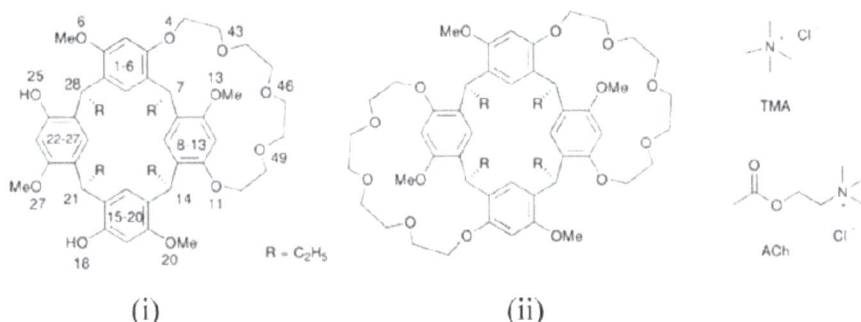

**Fig. (8).** Upper rim selective modification of mono-crown and bis-crown resorcinarene hosts (i) and (ii), quaternary ammonium guest acetylcholine (ACh) and tetramethylammonium (TMA) cation [42].

Tetramethoxy resorcinarene tribenzo-bis-crown ether has been developed by Nissinen *et al.* [43]. Two Equiv of 2-(2-hydroxyethoxy)phenol with m/p-bis(bromomethyl)benzene in the presence of $K_2CO_3$ in acetone yielded tetramethoxy resorcinarene tribenzo-bis-crown ether as shown in Fig. (9).

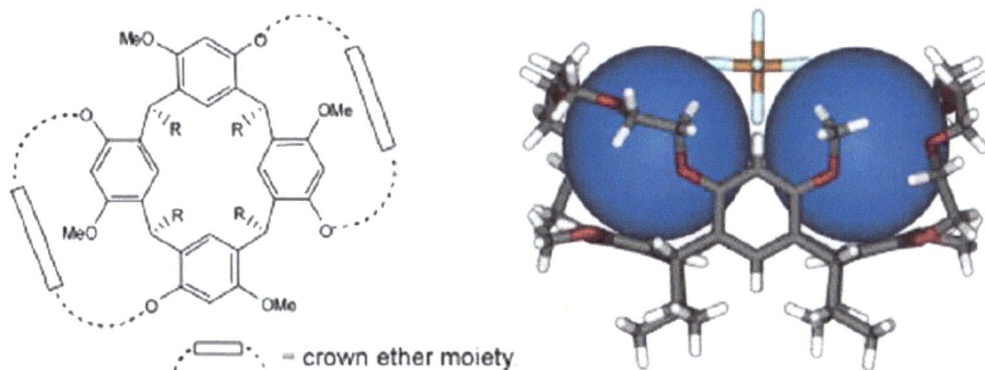

= crown ether moiety

**Fig. (9).** Upper rim selectively modified with tetramethoxy resorcinarene bis-crown [43].

## Selective functionalization of lower rim resorcinarene crowns

Selective functionalization of resorcinarene lower rim is introduced in the crown ether moieties. Hogberg [44] and Beer *et al.* [45] have reported lower rim resorcinol condensation reaction with 40-formylbenzo(15)crown-5 and 40-formylbenzo(18)crown-6 to give the lower rim resorcinarene crown derivatives of R = OH, n=1 and 2 and R = OAc, n=1 and 2 as shown in Fig. (**10**).

**Fig. (10).** Selectively lower rim modified crown ether resorcinarenes derivatives of –OH and –OAc [44, 45].

Poleska-Muchaldo *et al.* [46]. synthesized another two different types of lower rim crown derivatives in the same reaction condition. Two equivalent of resorcinol and one equivalent of azobenzo [15]crown-5 moiety in the presence of ethanol and HCl (as a catalyst) were first linked to a dimer followed by the second condensation reaction with octyl aldehyde to produce a bis-crown-substituted resorcin [4]arene as seen in Fig. (11).

**Fig. (11).** Lower rim selectively modified crown ether resorcinarenes derivatives [46].

## APPLICATIONS OF RESORCINARENE CROWNS

The resorcinarene derivatives find applications in the separation technique such as high performance liquid chromatography (HPLC), gas chromatography (GC) and ion chromatrography (IC), chemosensors, antibacterial and antioxidant activity, contrast agents, nanoparticles synthesis and catalytic systems.

## Chemical Separations

Silica particles surface covalently attached with lower rim modified resorcinarene derivatives were used as stationary phase in HPLC for effective separation of the compounds. Pietraszkiewicz *et al.* [47 - 49] developed resorcinarene with lower rim modification using four-$C_{11}H_{23}$ alkyl chains to increase the hydrophobicity of the ligand. The resulting C-tetra-n-undecylcalix [4]resorcinarene ligand was coated on surface of the reverse phase C18 (RP-18) column in acetonitrile, which proved a good stability. Ruderisch *et al.* [50] also developed HPLC stationary phase by silica substrate covalently attached with selective upper rim modified resorcinarene derivative of polar carbonate groups, used for high efficient stationary phase materials for polar analytes of low molecular weight acids in aqueous chromatographic conditions. This stationary phase column separated four compounds including uracil, phenol, naphthalene and anthracene. At first, phenol was separated strongly due to its high polar nature. Tan *et al.* [51] developed an HPLC separation column based on resorcinarenes 3-(C-methylcalix [4]resorcinarene)-2-hydroxypropoxy)-propylsilyl-appended silica particles (MCR-HPS). MCR-HPS column can specifically be used for the separation of benzene derivatives. For example, the elution order of nitrophenol is p- < o- < m-, which is due to a hydrophobic interaction between the resorcinarenes and nitrophenol derivative.

Resorcinarene derivatives can also be used for GC capillary column based on the thermal stability of resorcinarene derivatives. Zhang *et al.* [52] modified both lower and upper rim of resorcinarenes with phenyl and n-$C_5H_{11}$ groups, respectively. This column was utilized for the separation of alkanes, alkanols, phenols and aromatic hydrocarbons, especially chlorotoluene, cresol, and xylenol isomers. Schurig *et al.* [53 - 56] developed N-bromoacetyl-L-valine-tert--utyl-amide attached upper rim modified resorcinarene derivative as stationary phase in GC column. The performance of the column was checked and reported by injecting different amino acids.

Ion chromatography separation of transition metal ions such as $Mn^{2+}$, $Co^{2+}$, $Ni^{2+}$, $Cd^{2+}$, $Cu^{2+}$, and $Zn^{2+}$ was achieved using resorcinarene derivatives based on macrocyclic ligand functionalized with alanine and undecyl groups (AUA). Lamb *et al.* [57], developed AUA column containing four preorganized carboxylic acid groups, which is more selective for $Cu^{2+}$ as shown in Fig. (**12**). The chromatogram showed the order of elution as $Cu^{2+}$ > $Ni^{2+}$ > $Co^{2+}$ > $Zn^{2+}$ > $Mn^{2+}$ > $Cd^{2+}$ which is due to the stability constants of the formed complexes of metal ions with oxalic acid. AUA column showed selectivity for $Cu^{2+}$ because it has suitable coordination with macrocycle resorcinarene derivatives.

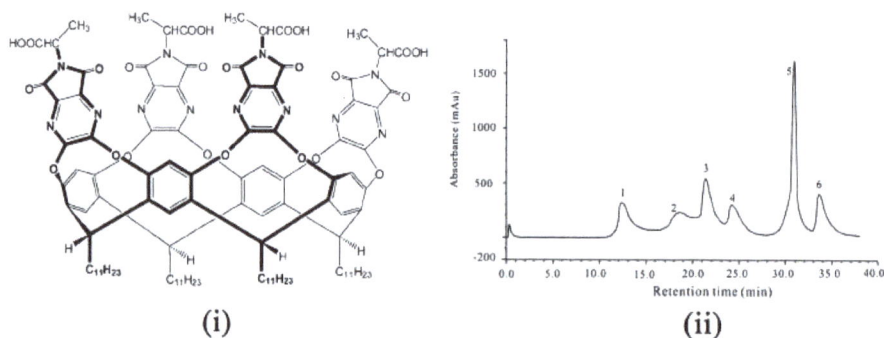

**Fig. (12).** (i) Structure of AUA. (ii) Chromatogram of separation of (1) $Cu^{2+}$, (2) $Ni^{2+}$, (3) $Zn^{2+}$, (4) $Co^{2+}$, (5) $Mn^{2+}$, and (6) $Cd^{2+}$ with oxalic acid gradient on AUA column [57].

## Chemical Sensors

Menon *et al.* [58], have developed a silver nanoprobe modified p-sulphonato-calix [4]resorcinarene (pSC4R–AgNP) as a powerful sensing tool for selective recognition of insecticide dimethoate in an aqueous medium. The pSC$_4$R can bind with amino residues and two thiol linkage groups of dimethoate *via* a host–guest interaction between the electron-rich cavity of pSC$_4$R and the electron-deficient dimethoate. The electrostatic interaction between the dimethoate and two pSC4R-modified AgNPs, resulted in the more favorable aggregation of pSC$_4$R–AgNPs in dimethoate than in other pesticides as shown in Fig. (**13**).

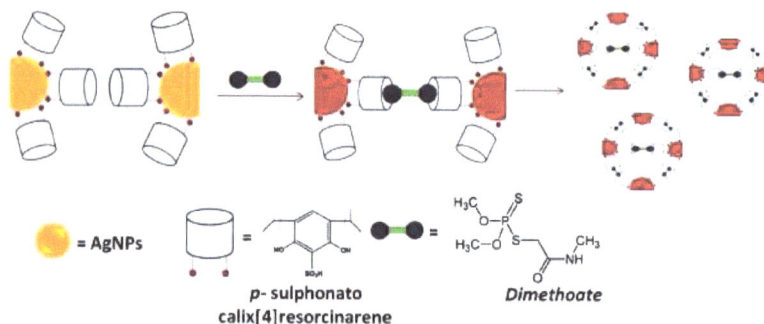

**Fig. (13).** Schematic diagram of dimethoate inclusion complex with pSC$_4$R–AgNPs [58].

Abdul Shaban *et al.* [59], have developed a calix [4]resorcinarene ionophores as chemical sensing tool for heavy metal ions in aqueous solutions based on quartz crystal microbalance with impedance analysis (QCM-I). The authors have synthesized four kinds of QCM-Calix based chemosensor ionophores, such as C-dec-9-en-1-ylcalix [4]resorcinarene (ionophore I), C-undecylcalix [4] resorcinarene (ionophore II), C-dec-9-enylcalix [4]resorcinarene-O-(S-

)-α-methylbenzylamine (ionophore III) and C-dec-9-enylcalix [4]resorcinarene---(R+)-α-methylbenzylamine (ionophore IV), which selectively detected $Pb^{2+}$, $Cd^{2+}$, $Hg^{2+}$, and $Cu^{2+}$, respectively in aqueous solutions. The detection mechanism involved the strong interaction of heavy metal ions ($Cd^{2+}$, $Cu^{2+}$, $Pb^{2+}$ and $Hg^{2+}$) with resorcinarene derivatives (Ionophores I, II, III, and IV) to form a host-guest complex.

## Catalytic Activity

Ziganshina *et al.* [60] developed a ferrocene-resorcinarene derivative as a template and reducing agent for the synthesis of silver nanoparticles (Ag-FcCA) inside the cavity as shown in Fig. (**14**) that was used as a catalyst for the reduction of p-nitrophenol.

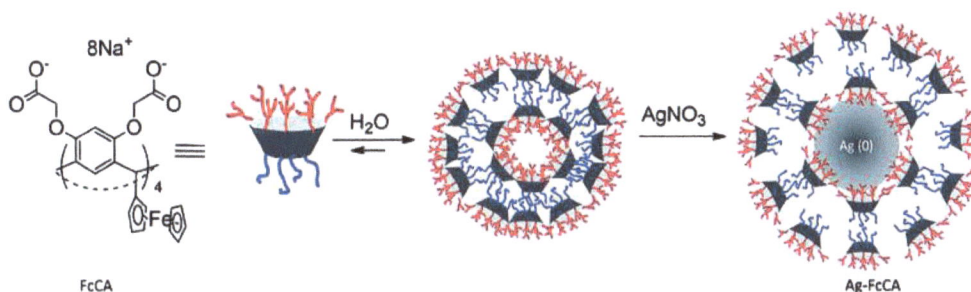

**Fig. (14).** Synthesis of AgNPs in self-assembled ferrocene-resorcinarene [60].

AgNPs-FcCA was utilized as a catalyst for the reduction of p-nitrophenol in the presence of sodium borohydride ($NaBH_4$). The reduction reaction of p-nitrophenol into p-aminophenol was completed within 15 min as seen in Fig. (**15**), indicating the prepared ferrocene-resorcinarene derivative silver nanocomposite (Ag-FcCA) showed good catalytic activity.

The catalytic epoxidation using Mn(III)-based resorcin [4]arene-metalloporphyrin and the size dependence of encapsulation ability of resorcin [4]arene conjugate in the epoxidation reactions of alkene was also reported.

Al-Azemi, *et al.* [61], have reported the catalytic epoxidation using Mn(III) based resorcin [4]arene–metalloporphyrin. Porphyrin conjugate with resorcinarene derivative showed higher catalytic activity in epoxidation reactions of styrene in the presence of pyridine as shown in Fig. (**16**). The binding studies have indicated that the rigid bowl-shaped cavitand, resorcin [4]arene was a better host for pyridine and styrene, increasing the catalytic efficiency in the presence of Mn(III) due to the encapsulation of styrene inside the cavity of resorcin [4]arene conjugate or Mn(III)-based metalloporphyrin system coordinated with pyridine axial ligand.

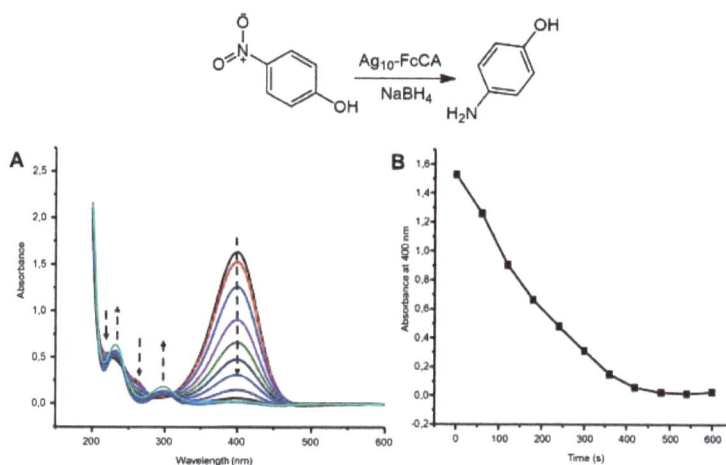

**Fig. (15).** (A) UV-Vis spectra of p-NPh reduction in the presence of Ag-FcCA, (B) Time-dependent changes in absorption at 400 nm [60].

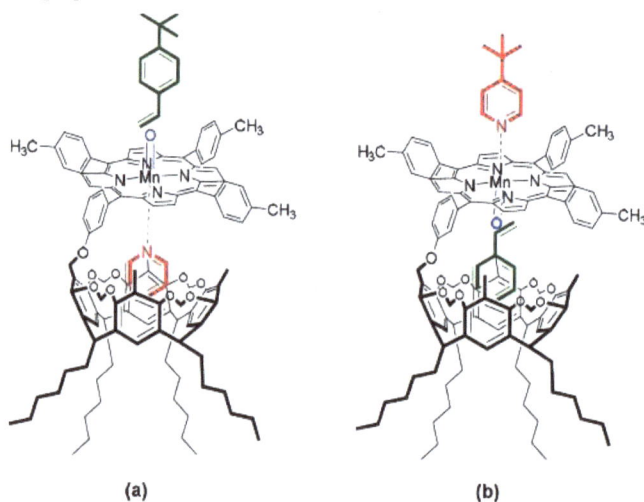

**Fig. (16).** Representation for the combinations of substrate and axial ligand utilized in the MnRCP-catalyzed epoxidation reactions: (a) pyridine–4-tert-butylstyrene; (b) 4-tert-butylpyridine–styrene [61].

## Lithographic Application

Synthesis, characterization and lithographic imaging of C-tetramethyl-calix [4]resorcinarene have been reported by Hiroshi Ito *et al.* [62]. C-tetramethyl-calix [4]resorcinarene was obtained by the condensation of resorcinol with aldehydes (acetaldehyde, benzaldehyde, and 4-isopropylbenzaldehyde) in the presence of con. HCl is followed by separation through fractional crystallization. Electron-beam patterning of a resist based on the IBM KRS containing C-tetramethyl-calix [4]resorcinarene protected with tBOCCH$_2$ provided dense and semidense 40 nm line/space patterns, indicating the electron beam chemical amplification resistance improves their contrast and performance as shown in Fig. (17).

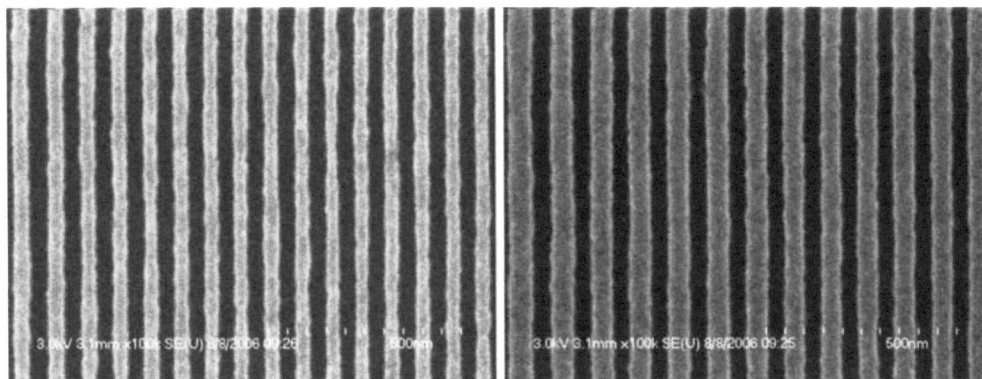

**Fig. (17).** Scanning electron micrographs of 40 nm dense and semidense patterns [62].

## Antioxidant and Antibacterial Applications

Macromolecules of C-methoxy calix [4]resorcinarene (CMPCR) have been used for medical field applications such as adsorbent for sunscreen and heavy metal [63], anti-HIV and HCV [64] and anti-tumour agents [65]. Budiana [66] successfully prepared and studied the antioxidant activity of C-Methoxyphenyl Calix [4] Resorcinarene and compared with vitamin C antioxidant activity. CMPCR shows the IC$_{50}$ value equal to 79 ppm when compared with vitamin C antioxidant activity (IC$_{50}$ - 20.96 ppm), indicating that the CMPCR compound has antioxidant activity equal to strong IC50 <100.

Resorcinarene bis-crown silver complexes have been synthesized by Nissinen *et al.* [67], and applied for antibacterial activity against E. coli. At first, a hydrophobic glass substrate was modified with resorcinarene bis-crown silver complexes upto four layers using the Langmuir–Blodgett approach. Bacterial (E. coli) carpet grown on agar gel, the silver was released from the bis-crown complexes in the presence of thiol groups on the bacterial surface, leads to bacterial death and inhibits the overall bacterial growth as observed in Fig. (18).

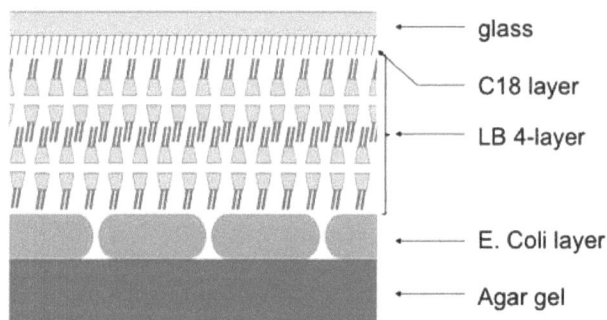

**Fig. (18).** Schematic representation of the experimental set-up used to test the antibacterial properties of C11BC5–Ag(I) complex [67].

## CONCLUSION

Remarkable progress was achieved in the development of resorcinarene crown derivatives, utilizing their host gust properties for many potential applications such as separation techniques using high performance liquid chromatography (HPLC), gas chromatography (GC) and ion chromatrography (IC). In addition, resorcinarenes were also used in chemosensors, antibacterial and antioxidant activity, contrast agents, nanoparticles synthesis and catalytic systems. Resorcinarene3-(C-methylcalix [4]resorcinarene)-2-hydroxypropoxy)-propylsi-yl-appended silica particles (MCR-HPS) showed excellent selective separations of benzene derivatives. Also, upper rim modified resorcinarene derivative of N-bromoacetyl-L-valine-tert-butyl-amide was used in the separation of tyrosine (Tyr) and ornithine (Orn). p-Sulphonato-calix [4]resorcinarene (pSC4R–AgNP) was used as a powerful sensing tool for insecticide detection. Also, the detection of heavy metal ions was achieved by using calix [4]resorcinarene ionophores. Ferrocene-resorcinarene derivative and Mn(III) based resorcin [4]arene–metalloporphyrin was mainly used for the catalytic activity of reduction of nitrophenol and epoxidation, respectively. C-Methoxy calix [4]resorcinarene (CMPCR) and bis-crown silver complexes have been used for antioxidant and antibacterial applications.

## CONSENT OF PUBLICATION

The author grants the publisher the sole and exclusive license of the full copyright of this material.

## CONFLICT OF INTEREST

The author declares no conflict of interest, financial or otherwise.

## ACKNOWLEDGEMENTS

I am very much thankful to SRMIST-Establishment programme for providing financial support for the post-doctoral fellowship.

## REFERENCES

[1]     Niederl, J.B.; Vogel, H.J. Aldehyde-Resorcinol Condensations. *J. Am. Chem. Soc.,* **1940**, *62*(9), 2512-2514.
        [http://dx.doi.org/10.1021/ja01866a067]

[2]     Erdtman, H.; Högberg, S.; Abrahamsson, S.; Nilsson, B. Cyclooligomeric phenol-aldehyde condensation products I. *Tetrahedron Lett.,* **1968**, *9*(14), 1679-1682.
        [http://dx.doi.org/10.1016/S0040-4039(01)99028-8]

[3]     Nilsson, B.; Liminga, R.; Olovsson, I.; Gronowitz, S.; Christiansen, H.; Rosén, U. The Crystal and Molecular Structure of the Synthetic Tetramer $C_{84}H_{84}Br_4O_{16}$. *Acta Chem. Scand.,* **1968**, *22*, 732-747.
        [http://dx.doi.org/10.3891/acta.chem.scand.22-0732]

[4]     Sliwa, W.; Deska, M. Cavitands. *Chem. Heterocycl. Compd.,* **2002**, *38*(6), 646-667.
        [http://dx.doi.org/10.1023/A:1019956900666]

[5]     Timmerman, P.; Verboom, W.; Reinhoudt, D.N. Resorcinarenes. *Tetrahedron,* **1996**, *52*(8), 2663-2704.
        [http://dx.doi.org/10.1016/0040-4020(95)00984-1]

[6]     Mandolini, L.; Ungaro, R. *Calixarenes in Action*; Imperial College Press: Singapore, **2000**, p. 271.
        [http://dx.doi.org/10.1142/p168]

[7]     Vicens, J.; Böhmer, V. *Topics in Inclusion Science*; Kluwer Academic Publishers: Netherlands, **1991**, p. 264.

[8]     Rodik, R.V.; Klymchenko, A.S.; Jain, N.; Miroshnichenko, S.I.; Richert, L.; Kalchenko, V.I.; Mély, Y. Virus-sized DNA nanoparticles for gene delivery based on micelles of cationic calixarenes. *Chemistry,* **2011**, *17*(20), 5526-5538.
        [http://dx.doi.org/10.1002/chem.201100154] [PMID: 21503994]

[9]     Aoyama, Y. Macrocyclic glycoclusters: from amphiphiles through nanoparticles to glycoviruses. *Chemistry,* **2004**, *10*(3), 588-593.
        [http://dx.doi.org/10.1002/chem.200305288] [PMID: 14767922]

[10]    Zakharova, L.Y.; Syakaev, V.V.; Voronin, M.A.; Valeeva, F.V.; Ibragimova, A.R.; Ablakova, Y.R.; Kazakova, E.K.; Latypov, S.K.; Konovalov, A.I. NMR and Spectrophotometry Study of the Supramolecular Catalytic System Based on Polyethyleneimine and Amphiphilic Sulfonatomethylated Calix[4]Resorcinarene. *J. Phys. Chem. C,* **2009**, *113*(15), 6182-6190.
        [http://dx.doi.org/10.1021/jp806541w]

[11]    Zakharova, L.Y.; Kudryashova, Y.R.; Selivanova, N.M.; Voronin, M.A.; Ibragimova, A.R.; Solovieva, S.E.; Gubaidullin, A.T.; Litvinov, A.I.; Nizameev, I.R.; Kadirov, M.K.; Galyametdinov, Y.G.; Antipin, I.S.; Konovalov, A.I. Novel membrane mimetic systems based on amphiphilic oxyethylated calix[4]arene: Aggregative and liquid crystalline behavior. *J. Membr. Sci.,* **2010**, *364*(1-2), 90-101.
        [http://dx.doi.org/10.1016/j.memsci.2010.08.005]

[12]    Holloway, A.F.; Nabok, A.; Hashim, A.A.; Penders, J. *Sens. Transducers J.,* **2010**, *113*, 71.

[13]    Lee, M.; Lee, S.J.; Jiang, L.H. Stimuli-responsive supramolecular nanocapsules from amphiphilic calixarene assembly. *J. Am. Chem. Soc.,* **2004**, *126*(40), 12724-12725.
        [http://dx.doi.org/10.1021/ja045918v] [PMID: 15469237]

[14]    Micali, N.; Villari, V.; Consoli, G.M.L.; Cunsolo, F.; Geraci, C. Vesicle-to-micelle transition in aqueous solutions of amphiphilic calixarene derivatives. *Phys. Rev. E Stat. Nonlin. Soft Matter Phys.,*

**2006**, *73*(5), 051904.
[http://dx.doi.org/10.1103/PhysRevE.73.051904] [PMID: 16802964]

[15]   Houmadi, S.; Coquière, D.; Legrand, L.; Fauré, M.C.; Goldmann, M.; Reinaud, O.; Rémita, S. Architecture-controlled "SMART" Calix[6]arene self-assemblies in aqueous solution. *Langmuir,* **2007**, *23*(9), 4849-4855.
[http://dx.doi.org/10.1021/la700271a] [PMID: 17397206]

[16]   Salorinne, K.; Nissinen, M. Novel tetramethoxy resorcinarene bis-crown ethers. *Org. Lett.,* **2006**, *8*(24), 5473-5476.
[http://dx.doi.org/10.1021/ol062138d] [PMID: 17107050]

[17]   Asfari, Z.; Lamare, V.; Dozol, J.F.; Vicens, J. A tribenzo modified 1,3-calix[4]-bis-crown-6: A highly selective receptor for cesium. *Tetrahedron Lett.,* **1999**, *40*(4), 691-694.
[http://dx.doi.org/10.1016/S0040-4039(98)02422-8]

[18]   Salorinne, K.; Lopez-Acevedo, O.; Nauha, E.; Häkkinen, H.; Nissinen, M. Solvent driven formation of silver embedded resorcinarene nanorods. *CrystEngComm,* **2012**, *14*(2), 347-350.
[http://dx.doi.org/10.1039/C1CE05737E]

[19]   Yi, Y.; Wang, Y.; Liu, H. Preparation of new crosslinked chitosan with crown ether and their adsorption for silver ion for antibacterial activities. *Carbohydr. Polym.,* **2003**, *53*(4), 425-430.
[http://dx.doi.org/10.1016/S0144-8617(03)00104-8]

[20]   Sliwa, W.; Kozlowski, C. *Calixarenes and Resorcinarenes*; Wiley-VCH: Weinheim, **2009**.

[21]   Gramage-Doria, R.; Armspach, D.; Matt, D. Metallated cavitands (calixarenes, resorcinarenes, cyclodextrins) with internal coordination sites. *Coord. Chem. Rev.,* **2013**, *257*(3-4), 776-816.
[http://dx.doi.org/10.1016/j.ccr.2012.10.006]

[22]   Egberink, R.J.M.; Cobben, P.L.H.M.; Vverboom, W.; Harkema, S.; Reinhoudt, D.N. Hügberg compounds with a functionalized box-like cavity. *Journal of Inclusion Phenomena and Molecular Recognition in Chemistry,* **1992**, *12*(1-4), 151-158.
[http://dx.doi.org/10.1007/BF01053858]

[23]   Reynolds, M.R.; Pick, F.S.; Hayward, J.J.; Trant, J.F. A Concise Synthesis of a Methyl Ester 2-Resorcinarene: A Chair-Conformation Macrocycle. *Symmetry (Basel),* **2021**, *13*(4), 627.
[http://dx.doi.org/10.3390/sym13040627]

[24]   Egberink, R.J.M.; Cobben, P.L.H.M.; Vverboom, W.; Harkema, S.; Reinhoudt, D.N. Hügberg compounds with a functionalized box-like cavity. *Journal of Inclusion Phenomena and Molecular Recognition in Chemistry,* **1992**, *12*(1-4), 151-158.
[http://dx.doi.org/10.1007/BF01053858]

[25]   Tunstad, L.M.; Tucker, J.A.; Dalcanale, E.; Weiser, J.; Bryant, J.A.; Sherman, J.C.; Helgeson, R.C.; Knobler, C.B.; Cram, D.J. Host-guest complexation. 48. Octol building blocks for cavitands and carcerands. *J. Org. Chem.,* **1989**, *54*(6), 1305-1312.
[http://dx.doi.org/10.1021/jo00267a015]

[26]   Thoden van Velzen, E.U.; Engbersen, J.F.J.; Reinhoudt, D.N. Self-Assembled Monolayers of Receptor Adsorbates on Gold: Preparation and Characterization. *J. Am. Chem. Soc.,* **1994**, *116*(8), 3597-3598.
[http://dx.doi.org/10.1021/ja00087a055]

[27]   Botta, B.; Di Giovanni, M.C.; Monache, G.D.; De Rosa, M.C.; Gacs-Baitz, E.; Botta, M.; Corelli, F.; Tafi, A.; Santini, A.; Benedetti, E.; Pedone, C.; Misiti, D. A Novel Route to Calix[4]arenes. 2. Solution- and Solid-State Structural Analyses and Molecular Modeling Studies. *J. Org. Chem.,* **1994**, *59*(6), 1532-1541.
[http://dx.doi.org/10.1021/jo00085a047]

[28]   Botta, B.; Iacomacci, P.; Di Giovanni, C.; Delle Monache, G.; Gacs-Baitz, E.; Botta, M.; Tafi, A.; Corelli, F.; Misiti, D. The tetramerization of 2,4-dimethoxycinnamates. A novel route to calixarenes. *J. Org. Chem.,* **1992**, *57*(12), 3259-3261.

[http://dx.doi.org/10.1021/jo00038a001]

[29]  Iwanek, W.; Syzdol, B. Lewis Acid-Induced Synthesis of Octamethoxyresorcarenes. *Synth. Commun.,* **1999**, *29*(7), 1209-1216.
[http://dx.doi.org/10.1080/00397919908086092]

[30]  Iwanek, W. The synthesis of octamethoxyresorc[4]arenes catalysed by Lewis acids. *Tetrahedron,* **1998**, *54*(46), 14089-14094.
[http://dx.doi.org/10.1016/S0040-4020(98)00859-X]

[31]  Morikawa, O.; Nagamatsu, Y.; Nishimura, A.; Kobayashi, K.; Konishi, H. Scandium triflate-catalyzed cyclocondensation of 1,3-dialkoxybenzenes with 1,3,5-trioxane. Formation of resorcin[4]arenes and confused resorcin[4]arenes. *Tetrahedron Lett.,* **2006**, *47*(24), 3991-3994.
[http://dx.doi.org/10.1016/j.tetlet.2006.04.015]

[32]  Moore, D.; Watson, G.W.; Gunnlaugsson, T.; Matthews, S.E. Selective formation of the rctt chair stereoisomers of octa-O-alkyl resorcin[4]arenes using Brønsted acid catalysis. *New J. Chem.,* **2008**, *32*(6), 994.
[http://dx.doi.org/10.1039/b714735j]

[33]  Deleersnyder, K.; Mehdi, H.; Horváth, I.T.; Binnemans, K.; Parac-Vogt, T.N. Lanthanide(III) nitrobenzenesulfonates and p-toluenesulfonate complexes of lanthanide(III), iron(III), and copper(II) as novel catalysts for the formation of calix[4]resorcinarene. *Tetrahedron,* **2007**, *63*(37), 9063-9070.
[http://dx.doi.org/10.1016/j.tet.2007.06.090]

[34]  Parac-Vogt, T.N.; Binnemans, K. Lanthanide(III) nosylates as new nitration catalysts. *Tetrahedron Lett.,* **2004**, *45*(15), 3137-3139.
[http://dx.doi.org/10.1016/j.tetlet.2004.02.084]

[35]  Parac-Vogt, T.N.; Deleersnyder, K.; Binnemans, K. Lanthanide(III) complexes of aromatic sulfonic acids as catalysts for the nitration of toluene. *J. Alloys Compd.,* **2004**, *374*(1-2), 46-49.
[http://dx.doi.org/10.1016/j.jallcom.2003.11.062]

[36]  Parac-Vogt, T.N.; Pachini, S.; Nockemann, P.; Van Hecke, K.; Van Meervelt, L.; Binnemans, K. Lanthanide( III ) Nitrobenzenesulfonates as New Nitration Catalysts: The Role of the Metal and of the Counterion in the Catalytic Efficiency. *Eur. J. Org. Chem.,* **2004**, *2004*(22), 4560-4566.
[http://dx.doi.org/10.1002/ejoc.200400475]

[37]  Barrett, A.G.M.; Braddock, D.C.; Henschke, J.P.; Walker, E.R. Ytterbium(III) triflate-catalysed preparation of calix[4]resorcinarenes: Lewis assisted Brønsted acidity. *J. Chem. Soc., Perkin Trans. 1,* **1999**, *1*(8), 873-878.
[http://dx.doi.org/10.1039/a809919g]

[38]  Peterson, K.E.; Smith, R.C.; Mohan, R.S. Bismuth compounds in organic synthesis. Synthesis of resorcinarenes using bismuth triflate. *Tetrahedron Lett.,* **2003**, *44*(42), 7723-7725.
[http://dx.doi.org/10.1016/j.tetlet.2003.08.093]

[39]  Vuano, B.; Pieroni, O.I. A one-step synthesis of O-functionalized resorcinarene under heterogeneous catalysis conditions. *Synthesis,* **1999**, *1999*(1), 72-73.
[http://dx.doi.org/10.1055/s-1999-3665]

[40]  Cram, D.J.; Jaeger, R.; Deshayes, K. Host-guest complexation. 65. Hemicarcerands that encapsulate hydrocarbons with molecular weights greater than two hundred. *J. Am. Chem. Soc.,* **1993**, *115*(22), 10111-10116.
[http://dx.doi.org/10.1021/ja00075a029]

[41]  Cai, Y.; Castro, P.P.; Gutierrez-Tunstad, L.M. 'Cloverleaf' crown ether resorcin[4]arenes. *Tetrahedron Lett.,* **2008**, *49*(13), 2146-2149.
[http://dx.doi.org/10.1016/j.tetlet.2008.01.092]

[42]  Salorinne, K.; Tero, T.R.; Riikonen, K.; Nissinen, M. Synthesis and structure of mono-bridged resorcinarene host: a ditopic receptor for ammonium guests. *Org. Biomol. Chem.,* **2009**, *7*(20), 4211-

4217.
[http://dx.doi.org/10.1039/b911389d] [PMID: 19795059]

[43]    Salorinne, K.; Nissinen, M. Alkali metal complexation properties of resorcinarene bis-crown ethers: effect of the crown ether functionality and preorganization on complexation. *Tetrahedron,* **2008**, *64*(8), 1798-1807.
[http://dx.doi.org/10.1016/j.tet.2007.11.103]

[44]    Hoegberg, A.G.S. Two stereoisomeric macrocyclic resorcinol-acetaldehyde condensation products. *J. Org. Chem.,* **1980**, *45*(22), 4498-4500.
[http://dx.doi.org/10.1021/jo01310a046]

[45]    Wright, A.J.; Matthews, S.E.; Fischer, W.B.; Beer, P.D. Novel resorcin[4]arenes as potassium-selective ion-channel and transporter mimics. *Chemistry,* **2001**, *7*(16), 3474-3481.
[http://dx.doi.org/10.1002/1521-3765(20010817)7:16<3474::AID-CHEM3474>3.0.CO;2-6] [PMID: 11560317]

[46]    Poleska-Muchlado, Z.; Luboch, E.; Biernat, J.F. Novel Calix[4]resorcinarenes with Side Azobenzo-15-crown-5 Residues. *Synth. Commun.,* **2008**, *38*(18), 3062-3067.
[http://dx.doi.org/10.1080/00397910802044298]

[47]    Pietraszkiewicz, O.; Pietraszkiewicz, M. Separation of pyrimidine bases on a HPLC stationary RP-18 phase coated with calix[4]resorcinarene. *J. Incl. Phenom. Macrocycl. Chem.,* **1999**, *35*(1/2), 261-270.
[http://dx.doi.org/10.1023/A:1008151100076]

[48]    Pietraszkiewicz, O.; Pietraszkiewicz, M. Separation of pyrimidine bases on HPLC stationary RP-18 phase coated with calix[4]resorcinarene. *Pol. J. Chem.,* **1998**, *72*, 2418.

[49]    Pietraszkiewicz, M.; Pietraszkiewicz, O.; Kozbial, M. Calix[4]resorcinarene as dynamic coating for modified stationary RP-18 phase for HPLC. *Pol. J. Chem.,* **1998**, *72*, 1963.

[50]    Ruderisch, A.; Iwanek, W.; Pfeiffer, J.; Fischer, G.; Albert, K.; Schurig, V. Synthesis and characterization of a novel resorcinarene-based stationary phase bearing polar headgroups for use in reversed-phase high-performance liquid chromatography. *J. Chromatogr. A,* **2005**, *1095*(1-2), 40-49.
[http://dx.doi.org/10.1016/j.chroma.2005.07.109] [PMID: 16275281]

[51]    Tan, H.M.; Soh, S.F.; Zhao, J.; Yong, E.L.; Gong, Y. Preparation and application of methylcalix[4]resorcinarene-bonded silica particles as chiral stationary phase in high-performance liquid chromatography. *Chirality,* **2011**, *23*(1E) Suppl. 1, E91-E97.
[http://dx.doi.org/10.1002/chir.20983] [PMID: 21837635]

[52]    Zhang, H.; Dai, R.; Ling, Y.; Wen, Y.; Zhang, S.; Fu, R.; Gu, J. Resorcarene derivative used as a new stationary phase for capillary gas chromatography. *J. Chromatogr. A,* **1997**, *787*(1-2), 161-169.
[http://dx.doi.org/10.1016/S0021-9673(97)00613-4]

[53]    Ruderisch, A.; Pfeiffer, J.; Schurig, V. Synthesis of an enantiomerically pure resorcinarene with pendant l-valine residues and its attachment to a polysiloxane (Chirasil-Calix). *Tetrahedron Asymmetry,* **2001**, *12*(14), 2025-2030.
[http://dx.doi.org/10.1016/S0957-4166(01)00352-4]

[54]    Pfeiffer, J.; Schurig, V. Enantiomer separation of amino acid derivatives on a new polymeric chiral resorc[4]arene stationary phase by capillary gas chromatography. *J. Chromatogr. A,* **1999**, *840*(1), 145-150.
[http://dx.doi.org/10.1016/S0021-9673(99)00224-1]

[55]    Ruderisch, A.; Pfeiffer, J.; Schurig, V. Mixed chiral stationary phase containing modified resorcinarene and β-cyclodextrin selectors bonded to a polysiloxane for enantioselective gas chromatography. *J. Chromatogr. A,* **2003**, *994*(1-2), 127-135.
[http://dx.doi.org/10.1016/S0021-9673(03)00423-0] [PMID: 12779224]

[56]    Levkin, P.A.; Ruderisch, A.; Schurig, V. Combining the enantioselectivity of a cyclodextrin and a diamide selector in a mixed binary gas-chromatographic chiral stationary phase. *Chirality,* **2006**,

*18*(1), 49-63.
[http://dx.doi.org/10.1002/chir.20219]

[57]   Li, N.; Allen, L.J.; Harrison, R.G.; Lamb, J.D. Transition metal cation separations with a resorcinarene-based amino acid stationary phase. *Analyst (Lond.),* **2013**, *138*(5), 1467-1474.
[http://dx.doi.org/10.1039/c2an36562f] [PMID: 23324944]

[58]   Menon, S.K.; Modi, N.R.; Pandya, A.; Lodha, A. Ultrasensitive and specific detection of dimethoate using a p-sulphonato-calix[4]resorcinarene functionalized silver nanoprobe in aqueous solution. *RSC Advances,* **2013**, *3*(27), 10623.
[http://dx.doi.org/10.1039/c3ra40762d]

[59]   Shaban, A.; Eddaif, L. Comparative Study of a Sensing Platform via Functionalized Calix[4]resorcinarene Ionophores on QCM Resonator as Sensing Materials for Detection of Heavy Metal Ions in Aqueous Environments. *Electroanalysis,* **2020**, *32*, 1.

[60]   Sergeeva, T.Y.; Samigullina, A.I.; Gubaidullin, A.T.; Nizameev, I.R.; Kadirov, M.K.; Mukhitova, R.K.; Ziganshina, A.Y.; Konovalov, A.I. Application of ferrocene-resorcinarene in silver nanoparticle synthesis. *RSC Advances,* **2016**, *6*(90), 87128-87133.
[http://dx.doi.org/10.1039/C6RA19961E]

[61]   Al-Azemi, T.F.; Vinodh, M.; Vinodh, M. Effect of the resorcin[4]arene host on the catalytic epoxidation of a Mn( III )-based resorcin[4]arene–metalloporphyrin conjugate. *RSC Advances,* **2015**, *5*(107), 88154-88159.
[http://dx.doi.org/10.1039/C5RA13767E]

[62]   Ito, H.; Nakayama, T.; Sherwood, M.; Miller, D.; Ueda, M. Characterization and Lithographic Application of Calix[4]resorcinarene Derivatives. *Chem. Mater.,* **2008**, *20*(1), 341-356.
[http://dx.doi.org/10.1021/cm7021483]

[63]   Budiana, I.G.M.N. Synthesis of Benzoic-Cinnamic Calyxseries[4]resorsinarena and Benzoil-Sinamoyl calix[4]resorcinarene and its Activity Test as Sunscreen and Adsorbent Cr (III), Pb (II) and Cd (II) In: *Dissertation*; Chemistry Department of Mathematics and Science Faculty UGM: Yogyakarta, **2015**.

[64]   Tsou, L.K.; Dutschman, G.E.; Gullen, E.A.; Telpoukhovskaia, M.; Cheng, Y.C.; Hamilton, A.D. Discovery of a synthetic dual inhibitor of HIV and HCV infection based on a tetrabutoxy-calix[4]arene scaffold. *Bioorg. Med. Chem. Lett.,* **2010**, *20*(7), 2137-2139.
[http://dx.doi.org/10.1016/j.bmcl.2010.02.043] [PMID: 20202840]

[65]   Kamada, R.; Yoshino, W.; Nomura, T. *et al.* Enhancement of transcriptional activity of mutant p53 tumor suppressor protein through stabilization of tetramer formation by calix[6]arene derivatives. *Bioorg. Med. Chem. Lett.,* **2010**, *20*(15), 4412-4415.
[http://dx.doi.org/10.1016/j.bmcl.2010.06.053] [PMID: 20605095]

[66]   Synthetic C-Methoxyphenyl Calix [4] Resorcinarene and Its Antioxidant Activity, *J Applied. Chem. Sci. (Camb.),* **2018**, *5*, 403.

[67]   Helttunen, K.; Moridi, N.; Shahgaldian, P.; Nissinen, M. Resorcinarene bis-crown silver complexes and their application as antibacterial Langmuir–Blodgett films. *Org. Biomol. Chem.,* **2012**, *10*(10), 2019-2025.
[http://dx.doi.org/10.1039/c2ob06920b] [PMID: 22290247]

# CHAPTER 4

# Pillararenes: Younger Luminescent Supramolecular Systems

**Palaniswamy Suresh**[1,*] and **Jeyaraj Belinda Asha**[1]

[1] *Supramolecular and Catalysis Lab, Dept. of Natural Products Chemistry, School of Chemistry, Madurai Kamaraj University, Madurai-625021, Tamilnadu, India*

**Abstract:** Since the development of supramolecular chemistry, synthetic macrocycles have also played an inevitable role in constructing the host-guest system. Among pillar[n]arenes, in short pillarenes, a decade-old younger member in the supramolecular family, after reported by Ogoshi *et al.* in 2008, has gained considerable attention. Due to the straightforward preparation methods, tunable cavity size, and symmetrical architecture makes it an ideal candidate in the supramolecular family. With this perspective, this chapter discusses a brief introduction to the synthesis, characterization, and structural features of different sizes of pillarenes. The presence of a confined hydrophobic and π-electron-rich cavity provided by a paraxyl ether or hydroquinone units offers a unique host-guest recognition capability towards positively charged and neutral molecules. Notably, the presence of a cavity with an aromatic wall provides a broad luminescent platform for various photophysical studies. This chapter elaborates on the contribution of pillarenes in tuning the photophysical properties of the small guest molecules and the formation of luminescent supramolecular materials. Further, the functionalization on the outer of the pillarenes has influenced the photophysical responses such as absorption and fluorescence, which paved a pathway for the development of supramolecular organic light-emitting functional material and novel sensor materials also discussed in this chapter. Finally, this chapter discusses all the progress and applications of luminescence pillarenes and their derivatives.

**Keywords:** Host, Host-Guest Chemistry, Luminescent materials, Pillararenes, Stimuli-Responsive.

## INTRODUCTION

The flourishing tailored interest from the biomolecules created a lot of attention with attraction in the development of various supramolecular systems, which are projected as bio-mimics [1]. The ever-growing attraction in supramolecular chemistry can be recognized from its versatile application in diverse platforms

---
[*] **Corresponding author Palaniswamy Suresh:** Supramolecular and Catalysis Lab, Dept. of Natural Products Chemistry, School of Chemistry, Madurai Kamaraj University, Madurai-625021, Tamilnadu, India; Tel: +919790296673; E-mail: suresh.chem@mkuniversity.ac.in

**Paulpandian Muthu Mareeswaran, Palaniswamy Suresh and Seenivasan Rajagopal (Eds.)**

that span a broad spectrum of chemistry, physics, material science, nanoscience and nanotechnology, biotechnology, biomaterials, and other fields. Because of the importance of the novel macrocyclic architecture in supramolecular science, the designing and preparation of such supramolecules have grown considerably. In the continuous evaluation of the chemistry of supramolecules, several supramolecules such as crown ethers, cyclodextrins, cucurbiturils, calixarenes and their structurally similar scaffolds have been explored extensively in different aspects from the past couple of decades. Nevertheless, owing to the development of modern synthetic strategies and updated modern characterization techniques triggered the exploration of novel macrocyclic and polymeric systems, which are considered additional feathers in the crown of supramolecular chemistry. Molecular recognition is one of the unique properties of the supramolecular systems that involve host-guest interactions and plays a vital role in the life-sustaining biological process. The non-covalent interactions originate from the molecular recognition that arises from the various bonding and nonbonding interactions such as hydrogen bonding, charge transfer, and π-π staking between molecules exhibiting molecular complementarity. To utilize such non-covalent interaction, understanding their role in the chemical reactions and exploring their applications in biomimetic systems, macrocyclic compounds provide a suitable platform. In the last century, after the Nobel award [2 - 5] Pederson, Lenn and Cram, the focus on the development of supramolecular chemistry has exponentially grown, and a spectrum of macrocyclic molecules have studied. However, owing to the remarkable molecular reorganization properties and mimicking the natural systems have drawn the attention of researchers and continuously focusing their engagement on the development of novel synthetic macrocyclic systems. In the continuous evolution of the synthetic supramolecular systems such as calixarene, cyclophane, crown ethers, and cucurbiturils, one more new type of columnar structural macrocyclic system with electron-rich cavities has been developed by Tomoki Ogoshi in 2008 [6] and named as 'pillararene'.

## Pillararenes

Pillararenes, a new type of macrocyclic synthetic supramolecule architectured molecules, consist of substituted hydroquinone units connected by methylene bridges at the 2 and 5 positions [6]. In the macrocyclic systems, the chemical structure of the pillararenes has resembled those of the calixarenes [7, 8]. To understand this younger macrocyclic system, structural features have been compared with the well-established structures of calixarenes. While comparing the chemical structure of pillararenes with calixarenes, the critical difference between both was found in the connecting position of the methylene bridges. In calixarenes, the phenolic units are joined by methylene bridges at the *meta* position, which causes the shape difference between pillararenes and calixarenes

(Fig. **1**). As a result, calixarenes show asymmetrical Calix-like structures, while pillararenes form symmetrical pillar-like structures. Indeed, the actual structure of pillarenes is different from that of typical calixarenes. Tomoki Ogoshi first reported about pillararenes as a *para*-bridge pillar-shaped novel host, pillar [5]arenes, formed by the condensation of 1,4-dimethoxybenze (DMB) with paraformaldehyde in the presence of an appropriate Lewis acid catalyst.

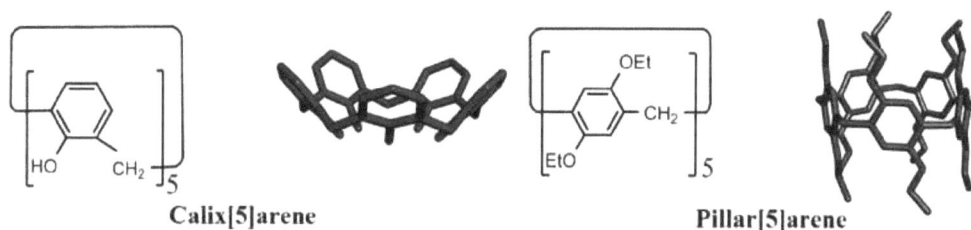

Calix[5]arene                                                       Pillar[5]arene

**Fig. (1).** Comparison of chemical structures of pillar [5]arene [P5A] and calix [5]arene. (Adopted with permission from ref. 8).

## Structure of Pillararenes

In the cyclodextrins and calixarenes, the size of the ring is determined based on the number of monomer units such as glucose and phenol, respectively. In a similar manner, the pillarenes ring size is determined by the number of hydroquinone units. In the coined name "pillar[n]arene, the letter 'n' means the presence of the number of the hydroquinone units. For example, pillar [5]arene means a cyclopentamer with five hydroquinone units. Commonly three types of pillararenes, such as pillar [5]arenes, pillar [6]arenes and pillar [7]arenes, are popular [9]. In supramolecular chemistry, while dealing with molecular recognition, understanding the structural features of the novel host molecule is extremely important because their structural features directly influence their host-

guest binding properties, being different from the basket-shaped structure of the *meta*-bridged calixarenes, pillar [5]arene (P [5]A) has a unique, symmetrical architecture [10] X-ray crystal structure of 1,4-dipropoxypillar [5]arene (DP [5]A) confirmed that it has a pentagon from the upper view and a pillar structure from the side view (Fig. **2**) [11]. The calculated average angle between the two-bridging carbon-carbon bonds is 108°, which is very close to the normal bond angle of the sp$^3$ carbon atom, 109°28'. Owing to the strain-free known bond angles P [5]A is conformationally stable. The diameter of the internal cavity of P [5]A was ~4.7 Å (Table **1**), which is close to that of curcubit [6]uril (~5.8 A) [12] and α-cyclodextrin (~4.7 Å) [13]. From another study, the crystal structure of P [6]A, has a hexagon-like cyclic structure, and the diameter of its internal cavity is ~6.7 Å, analogous to cucurbit [7]uril (~7.3 Å) [12] and β-cyclodextrin (~6.0 Å) (Fig. **3**)

[13]. Similarly, the structure of pillar [7]arene was reported with minimized energy structure. It is found that pillar [7]arene has a heptagonal pillar structure, and the diameter of its internal cavity is ~8.7 Å, analogous to the cucurbit [8]uril (~8.8 Å) and γ-cyclodextrin (~7.5 Å) [13].

P[5]A ; *n*=5
P[6]A ; *n*=6
P[7]A ; *n*=7

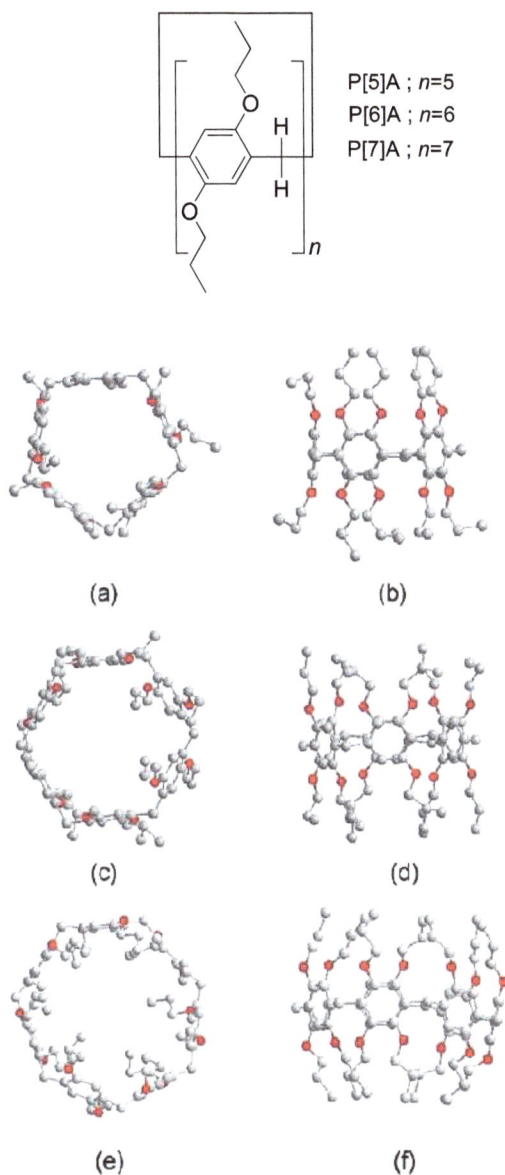

(a)  (b)

(c)  (d)

(e)  (f)

**Fig. (2).** Crystal structure of P [5]A (a,b) and P [6]A (c, d) and the minimized energy structure of P [7]A (e,f). Hydrogens were omitted for clarity. (Adopted with permission from ref. 9).

**Table 1. Calculated Structural Parameters for P [5]A, P [6]A and P [7]A based on van der Waals Radii of the Atoms**

| S. No. | | $A^c$ (Å) | $B^c$ (Å) | $H(\text{Å})^d$ | $V(\text{Å}^3)^e$ |
|:------:|:------:|:------:|:------:|:------:|:------:|
| 1. | P [5]A$^a$ | 4.7 | 13.5 | 7.8 | 152 |
| 2. | P [6]A$^a$ | 6.7 | 15.2 | 7.8 | 302 |
| 3. | P [7]A$^b$ | 8.7 | 16.9 | 7.8 | 493 |

$^a$ Based on X-ray crystal structures reported here. $^b$Based on the minimized energy model of P [7]A. $^c$Based on the diameter of the inscribed circle or the circumcircle of the regular pentagon, the regular hexagon, or the regular heptagon. $^d$Based on the distance between the two oxygen atoms on the same benzene ring. $^e$Based on the volume of the regular pentagonal pillar, hexagonal pillar, or heptagonal pillar.

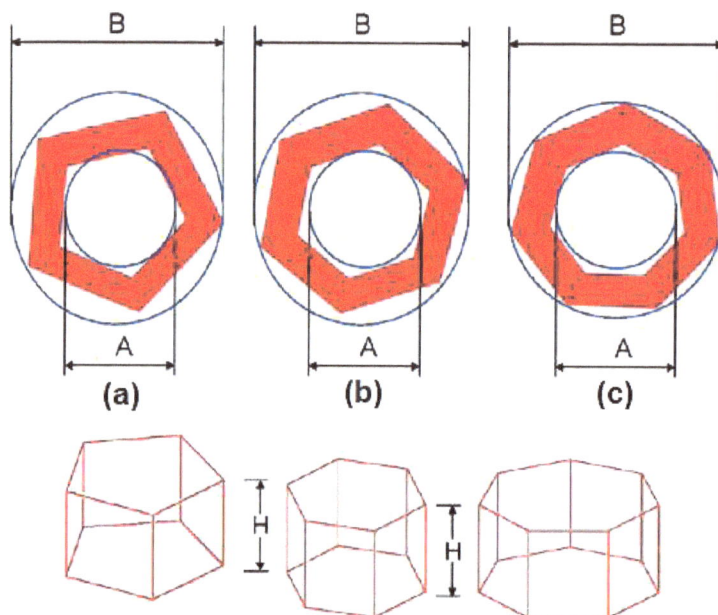

**Fig. (3).** Dimensional structural representation of a) P [5]A, b) P [6]A and c) P [7] A (Adopted with permission from ref. 9).

Though both the calixarene and pillararenes are made up of phenolic units, the conformational freedom remains in the cavities of pillararenes. In general, the symmetric pillar [5]arene exists in the two most stable confirmations, denoted as $p$S and $p$R and found in the exact equivalents in their crystal structures. Another exciting aspect is these two conformers are considered as enantiomers when they are conformationally fixed.

Understanding the factors affecting the rotation of the phenolic units helpful to recognize the interconversion between *p*S and *p*R form [14] (Fig. **4**) subsequently assists the separation of the enantiomers [15]. Further, the chemical functionalization of hydroquinone units with two different groups, the pillararene, should have conformational isomers and constitutional isomers. The presence of the hydroxyl group on the phenolic units influences the rotational behaviour of pillar [5]arene *via* the intramolecular H-bonding interactions, further other factors such as substituents, solvent, and addition of guest molecules are also controls.

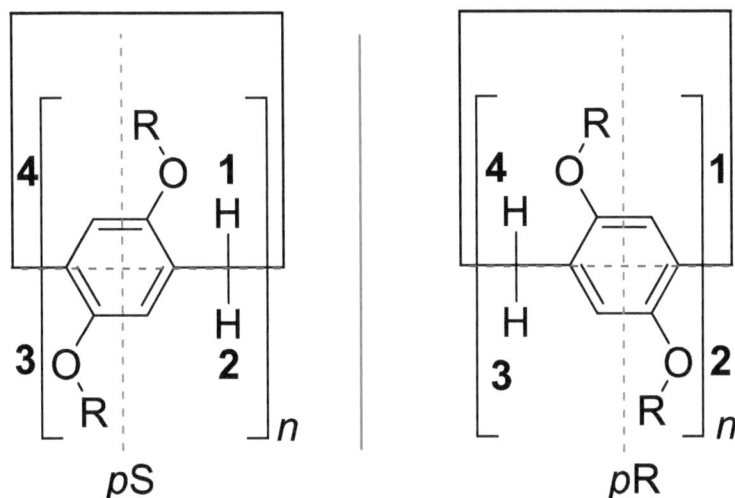

**Fig. (4).** The *p*S- and *p*R-conformations of pillar[n]arenes.

## Synthesis of Pillararenes

Compared to other supramolecular hosts, the synthesis of pillar[n]arenes are simple in terms of methodology and reaction conditions. Relatively, pillar[n]arenes are synthesized through a condensation reaction under a relatively moderate condition. For the synthesis of pillararenes, three common synthetic strategies are established (Scheme **1**) a) the Lewis acid-catalyzed condensation of 1,4-dialkoxybenzene and paraformaldehyde b) *p*-toluenesulfonic acid-catalyzed condensation of 1,4-dialkoxy-2,5-*bis*(alkoxymethyl)benzene, and c) appropriate Lewis's acid-catalyzed cyclooligomerization of 2,5-dialkoxybenzyl alcohol or 2,5-dialkoxybenzyl bromides. The common pillar[n]arene (n=5, 6 and 7) has been synthesized using these three methodologies. The first synthesis of pillar [5]arene was achieved by Ogoshi *et al.* in 2008 through the condensation of 1,4-dimethoxybenzene with paraformaldehyde in the presence of Lewis acid $BF_3.(C_5H_5)_2$ as a catalyst, reporting the corresponding crystal structure [6, 16]. Generally, cyclooligomerization of 1,4-dikeoxybenzene derivatives (monomers) and paraformaldehyde with a Lewis acid as the catalyst is the most reliable synthetic route to access pillararenes. The role of Lewis acids is inevitable in the synthesis of pillararenes, and available Lewis acids, boron trifluoride [6, 17], trifluoroacetic acid [18], trifluoromethanesulfonic acids [19], ferric chloride [20, 21], *p*-toluenesulfonic acid [22, 23], and sulfuric acid [24] were extensively employed as catalysts. On the continuous development of the other pillararenes, the primarily used monomer 1,4-disubstituted benzene was replaced with other substituted and functionalizable monomer, including 1,2,4,5-tetrasubstituted benzene derivatives, 1,4-dialkoxy-2,5-*bis*(alkoxymethyl)benzene [25, 26] 2,5-dialkoxylbenxyl alcohols [27] and 2,5-dialkoxybenzyl bromides [27] *via* condensation of monomers in the absence of paraformaldehyde. In another approach, pillararenes containing alternate repeating units were also synthesized through co-cyclooligomerization [28]. For example, by using 1,4-dimethoxybenzne and 1,4-dibutoxybenzne as comonomers, Hung's group successfully synthesized two different copillar [5]arene (Scheme **2**). The ratio of the repeating units in the final copillar [5]arenes could be tuned by carefully varying the starting ratio of the two monomers.

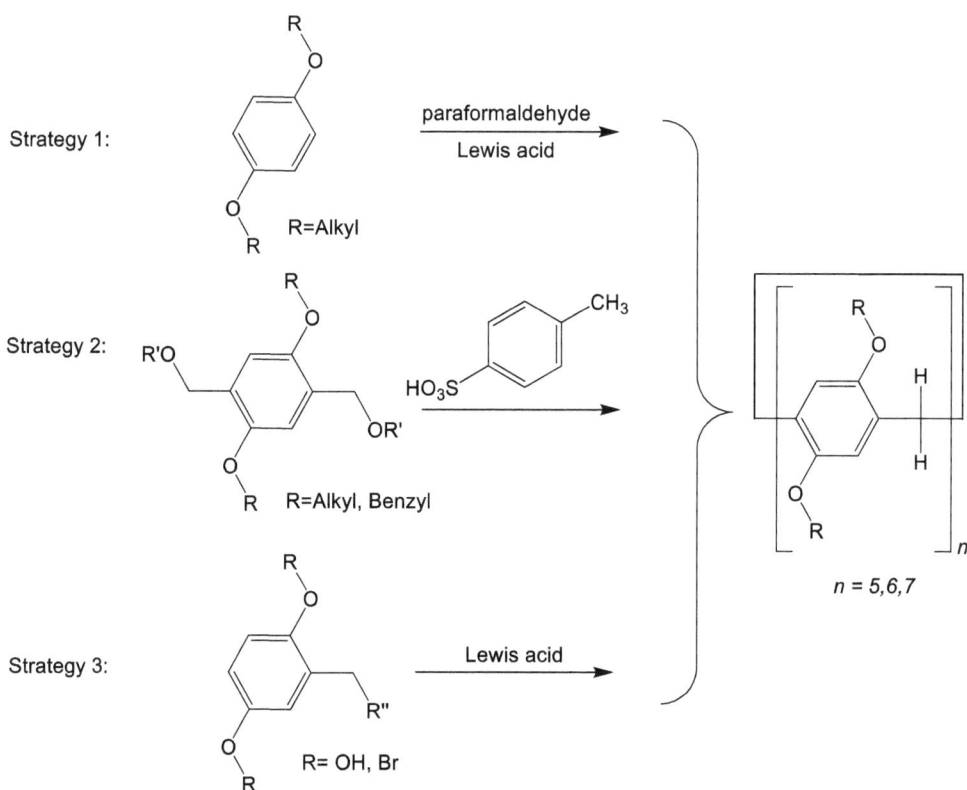

**Scheme (1).** Three general methods adopted for the synthesis of pillararenes.

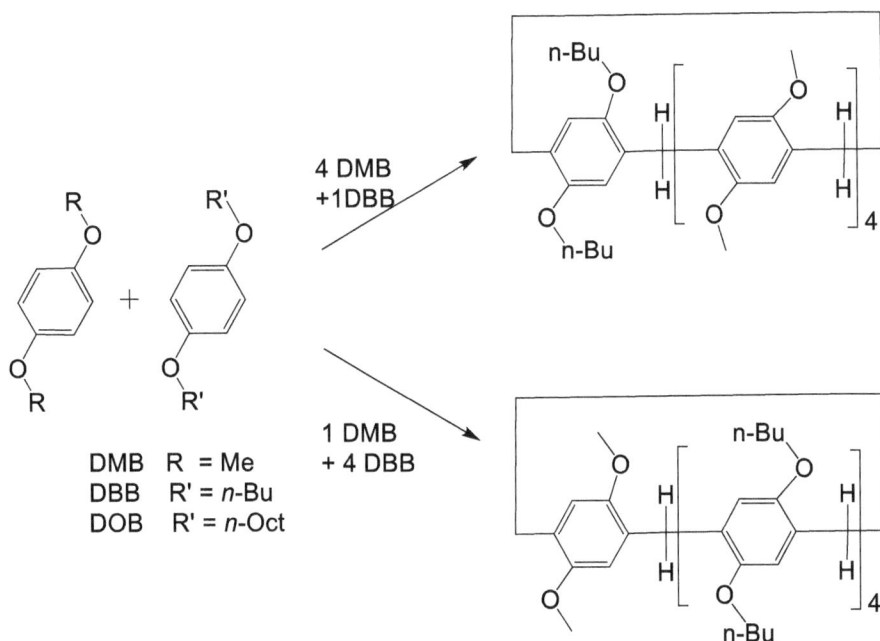

**Scheme (2).** Preparation of copillar [5]arenes by oligomerization of different hydroquinone diethers.

To expand the utility of the pillararenes in the diverse application, the functional groups present on the native pillararenes are not sufficient. Hence, native pillararenes are subjected to further functionalization through the available reactive functional groups present on the outer rims. Due to their easy accessibility for chemical modification, diverse pillararenes bearing desired functional group could be prepared conveniently. Such functionalized pillararenes could be accessed through the chemical modification of monomer or *post*-functionalization of parent pillararenes. In the macrocyclic parent pillar[n]arenes, reactive and functionalizable hydroxyl groups provide a unique platform for further chemical modification and functionalization. *E.g.*, Ogoshi *et al.* synthesized a modified DPhEpillar [5]arene from the native pillar [5]arene through the chemical modification of hydroxyl group with phenylethynyl groups (Scheme **3**), which showed new stimuli-responsive photoluminescent properties [29].

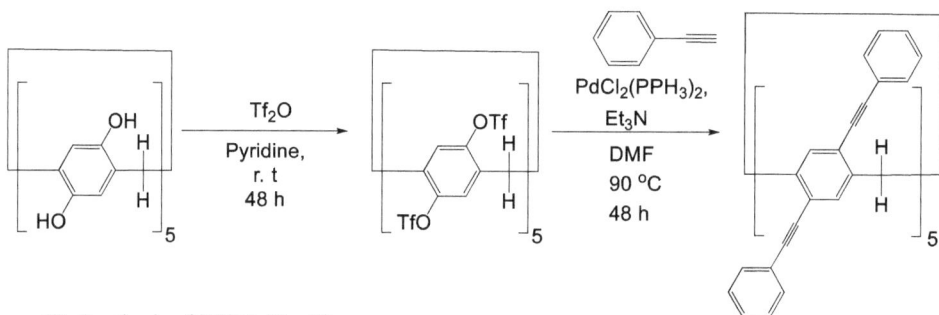

**Scheme (3).** Synthesis of DPhEpillar [5]arene.

Similarly, under alkaline condition, alkyl halide is functionalized on both sides of the pillar [5]arene, to yield the water-soluble pillararenes on further introduction

of carboxylate anions on the both upper and lower rims of pillar [5]arene [30]. Controlling the stoichiometry of the Lewis acid catalyst also yielded a selective demethylated pillar [5]arene, which supports the selective monofunctionalization. By decreasing the amount of Lewis acid catalyst, mono-de-O-methylation was carried out. Using monohydroxy pillar [5]arene as an intermediate, monofunctionalized pillar [5]arenes could be easily prepared [31]. The oxidation of alkoxy groups is another strategy for the post-functionalization of pillararenes. Ogoshi's group has devised a synthetic strategy based on oxidation catalyzed by [*bis*(trifluoroacetoxy)iodo]benzene and reduction of the pillararene units to prepare selective di- and tetrafunctionalized pillar[n]arenes (n=5,6) [32]. A simple chemical transformation such as oxidation, reduction, deprotection, and alkylation has made it possible to selectively modify pillararenes, making them suitable for advanced application.

## Host-Guest Binding Behaviours of Pillararenes

Compared to simple molecules, macromolecules are well known for their host-guest complexation behaviour. The formation of inclusion complexes, tuning or enhancing the properties of the guest molecules, is an inherent nature of the supramolecular system. In this context, pillar[n]arenes are established as novel synthetic hosts and have more advantages than traditional hosts. Compared to the other known synthetic host such as crown ethers and calixarenes, structurally pillararenes are highly symmetrical and rigid, facilitating selective binding to guest molecules. More added advantages of this synthetic host are facile access in excellent yield and pristine condition [32 - 35], which could not be possible for other supramolecules. Their pillar structure contains an interior cavity with a high π-electron density suitable for the capturing and binding of guest molecules.

CH---π interactions, hydrophobic and polar interactions inside and at the rim of the cavity are the general driving forces for the formation of host-guest inclusion complexes. Next, the selective and straightforward functionalization with desired substitutes on the benzene ring is another added advantage, enabling tweaking of their host-guest binding properties. Pillar[n]arenes without any modifications have been composed of the electron-donor hydroquinone and ionophores at both the ends. Thus, it exhibits interesting host-guest properties with electron-accepting molecules such as viologen and pyridine derivatives [36, 37] imidazolium cations [38] and *bis*(imidazolium) dications [39] in an organic medium. Another unique advantage of pillararenes are their aqueous solubility. Other hosts like cucurbiturils and cyclodextrins are usually insoluble in organic solvents. However, though the hydroxyl group is ornamented on the outer rims, their solubility in organic solvents is excellent, and it assists as necessary supplements to understand the host-guest binding behaviour in a non-aqueous medium. This unique host-guest complexation behaviour of pillararenes opened its application in all science disciplines, from medical to materials. Pillararene-based supramolecular polymers have many self-reparability, degradability, and self-adaptation characteristic features. Like other supramolecules, pillararenes cavities are more suitable for host-guest complexation [40], which are used for drug carrier studies catalytic application [41], adsorption studies [42], f-block element filters [43], also in bio-imaging applications [44]. Pillararene-based supramolecular polymers are also extensively explored in developing novel material applications due to many self-reparability, degradability, and self-adaptation characteristic features [45].

Though pillarenes and their derivatives find broad and diverse applications as a supramolecular host and as a potential host for molecular recognition, it is highly suitable for the development of reporter units. In supramolecular chemistry, the encapsulation of the guest molecule could be understood by any spectroscopic technique; in general, nuclear magnetic resonance spectroscopy (NMR) is most widely used. Indeed, it gave information about the inclusion behavior and mode of binding. Apart from complexation, the threading of guest molecules is combined with various alterations of their physical properties, which opened a gate for novel findings. The present book chapter focuses on the photophysical behaviour of the pillararenes and their corresponding inclusion complexes. In supramolecular chemistry, the fluorescence quantum yield $\Phi_F$ of a guest or a host molecule can dramatically change by host-guest complexations. The pathway of the fluorescence emission always competes with the radiationless deactivation processes of excited singlet states, and the complex formation may favour or disfavour each deactivation route. Observing and understanding the fluorescence behavior of a supramolecular host-guest system requires a fluorophore, typically in the host or in the guest molecule. In the case of the pillararenes, the presence of a cavity wall having aromatic rings makes itself highly fluorescent in nature; thus, pillararenes form versatile receptors and alter the guest molecules fluorescence behaviour upon complexation. Further, the simple and easily functionalizable outer rims are prompted to accommodate diverse fluorophores, allowing the opportunity to develop novel photoemitted supramolecular systems. Though the chemistry of pillarenes is a decade old, studies on the photophysical behaviour of pillararenes with guest molecules have extensively explored [46]. Furthermore, the tunable photophysical behaviour of the pillararenes and its inclusion complexes opened the door for the development of optical sensors and light-emitting smart materials. The successive titles provide the photophysical behaviour of pillararenes and resulting applications in emissive material applications.

## Photophysical Behaviours of Pillararenes

The literature clearly reveals that the pillararene exhibits two different types of the photoresponse. One is concerning the external conditions like temperature and pH; while performing a photophysical study, the confirmation of the structure totally varies [47]. Pillararene possess versatile phase of chemical and physical properties those already disclosed, even though to evolve its photophysical characteristic feature is interesting. Since most of the supramolecular molecules indulge in host-guest interactions [48], whereas the suitable element or organic molecule replace the guest and responsible for turn-on or turn-off response which is a fascinating as well as chemically and biologically important class of application. Herewith most of the luminescent motifs and sensor probes have designed to adapt with host-guest interactions, and few possess distinct behaviour. In most of the reports, the host-guest chemistry studied with the proton NMR and other techniques like electrospray ionization mass spectroscopy (ESI-MS) rather few reports predominantly explained the phenomenon using UV-vis. and emission spectroscopy. In the following subheadings, elaborate on the unleashed photophysical properties, behaviours, and applications of native and functionalized pillararenes.

## Pillararenes Based Luminescence Systems

Huang *et al.* studied photoresponsive host-guest complexation of pillararene. Where two types of pillararenes, pillar [5]arene and pillar [6]arene were taken as host moieties and the guest *trans* form of azobenzene could able to complex with pillar [6]arene, but it could not complex with pillar [5]arene derivatives owing to the cavity size difference between pillar [6]arene and pillar [5]arenes. The extemporaneous aggregation of its host-guest complex with the pillar [6]arene could be reversibly photo controlled by irradiation with UV and visible light and leading to a switch between irregular aggregates and vesicle-like aggregates (Scheme **4**). Hence this is a new strategy involved as a cavity size-selective and photoresponsive complex formation. A thorough investigation has been done with the selectivity of the guest; also, there exists a clear explanation for the selectivity concerning proton NMR [49].

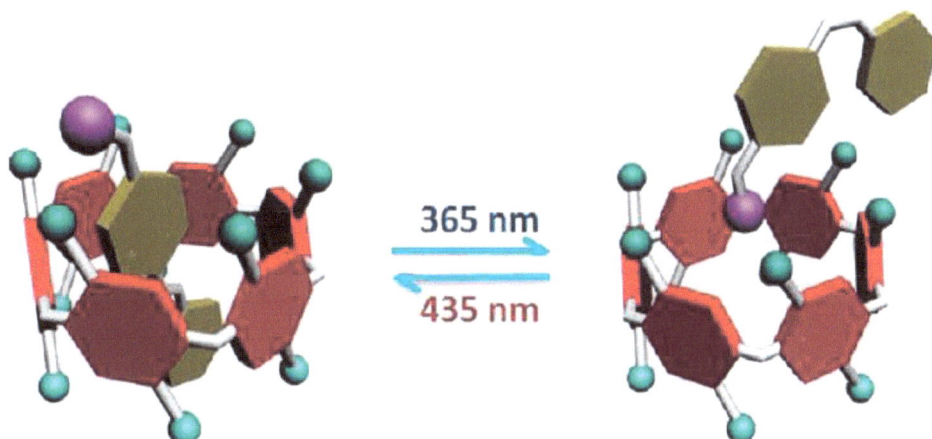

**Scheme (4).** Pillar [6]arene-based photoresponsive host-guest complexation (Adopted with permission from ref. 49).

Leyong Wang *et al.* synthesized supramolecular polymer-based pillar [5]arene dimers bridged with BODIPY and studied their application as a light-harvesting mimicking system of natural photosynthesis [50]. To develop an artificial light-harvesting system, a pillararene based system constructed with an antenna and a reaction centre. A boron-dipyrromethene (BODIPY)-bridged P [5]A dimer (**H** as donor) as an antenna and two BODIPY derivatives with mono-styryl/di-styryl group substituents (**G1** and **G2**) as a reaction centre, which resulted in an

AA/BB-type or A2/B3-type FRET-capable supramolecular polymer (Scheme **5**). In this architecture, two or three short alkyl chains with a triazole site and a cyano site at either end are employed as a neutral guest owing to their strong binding tendency towards P [5]A ($K_a = (1.2 \pm 0.2) \times 10^{-4}$ M$^{-1}$ in chloroform) [51, 52]. Here, the BODIPY dyes perfectly act as an ideal light-harvesting system due to high fluorescence yields and remarkable photostability [53, 54]. The host-guest complexation between the P [5]A and **G1** and **G2** were confirmed by $^1$H and DOSY-NMR studies. The two kinds of BODIPY bridged P [5]A supramolecular assemblies exhibited powerful absorption in a broad region from 300 to 700 nm and showed slightly different FRET effects. Due to the high complexation stability of the host-guest pair, they showed efficient FRET effects, and the energy transfer efficiencies were 51% for G1⊂ CH and 63% for G2⊂ CH. Here too high fluorescence yields and remarkable photostability of the BODIPY dyes made them an ideal light-harvesting system. This work demonstrated the light-harvesting capability of pillararene based system and extended the potential applications of pillarenes in the field of optoelectronics materials.

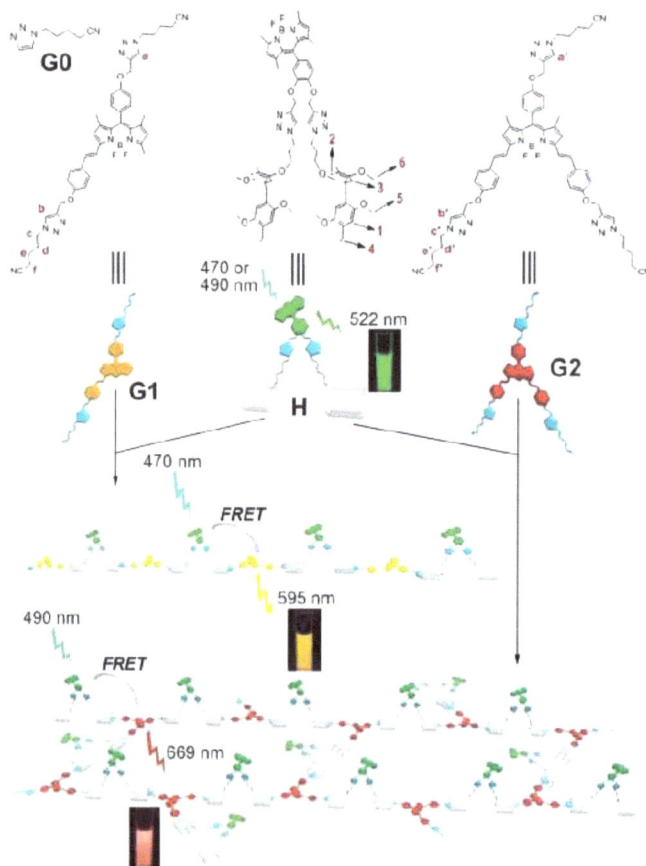

**Scheme (5).** The cartoon representation of the construction of pillar [5]arene-BODIPY based two kinds of FRET-capable light -harvesting mimics. (Adopted with permission from ref. 50).

Using water-soluble pillar [5]arene derivative (WP5), a bolo-type supra-amphiphile was prepared and explored it properties as a double fluorescence sensor [55]. Studied complexation the behaviour between WP5 and imidazolium derivatives utilized for the development of host-guest based fluorescence process. In which the host-guest interaction originates the photophysical behaviour of the WP5 with imidazolium functionalized rod-coil (**1**). From the NMR studies, dynamic light scattering, and transmission electron microscope images confirmed the self-assembly behaviour of **1** into nano-sheets in water. But the bola-type supra self-assembled into vesicles, the nano-sheet nature was disturbed and exhibited very strong fluorescence due to the influence of two bulky WP5 rings at its two ends. It is understood from suppressing the electronic coupling of the quinquephenyl aromatic rings, leading to the enhanced fluorescence. The intense fluorescence response of the bola type supra amphiphile (Scheme **6**) was weakened by two types of external stimuli. The addition of paraquat and lowering the pH decreases the fluorescence intensity. Hence, this water soluble pillarenes derived bolo-type supra-amphiphile could serve as a paraquat sensor and measure the pH changes. Also, this type of stimuli dependent response of pillarenes used for the fabrication of new kinds of smart sensing materials.

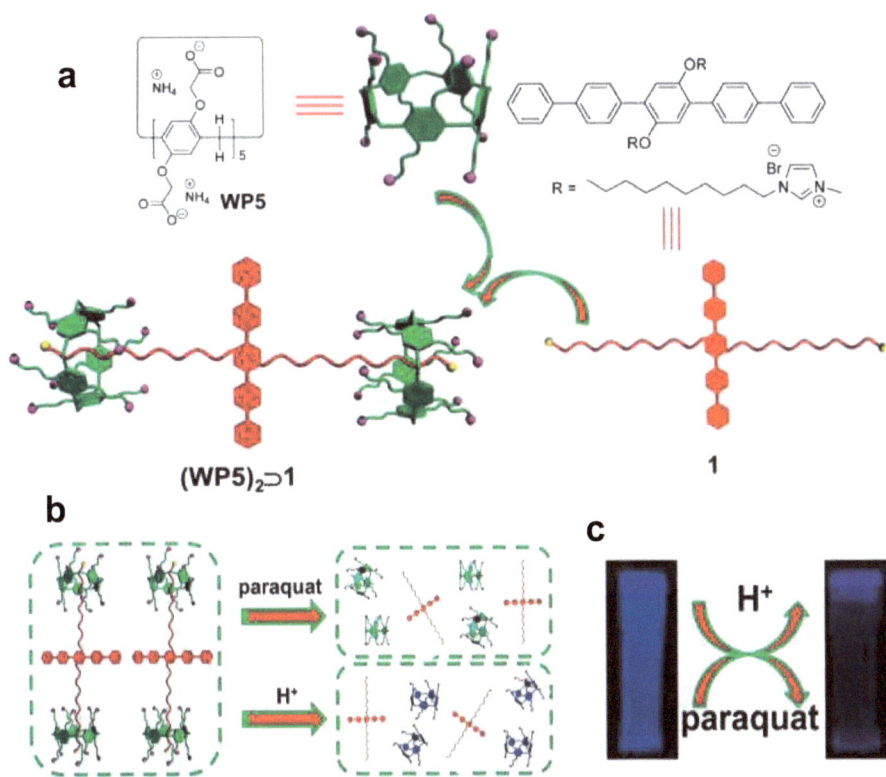

**Scheme (6).** a) Chemical structures and cartoon representations of WP5H, WP5, rod–coil molecule **1**, and bola-type supra-amphiphile (WP5)2⊃ 1b) Schematic representation of the bola-type supra-amphiphile treatment by paraquat or H$^+$ c). Fluorescent photographs of the bola-type supra-amphiphile aqueous solutions upon irradiation with a 365 nm light source before (left) and after (right) treatment with paraquat or H$^+$ (Adopted with permission from ref. 55).

## Pillararenes in Sensors Applications

Pillarenes and derivatives are used as a stabilizing agent in the synthesis of nanoparticles. The presence of polar and ionizable functional groups on the outer rims stabilize and control the size of the nanoparticles. Gold nanoparticles were synthesized using ammonium pillar [5]arene with quasi-spherical morphology and size control (up to 120 nm) through seeding growth. The surface enhanced Raman spectroscopy (SERS) studies demonstrated the role of AP [5]A, and it avoided the further nucleation of gold salts through perpendicularly adsorbed onto the gold nanoparticle surface. Further, the application of the AP [5]A stabilized Au nanoparticles was successfully tested for the detection of guest molecules such as 2-naphthoic acid and pyrene through the SERS technique [56]. In another work, Yang *et al.*, reported the synthesis of amino-pillar [5]arene through functionalization of the ethylenediamine on both rims, which selectively recolonizes $Au^{3+}$ ions [57]. This amine-functionalized pillar [5]arene showed excellent luminescence under 365 nm, which is used as a selective fluorescent probe for the selective detection of $Au^{3+}$ ion over screened twenty-four metal ions. Studies revealed that $Au^{3+}$ selectively quenches the emission intensity of the amino pillarene over other coexisting twenty-four other metal ions worked in the range of pH 1 to 13.5 and do not cause any marked interference. The detection limit is as low as $7.59 \times 10^{-8}$ mol. $L^{-1}$ and the binding ratio of the probe and $Au^{3+}$ ions calculated as 2:1.

A water-soluble carboxylatopillar [5]arene (CP [5]A), was used as a macrocyclic synthetic receptor for the *in situ* preparation of gold nanoparticles (AuNPs) through supramolecular host-AuNP interactions [58]. The CP [5]A-modified AuNPs showed good dispersion and narrow sized distributions (3.1 ± 0.5 nm) in an aqueous solution, where the presence of the five carboxylate groups on both rims of CP [5]A serving as a stabilizer for AuNPs. CP [5]A-modified AuNPs forms a strong binding guest aromatic molecule viologen I and viologen II (dimer of viologen I), and its complexation nature understood from binding and transmission electron microscope. Viologen II favours the formation of 1D and 3D assembly of CP [5]A-modified Au nanoparticles, meantime, the presence of a small amount of viologen I caused the aggregation of Au nanoparticles (Scheme 7). This supramolecular self-assembly of AuNPs with CP [5]A and stable complexation with guest molecules utilized as an optical probe for real-time detection of the herbicide, paraquat (common name of viologen I) through simple absorption spectra and naked-eye detection.

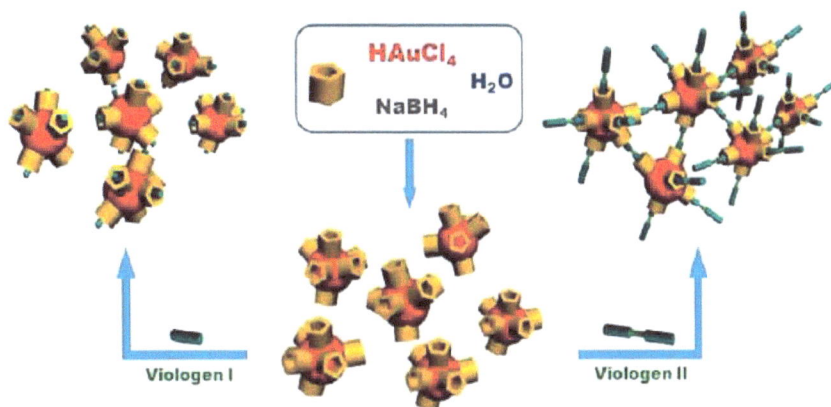

**Scheme (7).** CP [5]A-modified AuNPs and their supramolecular self-assembly upon addition of guest viologen molecules (I and II) (Adopted with permission from ref. 58).

Based on host-guest complexation behaviour of pillararene, a sidechain polypseudorotaxane was formed from modified conjugated polymeric pillar [5]arene with n-octylpyrazinium hexafluorophosphate [59]. Due to the efficient electron transfer from the conjugated backbone to the guest molecule n-octylpyrazinium cation, the sidechain polypseudorotaxanes actual fluorescence intensity got much weaker than that of the conjugated polymer (Scheme **8**). The change in the difference in the fluorescence intensity owing to the complexation behaviour could easily be distinguished by the naked eye under UV light (365 nm) and the decreased fluorescence intensity.

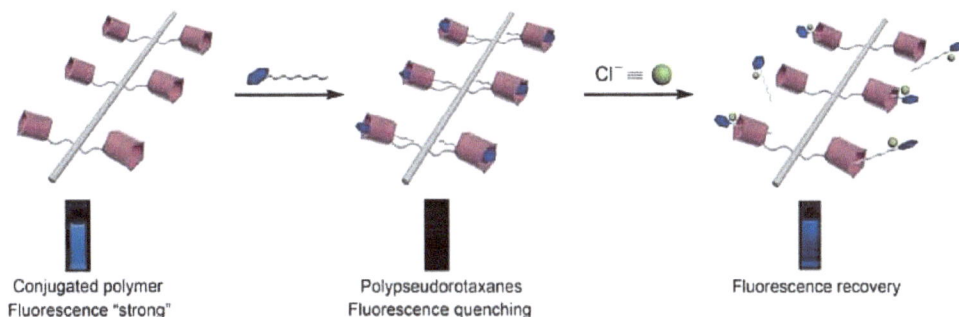

**Scheme (8).** Pillar [5]arene based sidechain polypseudorotaxanes as a chloride anion-responsive fluorescent sensor. (Adopted with permission from ref. 59).

The fluorescence intensity of the polypseudorotaxanes has been restored by adding tetrabutylammonium chloride, which leads to the disassembly process between the host-guest interaction of the conjugated polymer and n-octylpyrazinium salt. Controlling the fluorescence intensity of the pillararene derived polypseudorotaxanes in the presence of an external chloride anion through the host-guest interaction identified as an efficient anion-responsive fluorescence sensor.

Another finding by Wang *et al.* reported the application of polypseudorotaxanes constructed from pillar [5]arene moieties. Where n-octylpyrazinium cations acting as a guest and the resulting emissive nature used as a fluorescent sensor for the selective detection of halogen ions. The addition of $Cl^-$, $Br^-$ and $I^-$ as corresponding tretrabutylammonium salt collapses the pillar [5]arene-based pseudorotaxanes and polypseudorotaxanes, which recover the original emissive properties and enhance the fluorescence intensity. The fluorescent enhancement of the pseudorotaxane and the polypseudorotaxanes increases in the order of $I^- < Br^- < Cl^-$, and the differences in fluorescence intensity could be easily distinguished by naked eyes under UV light illumination, which resulted in the development of simple pillarenes based sensors for halide ions [60].

Pangkuan Chen *et al.* reported the unusual light-emitting behaviour of the homo pillarenes (MeP5, EtP5, PrP5, BuP5, DBuP5 and DBP5) and copillar [5]arenes (DBuP5 and DBP5) [6, 28, 61, 62] with combinations of alkoxy functionalities. Usually, these pillararene derivatives weakly emittive in dilute solutions. From the photophysical studies, the emission intensity of pillararene derivatives is highly enhanced while changing the solvent (Table **2**). As a representative example, a pillararene, MeP5, does not show shifts in the absorption peak position in $EtOH/CH_2Cl_2$ and poor emission with an EtOH content $f_{EtOH} < 80\%$ was observed. While further increased the EtOH content up to 98%, the MeP5 solution turned became heterogeneous in the highly polar medium of $EtOH/CH_2Cl_2$, resulted in a dramatic enhancement of luminescence intensity at $\lambda_{em} = 326$ nm. As a result, the fluorescence quantum yields were enhanced from 0.1% (in 100% $CH_2Cl_2$) to 13.2%

**Table 2. Photophysical data of pillar [5]arenes in the solid state (S) and in dilute solutions (L) in $CH_2Cl_2$.**

| Compounds | $\lambda_{em}/\lambda_{ex}$ (S) (nm) | $\Phi_S$ (%) | $\Phi_L{}^a$ (%) | $\Phi_L{}^b$ (%) |
|-----------|-------------------|--------|----------|----------|
| MEP5 | 327/271 | 19 | 0.1 | 0.5 |
| EtP5 | 325/285 | 9 | 0.1 | 0.6 |
| PrP5 | 327/280 | 5 | 0.1 | 0.6 |
| BuP5 | 353/275 | 10 | 0.1 | 0.7 |
| DBuP5 | 329/285 | 10 | 0.2 | 0.9 |
| DBP5 | 326/285 | 4 | 0.1 | 0.4 |

*(Table 2) cont.....*

| Compounds | $\lambda_{em}/\lambda_{ex}$ (S) (nm) | $\Phi_s$ (%) | $\Phi_L{}^a$ (%) | $\Phi_L{}^b$ (%) |
|---|---|---|---|---|
| [a] Fluorescence quantum efficiencies ($\Phi_L$) measured in $CH_2Cl_2$ (c = 5.0 x10$^{-4}$ M). [b]$\Phi_L$ measured in $CH_2Cl_2$ (c = 2.0 x 10$^{-5}$ M) | | | | |

(in EtOH/$CH_2Cl_2$, $f_{EtOH}$ = 98%) in response to a gradual increase of the solvent polarity. The switching of the strong luminescence rationalized through the aggregation induced emission enhancement (AIE) mechanism. These AIE turn-on fluorescence behaviours explained through their intramolecular rotations of Mep5 monomer units, which are restricted in aggregated states of their unique ring-constrained conformations (Fig. **5**). The formation of the aggregates and the respective particle size were evaluated *via* SEM images and the dynamic light scattering (DLS) technique. Added to that, the molecular packing of single crystals of MeP5 endorses the AIE behaviour. This AIEgen luminescence from the macrocyclic pillararene is used for the detection of $Fe^{3+}$ by fluorescence quenching. The major emission intensity rapidly decreased and with a quenching efficiency >90% with 1.0 equiv. of $Fe^{3+}$ ion. The limit of detection (LOD) was calculated to be 24.6 mM. This AIEgen behaviour disclosed fluorescence behaviour of weakly emissive pillararene, nevertheless, provides alternative AIEgens for luminescence sensing of guest molecules [63].

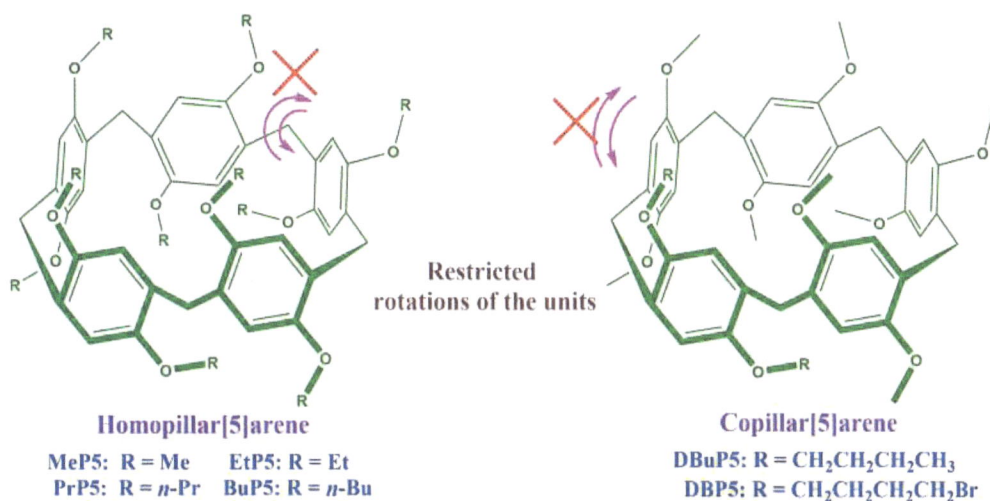

**Restricted rotations of the units**

**Homopillar[5]arene**

**MeP5: R = Me    EtP5: R = Et**
**PrP5: R = *n*-Pr    BuP5: R = *n*-Bu**

**Copillar[5]arene**

**DBuP5: R = $CH_2CH_2CH_2CH_3$**
**DBP5: R = $CH_2CH_2CH_2CH_2Br$**

**Fig. (5).** Proposed AIE mechanism for pillar [5]arenes in aggregates and solid-state emission (Adopted with permission from ref. 63).

Bingbing Shi *et al.* developed a copper(II) ion-selective self-assembly using a water-soluble pillar [5]arene containing rhodamine-B amphiphile [64] The self-assembly of rhodamine B triethylammonium bromide acting as a guest that forms inclusion complex with P [5]A, the resulting host-guest complex acts as a supra-amphiphile and self-assembled into vesicles. While adding the copper(II) chloride to the self-assembled vesicles, it turned into solid nanoparticles, accompanied by the colour change from colourless to red (Scheme **9**). The changing process from self-assembled vesicles to stable nanoparticles could easily be detected by fluorescence emission spectroscopy and naked-eye detection and explored as the tool for the detection of copper(II) ions.

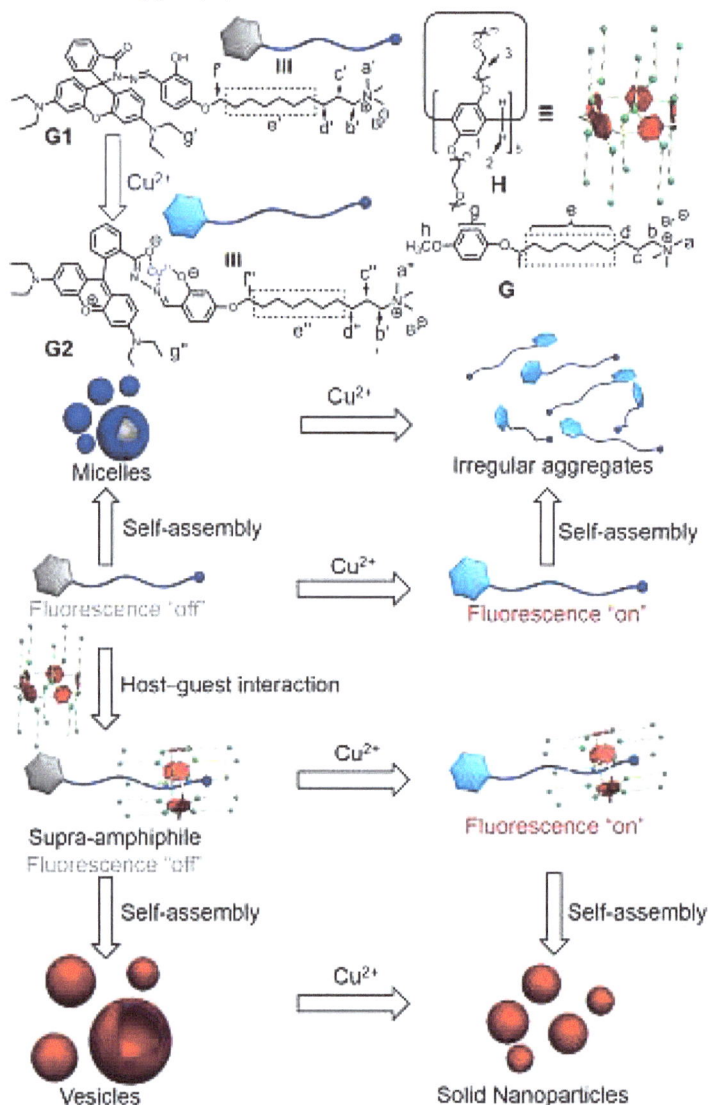

**Scheme (9).** Cartoon representation of the pillararene-based host-guest interaction and Cu(II) ion-responsive self-assembly and morphology transformation (Adopted with permission from ref. 64).

Wang *et al.* developed AIEgen supramolecular polymer from pillararenes which are having reversible fluorescence quenching properties. The supramolecular polymer was fabricated from monofunctionalized pillar [5]arene and salicylaldehyde azine linked through the C=N bonding [65]. It showed AIE behaviour in chloroform through host-guest interaction between salicylaldehyde azine-containing pillar [5]arene dimer and a homotopic (BB-type) guest (G), bearing two neutral guests at its two ends. The linear supramolecular polymer displayed a strong fluorescence at a high concentration. The linear polymer structure changed into crossed liked polymer after the addition of Cu(II), which quenched fluorescence and further addition of $CN^-$, the structure resorted as linear polymer, and AIE is also recovered. This host-guest structural dependent AIE properties of the pillararenes suggested as a promising candidate for the development of advanced sensor materials for copper (II) and cyanide ion. The self-assembly and dissociation behaviour from the pillararene host-guest interaction and corresponding emissive property were used to detect copper ions.

Mechanically interlocked molecules (MIMs) have attracted the wide attention of supramolecular chemists. Presence of rigid pillar-shaped architecture, unique host-guest properties, and easy functionalization properties [66], made pillararenes are versatile candidates for MIMs. Pillararene based MIMS has been developed and studied for optical ions sensing-based in ion-induced reaction and ion-dipole interactions [67]. Based on the host-guest chemistry of pillar [5]arene, a mechanically interlocked molecular rotaxane sensor has been constructed to detect fluoride ions. Based on a new concept of "self-immolative rotaxane sensor" the MIM has been developed using highly fluorescent anthracene-bearing pillar [5]arene as a "wheel, dinitrobenzene-containing alkyl chain as an "axle", and tertbutyldiphenylchlorosilane as a "stopper". In the mechanically interlocked rotaxane, the fluorescence nature is quenched owing to the presence of dinitrobenzene [68]. Conversely, owing to the weak binding affinity of the alkyl chain with pillararene, when the guest escaped from the host, led to in a strong fluorescence enhancement. In the presence of fluoride anion, the selective fluoride-promoted Si-O cleavage breaks the interlocked structure through the dethreading of the axle with a weak association in DMSO/THF = (1:1 mixture) polar solvents (Fig. **6**). With the self-immolative characteristic of the anthracene-bearing pillar [5]arene, the fluorescence of the system is regained with a 48-fold enhancement, making the "turn-on" sensing for fluoride ion feasible. The anthracene-bearing Pillar [5]arene showed excellent selectivity from high emission and weak-guest recognition, and the detection limit is determined as 3.8 x $10^{-7}$ M. The sensing mechanism was proved by NMR and MALDI-TOF. The selectivity of the rotaxane tested with other anions including $Cl^-$, $Br^-$, $I^-$, $ClO^{4-}$, $H_2PO^{4-}$, $HSO_4^-$, $NO_3^-$, $OAc^-$, and $PF_6^-$ none of them was able to resume the fluorescence effectively. Based on the reliable result from the rotaxane sensor, a

Based on the reliable result from the rotaxane sensor, a paper-based fluoride test kit was developed.

Another rhodamine-appended copillar [5]arene has been synthesized through an amide bond (Fig. 7). This rhodamine-modified copillararene selectively reorganizes $Hg^{2+}$ over a series of other metal ions was studied in acetonitrile medium. The sensing behaviour has been revealed by a change in colour from colourless to pink *via* spirolactam ring-opening, and the visual sensing was validated with a detection limit of 2.85 x $10^{-8}$ M. In the fluorescence analysis, turn-on response indicates the selective detection of the $Hg^{2+}$. The same rhodamine modified copillararene showed the selectivity towards $Cu^{2+}$, understood by the colour change from colourless to reddish-violet colour. On the other hand, the selectivity between $Hg^{2+}$ and $Cu^{2+}$ is discriminated using tetrabutylammonium iodide (TBAI). In the presence of TBAI, the reddish-violet colour of the (RhP) $Cu^{2+}$ complex becomes colourless with the regeneration of the original electronic spectrum of RhP. Under identical conditions, the pink colour of the RhP-$Hg^{2+}$ complex is not changed in the presence of TBAI. This study revealed the role of simply modified copillararenes in the selective sensing of metal ion in the in organic medium. The more notable advantage of the present RhP is that the $Hg^{2+}$ detection is purely organic medium, which could not be possible with other reported methods [69].

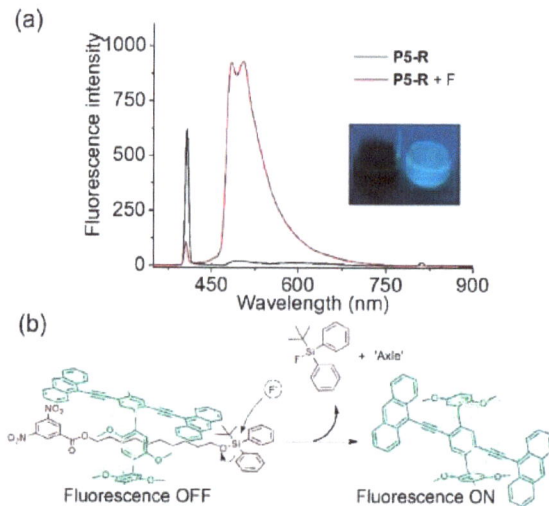

**Fig. (6).** (a) Fluorescence spectra of P5-R (0.10 mM) in DMSO/THF = 1:1 mixture at room temperature. $\lambda_{ex}$ = 408 nm. Inset photograph: emission of P5-R before (left) and after (right) addition of TBAF under 365 nm UV light. The emission appears to be blue due to the bright UV light. (b) Illustration of the mechanism of fluoride sensing. (Adopted with permission from ref. 65).

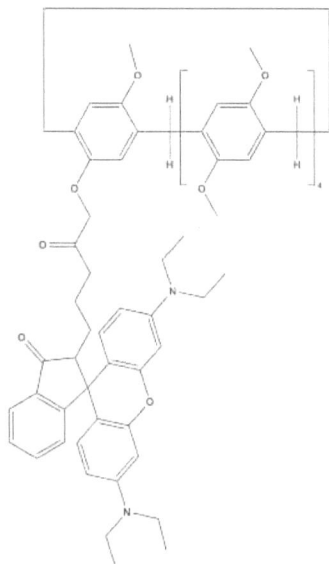

**Fig. (7).** Structures of rhodamine functionalized copillar [5]arene.

## Pillararenes in Metal Ions Scavenging

Pollution caused by the metal ions is a significant challenge from the effluent and discarded waste. In sensing and removing hazardous and toxic metals, supramolecular systems are rarely studied or yet not used in toxic heavy metal ion removal applications. Indeed, a supramolecular based sensing system has more advantages than simple molecule-based chemosensors. Supramolecular assemblies possess dual application like sensing, capture and removal of metal ions. In this context, a pillarenes based supramolecular architecture is constructed *via* assembly of a thymine-appended copillar [5]arene (TCP) (as a catcher) and a tetraphenylethylene (TPE) derivative (as an indicator) together with $Hg^{2+}$ (Scheme **10**) [70]. In this supramolecular assembly, the catcher TCP coordinate with $Hg^{2+}$ through appended thiamine and produced a liner wire; meantime, it binds with the indicator TPE through host-guest interactions. When the non-fluorescent TPE is intertwined with the complex TPE⊂TCP @ $Hg^{2+}$ formed as nanoparticles, it exhibits strong aggregation-induced emission (AIE) fluorescence. It enables convenient detection of this aggregation process and facile monitoring of the metal ion removal procedure. More importantly, the host-guest complex TPE⊂ TCP can be recycled by a simple treatment with sodium sulfide in acetone/dichloromethane. Reuse of the recycled pseudorotaxane complex showed almost no loss of activities in the sensing and capturing of $Hg^{2+}$. This feature paves the way for the practical application of the supramolecular polymer in sensing and removing heavy metals using the fluorescence detection method.

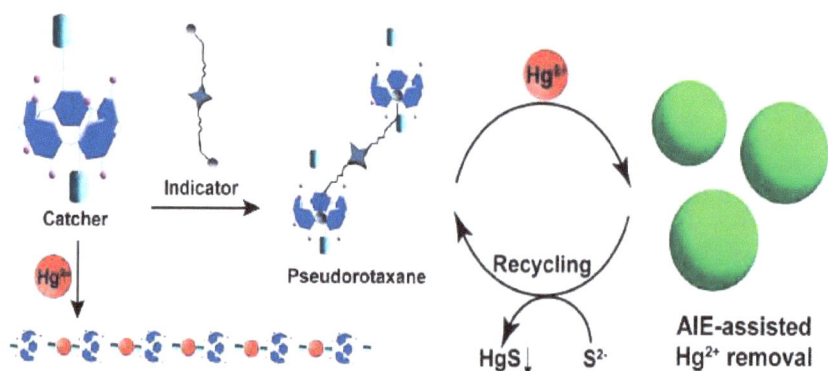

**Scheme (10).** Thymine-appended copillar [5]arene (TCP) derived supramolecular polymers for the AIE sensing and removal of $Hg^{2+}$. (Adopted with permission from ref. 70).

## Pillararenes Derived Stimuli Luminescent Materials

The demand for fluorescent materials has continuously increased in the expanding materials world and the growing demand for sensing and lighting materials. The self-assembled fluorescent materials and supramolecular derived host-guest system with emissive properties play an essential role in the development of smart materials in optoelectronics, fluorescent sensors, memory chips, and security inks. In developing such luminescent materials, aromatic rich pillarenes based supramolecular assemblies have also been explored. The presence of an electron-rich aromatic host system, the formation of strong host-guest interaction even in the organic medium and energy and electron transfer between pillarenes guest molecules make pillarenes a promising candidate in the design and development of novel luminescent material. Yong Yao *et al.* developed fluorescent supramolecular polymetric pillarenes materials based on the stable coordination between terpyridyl groups and zinc ion. A zinc ion responsive fluorescent terpyridyl group functionalized pillar [5]arene-based supramolecular polymer was constructed. When a neutral guest is possessing two triazole binding sites with a long alkyl chain, it forms a host-guest inclusion complex with pillar [5]arene cavities and the coordination between zinc ions and terpyridyl groups. The resulting linear supramolecular ternary polymer acts as an excellent fluorescent supramolecular polymer depending on the phase transition through heating and cooling. The high concentration supramolecular polymers showed reversible glue-sol phase transition upon heating and cooling in $CH_3CN/CHCl_3$ (v/v=1/1). In addition, the fluorescent supramolecular polymer showed the base-stimulus responsive property. The prepared pillarenes supramolecular polymer was fabricated as a thin film and demonstrated as a convenient test kit for detecting hydroxide (OH⁻) anions [71].

Tetraphenylethene (TPE) [72] and/or 9,10-distyrylanthracene (DSA) [73] functionalized or complexed several pillararene-based fluorescent supramolecules has reported. Due to the extended conjugation and inherent photoluminescence properties of TPE and DSA, when they are functionalized or bound with pillarenes, their enhanced fluorescence response paves a path for diverse applications. Yang *et al.* developed a pillar [5]arene-based supramolecular solid-state material with tunable luminescent properties that arise from host-guest complexation between TPE and DSA-linked pillararene derivatives with neural guest compounds [74]. The fluorescence emission of TPE and DSA-linked pillararene was strongly enhanced. Their colour was changed upon the host-guest assembly, which can be ascribed to the supramolecular assembly-induced enhanced emission and FRET between the TPE appended P [5]A and DSA. This supramolecular assembly showed different lengths of alkyl ether chains between P [5]A and TPE also tune the fluorescence emission during the DSA-guest binding and exhibit thermo and solvent dual-responsive feature.

Such types of pillararene derived tunable fluorescence supramolecular assemblies have found potential as smart optical materials. Yang *et al.* envisioned a new material from diatopic pillarenes to fabricate multi-responsive fluorescent sensors and smart supramolecular materials. Using anthracene functionalized pillar [5]arenes (**M1** and **M2**) supramolecular fluorescent system was studied based on the effect of host-guest complexation on the various neural and charged guest molecules [75]. The fluorescence enhancement was inferred while forming the cyclic dimeric daisy chains ([c2]daisy chains). Initial studies reveal that at the excitation wavelength of 370 nm, **M1** was strongly emittive than **M2**, which was 10 nm red-shifted from that of the unsubstituted anthracene. When a guest molecule added with **M1,** the fluorescence intensity increased owing to the formation of pseudorotaxane of **M1** with the guest (Scheme **11**). The fluorescence enhancement mechanism of **M1** has thoroughly investigated employing theoretical calculation and time-resolved fluorescent measurements. The motions of the anthracene unit are exceptionally restricted in **M1** due to the formation of a [c2]daisy chain structure, which blocks the non-radiative process and accelerates the radiative decay pathways. Therefore, the fluorescence intensity of **M1** is dramatically enhanced. However, an external stimulus such as temperature, solvent composition, pH and counter anions could switch the formation and deformation of [c2]daisy chains of **M1** in solution, resulting in the on-off switching of its fluorescence emission. The strong binding and the construction of [c2]daisy with guest realized from Job's plot fluorescence titration and demonstrated the 1:1 complexation of **M1** with guest and the association constant calculated as $4.72 \times 10^3$ M$^{-1}$. Thus, this simple pillararene derived material can act as a multi-responsive supramolecular sensor, exhibiting the on-off switching of its fluorescence emission.

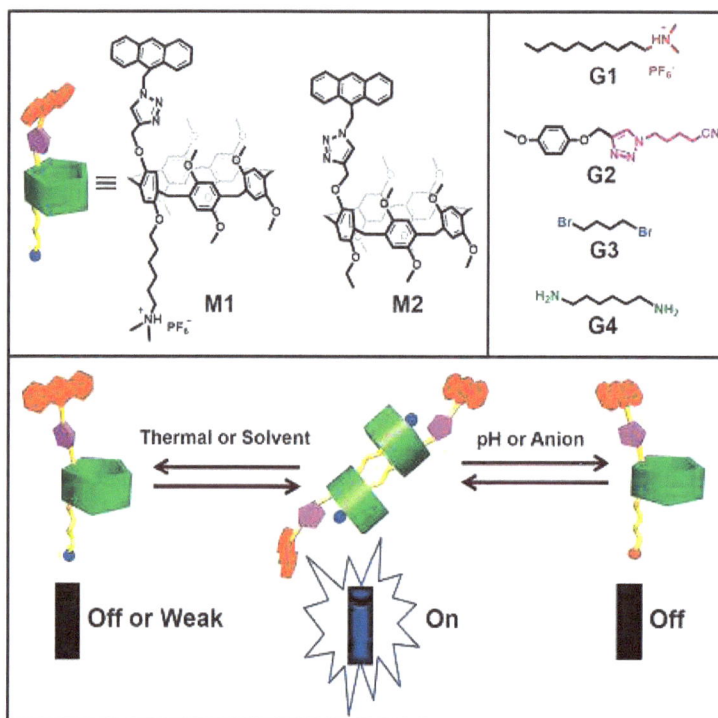

**Scheme (11).** a) Structure of anthracene functionalized pillar [5]arenes (**M1** and **M2**) and guest molecules. b) Switchable emissions of M1 under different external stimuli: thermal, solvent composition, pH and anion. (Adopted with permission from ref. 75).

Zhao *et al.* prepared a thermo-responsive fluorescent vesicle from pillarenes for temperature-dependent fluorescent sensors [76]. The out rim of the pillar [5]arene is mono-functionalized with fluorescein (FITC) and formed fluorescent vesicles through self-assembly. The formation of the fluorescent vesicles was confirmed in acetone medium using DLS and developed as a larger aggregate with a diameter of around 50 nm at a CAC value of 0.008 mM. The self-assembly and the formation of vesicles were further proved by the TEM and SEM analysis. The possibility of self-inclusion of FITC into P [5] is ruled out owing to the size of the FITC molecule being bigger than the cavity of pillar [5]arene (D = 1.35 nm). Under this circumstance, the fluorescein modified P [5]A formed fluorescent vesicles *via* the self-assembly process. The fluorescent behaviours and the formation of self-assembly are understood through intermolecular π-π interaction between the FITC. The fluorescence emission intensity gradually enhanced up to the CAC concentration, and a further increase in the concentration beyond the CAC leads to a drop in fluorescence intensity. This behaviour is inferred from DLS studies and electron microscopic images.

Similarly, the formation of self-assembly vesicles is directly influenced by temperature. The vesicles are stable under lower temperatures (10 °C) and exhibit a good fluorescence response. Meanwhile, the rise in the temperature (up to 50 °C) leads to a gradual decrease in fluorescence intensity due to the dissociations of self-assembly. These thermo-responsive assembly/disassembly processes and corresponding fluorescence emittive response of the compound fluorescein modified P [5]A are reversible, which could be used as thermo-responsive fluorescent sensors.

Functional soft materials attracted broad research interest in the evolving materials world because of their controllable mobility and flexibility. Among gel types, materials have widely been considered for diverse applications. In supramolecular chemistry, the host-guest complexation through non-covalent interaction in macrocyclic moieties has been utilized to form supramolecular gels. Pillararenes have also been employed as a suitable supramolecular host for the preparation of various gel materials through host-guest reorganization [77]. Such chemically cross-linked pillararene derived polymeric gels showed stability and flexibility with excellent stimuli-responsive behaviours, which find broad sensing and material applications [78]. Similarly, a novel pillar [5]arene-based fluorescent supramolecular gel was reported using the two cooperative interactions [79]. Using acrylate-modified pillar [5]arene (ATP5) as a host and biquaternary ammonium salt (G) as a guest molecule, a cross-linked host-guest supramolecule (HGSM) was formed. Further copolymerization with the acrylate functionalized monomer under UV light (365 nm) resulted in a supramolecular gel (Scheme **12**). The synthesized chemically cross-linked supramolecular gel showed a remarkable self-healing capacity due to the dynamic self-assembly induced by non-covalent host-guest interactions. The covalent bonding formed *via* UV-initiated polymerization provides a strongly cross-linked network with compression resistance and anti-sliding ability. This pillarenes based soft gel showed high elastic and strong fatigue-resistant capacity; further, the host-guest recognition enables sufficient self-heal properties. This gel offered a dual response to the temperature and pH changes. It showed reversible gel-sol phase transitions on several heating-cooling cycles. This temperature response of the gel was further inferred from emission properties. While increasing the temperature, the emission intensity of the gel decreased because of the traditional mechanism of thermal quenching of luminance.

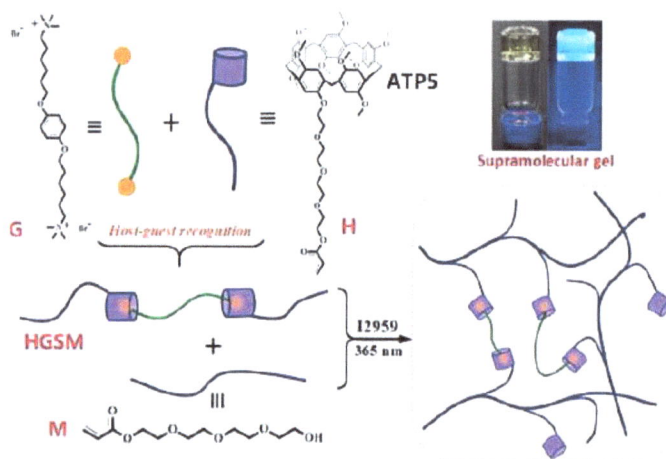

**Scheme (12).** Acrylate-modified pillar [5]arene derived gelation induced by copolymerization. (Adopted with permission from ref. 79).

Yang *et al.* reported a blue fluorescent supramolecular polymer based on pillar [5]arene tetramer [80]. A new type of pillararene derivative is prepared from functionalizing TPE with four pillar [5]arene. This tetraphenylethene-bridged pillarenes tetramer form an inclusion complex with cyano sites containing triazole-based neutral linker containing (G2) and resulted in a pillarenes-based stimuli-responsive supramolecular gel with strong blue fluorescence. The presence of TPE promotes the AIE; however, noticeable fluorescence enhancement is due to the supramolecular polymerization attributed to restriction of internal rotation (RIR) of the TPE cores pillarenes functionalized tetramers. Furthermore, blue supramolecular gels formed at a host concentration of 70 mmol $L^{-1}$ with a host-guest molar ratio of 1:2. The supramolecular polymerization and depolymerization and the resulting on-off switchable fluorescence properties and solvent (chloroform) induced assembly and disassembly of the supramolecular system facilitate this blue emittive nature of these pillararene tetramers. Also, the fluorescent intensity of pillararene tetramers with guest assemblies in chloroform showed suitable temperature and solvent responsiveness.

## CONCLUSIONS

In this chapter, we discussed the younger, attractive, and reactive synthetic supramolecular host pillararenes. In the first part, the origin and structural features of different types of pillararenes are discussed. In which the size, crystal structure and possible confirmations are presented. Next, the synthesis of a decade-old macromolecular host is addressed in brief. Where the general synthesis of native, copillararenes and functionalized pillararenes are discussed. In the synthesis, for native and functionalized derivatives, a representative synthesis is discussed with limited examples. Further, the host-guest complexation with guest molecules and complexation behaviour are deliberated. Additionally, the complexation behaviour, change in the guest's photophysical behaviour, and the resulting novel luminescent supramolecular materials are reviewed in detail. The preparation of luminescent both native and diverse functionalized pillararenes and their application in the development of sensors, artificial light-harvesting system and stimuli-responsive signaling material are discussed with appropriate examples. Overall content discussed in this chapter reveals the straightforward synthesis with high yield and corresponding inherent photophysical properties of the pillarenes will open the opportunity to develop novel and reliable light-emitting and optoelectronic materials in the near future.

## CONSENT OF PUBLICATION

## CONFLICT OF INTEREST

The author declares no conflict of interest, financial or otherwise.

## ACKNOWLEDGEMENT

Dr. P. S. gratefully acknowledges the financial support from Science and Engineering Research Board (SERB), New Delhi, India University Grants Commission (UGC), New Delhi, India and Council for Scientific and Industrial Research CSIR, New Delhi, India.

## REFERENCES

[1]    Williams, G.T.; Haynes, C.J.E.; Fares, M.; Caltagirone, C.; Hiscock, J.R.; Gale, P.A. Advances in applied supramolecular technologies. *Chem. Soc. Rev.,* **2021**, *50*(4), 2737-2763.
[http://dx.doi.org/10.1039/D0CS00948B] [PMID: 33438685]

[2]    Nobel Prize in Chemistry: Jean-Marie Lehn with Donald J. Cram and Charles J. Pedersen in **1987**, La lettre du Collège de France.

[3]    Lehn, J.M. Supramolecular chemistry - scope and perspectives molecules, supermolecules and molecular devices. *Angew. Chem. Int. Ed. Engl.,* **1988**, *27*(1), 89-112.
[http://dx.doi.org/10.1002/anie.198800891]

[4]    Cram, D.J. The Design of Molecular Hosts, Guests, and Their Complexes (Nobel Lecture). *Angew. Chem. Int. Ed. Engl.,* **1988**, *27*(8), 1009-1020.
[http://dx.doi.org/10.1002/anie.198810093]

[5]    Lehn, J.M. Perspectives in supramolecular chemistry - from molecular recognition towards molecular information-processing and self-organization. *Angew. Chem. Int. Ed. Engl.,* **1990**, *29*(11), 1304-1319.
[http://dx.doi.org/10.1002/anie.199013041]

[6] Ogoshi, T.; Kanai, S.; Fujinami, S.; Yamagishi, T.; Nakamoto, Y. *para*-Bridged symmetrical pillar[5]arenes: their Lewis acid catalyzed synthesis and host-guest property. *J. Am. Chem. Soc.,* **2008**, *130*(15), 5022-5023.
[http://dx.doi.org/10.1021/ja711260m] [PMID: 18357989]

[7] Guo, D.S.; Liu, Y. Calixarene-based supramolecular polymerization in solution. *Chem. Soc. Rev.,* **2012**, *41*(18), 5907-5921.
[http://dx.doi.org/10.1039/c2cs35075k] [PMID: 22617955]

[8] Song, N.; Kakuta, T.; Yamagishi, T.; Yang, Y.W.; Ogoshi, T. Molecular-Scale Porous Materials Based on Pillar[n]arenes. *Chem,* **2018**, *4*(9), 2029-2053.
[http://dx.doi.org/10.1016/j.chempr.2018.05.015]

[9] Xue, M.; Yang, Y.; Chi, X.; Zhang, Z.; Huang, F. Pillararenes, a new class of macrocycles for supramolecular chemistry. *Acc. Chem. Res.,* **2012**, *45*(8), 1294-1308.
[http://dx.doi.org/10.1021/ar2003418] [PMID: 22551015]

[10] Han, C.; Ma, F.; Zhang, Z.; Xia, B.; Yu, Y.; Huang, F. DIBPillar[n]arenes (*n* = 5, 6): syntheses, X-ray crystal structures, and complexation with *n*-octyltriethyl ammonium hexafluorophosphate. *Org. Lett.,* **2010**, *12*(19), 4360-4363.
[http://dx.doi.org/10.1021/ol1018344] [PMID: 20831166]

[11] Yu, G.; Xue, M.; Zhang, Z.; Li, J.; Han, C.; Huang, F. A water-soluble pillar[6]arene: synthesis, host-guest chemistry, and its application in dispersion of multiwalled carbon nanotubes in water. *J. Am. Chem. Soc.,* **2012**, *134*(32), 13248-13251.
[http://dx.doi.org/10.1021/ja306399f] [PMID: 22827832]

[12] Lee, J.W.; Samal, S.; Selvapalam, N.; Kim, H.J.; Kim, K. Cucurbituril homologues and derivatives: new opportunities in supramolecular chemistry. *Acc. Chem. Res.,* **2003**, *36*(8), 621-630.
[http://dx.doi.org/10.1021/ar020254k] [PMID: 12924959]

[13] Hapiot, F.; Tilloy, S.; Monflier, E. Cyclodextrins as supramolecular hosts for organometallic complexes. *Chem. Rev.,* **2006**, *106*(3), 767-781.
[http://dx.doi.org/10.1021/cr050576c] [PMID: 16522008]

[14] Ogoshi, T.; Shiga, R.; Yamagishi, T.; Nakamoto, Y. Planar-chiral pillar[5]arene: chiral switches induced by multiexternal stimulus of temperature, solvents, and addition of achiral guest molecule. *J. Org. Chem.,* **2011**, *76*(2), 618-622.
[http://dx.doi.org/10.1021/jo1021508] [PMID: 21190363]

[15] Ogoshi, T.; Masaki, K.; Shiga, R.; Kitajima, K.; Yamagishi, T. Planar-chiral macrocyclic host pillar[5]arene: no rotation of units and isolation of enantiomers by introducing bulky substituents. *Org. Lett.,* **2011**, *13*(5), 1264-1266.
[http://dx.doi.org/10.1021/ol200062j] [PMID: 21288006]

[16] Ogoshi, T.; Aoki, T.; Kitajima, K.; Fujinami, S.; Yamagishi, T.; Nakamoto, Y. Facile, rapid, and high-yield synthesis of pillar[5]arene from commercially available reagents and its X-ray crystal structure. *J. Org. Chem.,* **2011**, *76*(1), 328-331.
[http://dx.doi.org/10.1021/jo1020823] [PMID: 21142202]

[17] Hu, X.B.; Chen, Z.; Chen, L.; Zhang, L.; Hou, J.L.; Li, Z.T. Pillar[n]arenes (n = 8–10) with two cavities: synthesis, structures and complexing properties. *Chem. Commun. (Camb.),* **2012**, *48*(89), 10999-11001.
[http://dx.doi.org/10.1039/c2cc36027f] [PMID: 23038422]

[18] Boinski, T.; Szumna, A. A facile, moisture-insensitive method for synthesis of pillar[5]arenes—the solvent templation by halogen bonds. *Tetrahedron,* **2012**, *68*(46), 9419-9422.
[http://dx.doi.org/10.1016/j.tet.2012.09.006]

[19] Wang, K.; Tan, L.L.; Chen, D.X. *et al.* One-pot synthesis of pillar[n]arenes catalyzed by a minimum amount of TfOH and a solution-phase mechanistic study. *Org. Biomol. Chem.,* **2012**, *10*(47), 9405-

9409.
[http://dx.doi.org/10.1039/c2ob26635k] [PMID: 23108705]

[20]   Tao, H.; Cao, D.; Liu, L.; Kou, Y.; Wang, L.; Meier, H. Synthesis and host-guest properties of pillar[6]arenes. *Sci. China Chem.,* **2012**, *55*(2), 223-228.
[http://dx.doi.org/10.1007/s11426-011-4427-3]

[21]   Wang, X.; Han, K.; Li, J.; Jia, X.; Li, C. Pillar[5]arene–neutral guest recognition based supramolecular alternating copolymer containing [c2]daisy chain and *bis*-pillar[5]arene units. *Polym. Chem.,* **2013**, *4*(14), 3998-4003.
[http://dx.doi.org/10.1039/c3py00462g]

[22]   Kou, Y.; Cao, D.; Tao, H.; Wang, L.; Liang, J.; Chen, Z.; Meier, H. Synthesis and inclusion properties of pillar[n]arenes. *J. Incl. Phenom. Macrocycl. Chem.,* **2013**, *77*(1-4), 279-289.
[http://dx.doi.org/10.1007/s10847-012-0242-5]

[23]   Han, C.; Zhang, Z.; Chi, X.; Zhang, M.; Yu, G.; Huang, F. Synthesis of 1,4-Bis( *n* - propoxy)pillar[7]arene and Its Host-guest Chemistry. *Huaxue Xuebao,* **2012**, *70*(17), 1775-1778.
[http://dx.doi.org/10.6023/A12060296]

[24]   Santra, S.; Kopchuk, D.S.; Kovalev, I.S.; Zyryanov, G.V.; Majee, A.; Charushin, V.N.; Chupakhin, O.N. Solvent-free synthesis of pillar[6]arenes. *Green Chem.,* **2016**, *18*(2), 423-426.
[http://dx.doi.org/10.1039/C5GC01505G]

[25]   Cao, D.; Kou, Y.; Liang, J.; Chen, Z.; Wang, L.; Meier, H. A facile and efficient preparation of pillararenes and a pillarquinone. *Angew. Chem. Int. Ed.,* **2009**, *48*(51), 9721-9723.
[http://dx.doi.org/10.1002/anie.200904765] [PMID: 19924749]

[26]   Kou, Y.; Tao, H.; Cao, D.; Fu, Z.; Schollmeyer, D.; Meier, H. Synthesis and Conformational Properties of Nonsymmetric Pillar[5]arenes and Their Acetonitrile Inclusion Compounds. *Eur. J. Org. Chem.,* **2010**, *2010*(33), 6464-6470.
[http://dx.doi.org/10.1002/ejoc.201000718]

[27]   Ma, Y.; Zhang, Z.; Ji, X.; Han, C.; He, J.; Abliz, Z.; Chen, W.; Huang, F. Preparation of Pillar[*n*]arenes by Cyclooligomerization of 2,5-Dialkoxybenzyl Alcohols or 2,5-Dialkoxybenzyl Bromides. *Eur. J. Org. Chem.,* **2011**, *2011*(27), 5331-5335.
[http://dx.doi.org/10.1002/ejoc.201100698]

[28]   Zhang, Z.; Xia, B.; Han, C.; Yu, Y.; Huang, F. Syntheses of copillar[5]arenes by co-oligomerization of different monomers. *Org. Lett.,* **2010**, *12*(15), 3285-3287.
[http://dx.doi.org/10.1021/ol100883k] [PMID: 20583777]

[29]   Ogoshi, T.; Umeda, K.; Yamagishi, T.; Nakamoto, Y. Through-space π-delocalized Pillar[5]arene. *Chem. Commun. (Camb.),* **2009**, (32), 4874-4876.
[http://dx.doi.org/10.1039/b907894k] [PMID: 19652810]

[30]   Ogoshi, T.; Hashizume, M.; Yamagishi, T.; Nakamoto, Y. Synthesis, conformational and host–guest properties of water-soluble pillar[5]arene. *Chem. Commun. (Camb.),* **2010**, *46*(21), 3708-3710.
[http://dx.doi.org/10.1039/c0cc00348d] [PMID: 20390156]

[31]   Ogoshi, T.; Demachi, K.; Kitajima, K.; Yamagishi, T. Monofunctionalized pillar[5]arenes: synthesis and supramolecular structure. *Chem. Commun. (Camb.),* **2011**, *47*(25), 7164-7166.
[http://dx.doi.org/10.1039/c1cc12333e] [PMID: 21607270]

[32]   Ogoshi, T.; Yamafuji, D.; Kotera, D.; Aoki, T.; Fujinami, S.; Yamagishi, T. Clickable di- and tetrafunctionalized pillar[n]arenes (n = 5, 6) by oxidation-reduction of pillar[n]arene units. *J. Org. Chem.,* **2012**, *77*(24), 11146-11152.
[http://dx.doi.org/10.1021/jo302283n] [PMID: 23198965]

[33]   Cao, D.; Meier, H. Pillar[ *n* ]arenes-a Novel, Highly Promising Class of Macrocyclic Host Molecules. *Asian J. Org. Chem.,* **2014**, *3*(3), 244-262.
[http://dx.doi.org/10.1002/ajoc.201300224]

[34] Wang, M.; Du, X.; Tian, H.; Jia, Q.; Deng, R.; Cui, Y.; Wang, C.; Meguellati, K. Design and synthesis of self-included pillar[5]arene-based bis-[1]rotaxanes. *Chin. Chem. Lett.,* **2019**, *30*(2), 345-348.
[http://dx.doi.org/10.1016/j.cclet.2018.10.014]

[35] Cao, D.; Meier, H. Synthesis of Pillar[6]arenes and Their Host–Guest Complexes. *Synthesis,* **2015**, *47*(8), 1041-1056.
[http://dx.doi.org/10.1055/s-0034-1378688]

[36] Li, C.; Xu, Q.; Li, J.; Feina Yao, ; Jia, X. Complex interactions of pillar[5]arene with paraquats and *bis*(pyridinium) derivatives. *Org. Biomol. Chem.,* **2010**, *8*(7), 1568-1576.
[http://dx.doi.org/10.1039/b920146g] [PMID: 20336850]

[37] Ogoshi, T.; Yamafuji, D.; Aoki, T.; Yamagishi, T. Photoreversible transformation between seconds and hours time-scales: threading of pillar[5]arene onto the azobenzene-end of a viologen derivative. *J. Org. Chem.,* **2011**, *76*(22), 9497-9503.
[http://dx.doi.org/10.1021/jo202040p] [PMID: 22004422]

[38] Ogoshi, T.; Tanaka, S.; Yamagishi, T.; Nakamoto, Y. Ionic Liquid Molecules (ILs) as Novel Guests for Pillar[5]arene: 1:2 Host–Guest Complexes between Pillar[5]arene and ILs in Organic Media. *Chem. Lett.,* **2011**, *40*(1), 96-98.
[http://dx.doi.org/10.1246/cl.2011.96]

[39] Li, C.; Zhao, L.; Li, J.; Ding, X.; Chen, S.; Zhang, Q.; Yu, Y.; Jia, X. Self-assembly of [2]pseudorotaxanes based on pillar[5]arene and *bis*(imidazolium) cations. *Chem. Commun. (Camb.),* **2010**, *46*(47), 9016-9018.
[http://dx.doi.org/10.1039/c0cc03575k] [PMID: 21057678]

[40] Ma, Y.; Ji, X.; Xiang, F.; Chi, X.; Han, C.; He, J.; Abliz, Z.; Chen, W.; Huang, F. A cationic water-soluble pillar[5]arene: synthesis and host–guest complexation with sodium 1-octanesulfonate. *Chem. Commun. (Camb.),* **2011**, *47*(45), 12340-12342.
[http://dx.doi.org/10.1039/c1cc15660h] [PMID: 22011730]

[41] Ogoshi, T.; Ueshima, N.; Yamagishi, T. An amphiphilic pillar[5]arene as efficient and substrate-selective phase-transfer catalyst. *Org. Lett.,* **2013**, *15*(14), 3742-3745.
[http://dx.doi.org/10.1021/ol4016546] [PMID: 23815706]

[42] Ogoshi, T.; Sueto, R.; Yoshikoshi, K.; Yamagishi, T. One-dimensional channels constructed from per-hydroxylated pillar[6]arene molecules for gas and vapour adsorption. *Chem. Commun. (Camb.),* **2014**, *50*(96), 15209-15211.
[http://dx.doi.org/10.1039/C4CC06591C] [PMID: 25339195]

[43] Fang, Y.; Wu, L.; Liao, J.; Chen, L.; Yang, Y.; Liu, N.; He, L.; Zou, S.; Feng, W.; Yuan, L. Pillar[5]arene-based phosphine oxides: novel ionophores for solvent extraction separation of f-block elements from acidic media. *RSC Advances,* **2013**, *3*(30), 12376-12383.
[http://dx.doi.org/10.1039/c3ra41251b]

[44] Chen, J.Y.; Li, X.Y.; Wu, J.; Wu, Y.; Kuang, G.C. Pillar[5]arene-BODIPY host-guest interaction induced fluorescence enhancement and lysosomes targetable bioimaging in dilute solution. *Tetrahedron,* **2020**, *76*(48), 131698-131703.
[http://dx.doi.org/10.1016/j.tet.2020.131698]

[45] Li, H.; Yang, Y.; Xu, F.; Liang, T.; Wen, H.; Tian, W. Pillararene-based supramolecular polymers. *Chem. Commun. (Camb.),* **2019**, *55*(3), 271-285.
[http://dx.doi.org/10.1039/C8CC08085B] [PMID: 30418439]

[46] Cao, D.; Meier, H. Pillararene-based fluorescent sensors for the tracking of organic compounds. *Chin. Chem. Lett.,* **2019**, *30*(10), 1758-1766.
[http://dx.doi.org/10.1016/j.cclet.2019.06.026]

[47] Yang, J.; Shao, L.; Yu, G. Construction of pillar[6]arene-based $CO_2$ and UV dual-responsive supra-amphiphile and application in controlled self-assembly. *Chem. Commun. (Camb.),* **2016**, *52*(15),

3211-3214.
[http://dx.doi.org/10.1039/C5CC10617F] [PMID: 26813155]

[48]    Shi, B.; Shangguan, L.; Wang, H. *et al.* Pillar[5]arene-Based Molecular Recognition Induced Crystal-to-Crystal Transformation and Its Application in Adsorption of Adiponitrile in Water. *ACS Materials Letters,* **2019**, *1*(1), 111-115.
[http://dx.doi.org/10.1021/acsmaterialslett.9b00163]

[49]    Yu, G.; Han, C.; Zhang, Z.; Chen, J.; Yan, X.; Zheng, B.; Liu, S.; Huang, F. Pillar[6]arene-based photoresponsive host-guest complexation. *J. Am. Chem. Soc.,* **2012**, *134*(20), 8711-8717.
[http://dx.doi.org/10.1021/ja302998q] [PMID: 22540829]

[50]    Meng, L.B.; Li, D.; Xiong, S.; Hu, X.Y.; Wang, L.; Li, G. FRET-capable supramolecular polymers based on a BODIPY-bridged pillar[5]arene dimer with BODIPY guests for mimicking the light-harvesting system of natural photosynthesis. *Chem. Commun. (Camb.),* **2015**, *51*(22), 4643-4646.
[http://dx.doi.org/10.1039/C5CC00398A] [PMID: 25690934]

[51]    Li, C.; Han, K.; Li, J. *et al.* Supramolecular Polymers Based on Efficient Pillar[5]arene-Neutral Guest Motifs. *Chemistry,* **2013**, *19*(36), 11892-11897.
[http://dx.doi.org/10.1002/chem.201301022] [PMID: 23922310]

[52]    Wang, X.; Deng, H.; Li, J.; Zheng, K.; Jia, X.; Li, C. A neutral supramolecular hyperbranched polymer fabricated from an AB2 -type copillar[5]arene. *Macromol. Rapid Commun.,* **2013**, *34*(23-24), 1856-1862.
[http://dx.doi.org/10.1002/marc.201300731] [PMID: 24285568]

[53]    Ulrich, G.; Ziessel, R.; Harriman, A. The chemistry of fluorescent bodipy dyes: versatility unsurpassed. *Angew. Chem. Int. Ed.,* **2008**, *47*(7), 1184-1201.
[http://dx.doi.org/10.1002/anie.200702070] [PMID: 18092309]

[54]    Coskun, A.; Akkaya, E.U. Ion sensing coupled to resonance energy transfer: a highly selective and sensitive ratiometric fluorescent chemosensor for Ag(I) by a modular approach. *J. Am. Chem. Soc.,* **2005**, *127*(30), 10464-10465.
[http://dx.doi.org/10.1021/ja052574f] [PMID: 16045314]

[55]    Yao, Y.; Chi, X.; Zhou, Y.; Huang, F. A bola-type supra-amphiphile constructed from a water-soluble pillar[5]arene and a rod–coil molecule for dual fluorescent sensing. *Chem. Sci. (Camb.),* **2014**, *5*(7), 2778-2782.
[http://dx.doi.org/10.1039/c4sc00585f]

[56]    Montes-García, V.; Fernández-López, C.; Gómez, B.; Pérez-Juste, I.; García-Río, L.; Liz-Marzán, L.M.; Pérez-Juste, J.; Pastoriza-Santos, I. Pillar[5]arene-mediated synthesis of gold nanoparticles: size control and sensing capabilities. *Chemistry,* **2014**, *20*(27), 8404-8409.
[http://dx.doi.org/10.1002/chem.201402073] [PMID: 24888988]

[57]    Yang, J.L.; Yang, Y.H.; Xun, Y.P.; Wei, K.K.; Gu, J.; Chen, M.; Yang, L.J. Novel Amino-pillar[5]arene as a Fluorescent Probe for Highly Selective Detection of Au [3+] Ions. *ACS Omega,* **2019**, *4*(18), 17903-17909.
[http://dx.doi.org/10.1021/acsomega.9b02951] [PMID: 31681900]

[58]    Li, H.; Chen, D.X.; Sun, Y.L.; Zheng, Y.B.; Tan, L.L.; Weiss, P.S.; Yang, Y.W. Viologen-mediated assembly of and sensing with carboxylatopillar[5]arene-modified gold nanoparticles. *J. Am. Chem. Soc.,* **2013**, *135*(4), 1570-1576.
[http://dx.doi.org/10.1021/ja3115168] [PMID: 23256789]

[59]    Sun, S.; Hu, X.Y.; Chen, D.; Shi, J.; Dong, Y.; Lin, C.; Pan, Y.; Wang, L. Pillar[5]arene-based side-chain polypseudorotaxanes as an anion-responsive fluorescent sensor. *Polym. Chem.,* **2013**, *4*(7), 2224-2229.
[http://dx.doi.org/10.1039/c3py00162h]

[60]    Sun, S.; Shi, J.B.; Dong, Y.P.; Lin, C.; Hu, X.Y.; Wang, L.Y.; Pillar, A. A pillar[5]arene-based side-chain pseudorotaxanes and polypseudorotaxanes as novel fluorescent sensors for the selective

detection of halogen ions. *Chin. Chem. Lett.,* **2013**, *24*(11), 987-992.
[http://dx.doi.org/10.1016/j.cclet.2013.07.014]

[61] Chen, J.F.; Lin, Q.; Yao, H.; Zhang, Y.M.; Wei, T.B. Pillar[5]arene-based multifunctional supramolecular hydrogel: multistimuli responsiveness, self-healing, fluorescence sensing, and conductivity. *Mater. Chem. Front.,* **2018**, *2*(5), 999-1003.
[http://dx.doi.org/10.1039/C8QM00065D]

[62] Li, Z.; Li, X.; Yang, Y.W. Conjugated Macrocycle Polymer Nanoparticles with Alternating Pillarenes and Porphyrins as Struts and Cyclic Nodes. *Small,* **2019**, *15*(12), 1805509.
[http://dx.doi.org/10.1002/smll.201805509] [PMID: 30735309]

[63] Chen, J.F.; Meng, G.; Zhu, Q.; Zhang, S.; Chen, P. Pillar[5]arenes: a new class of *AIE* gen macrocycles used for luminescence sensing of Fe$^{3+}$ ions. *J. Mater. Chem. C Mater. Opt. Electron. Devices,* **2019**, *7*(38), 11747-11751.
[http://dx.doi.org/10.1039/C9TC03831K]

[64] Xia, D.; Wang, P.; Shi, B. Cu(II) Ion-Responsive Self-Assembly Based on a Water-Soluble Pillar[5]arene and a Rhodamine B-Containing Amphiphile in Aqueous Media. *Org. Lett.,* **2017**, *19*(1), 202-205.
[http://dx.doi.org/10.1021/acs.orglett.6b03486] [PMID: 28029260]

[65] Wang, P.; Liang, B.; Xia, D. A Linear AIE Supramolecular Polymer Based on a Salicylaldehyde Azine-Containing Pillararene and Its Reversible Cross-Linking by Cu$^{II}$ and Cyanide. *Inorg. Chem.,* **2019**, *58*(4), 2252-2256.
[http://dx.doi.org/10.1021/acs.inorgchem.8b02896] [PMID: 30694053]

[66] Zhu, H.; Shangguan, L.; Shi, B.; Yu, G.; Huang, F. Recent progress in macrocyclic amphiphiles and macrocyclic host-based supra-amphiphiles. *Mater. Chem. Front.,* **2018**, *2*(12), 2152-2174.
[http://dx.doi.org/10.1039/C8QM00314A]

[67] Yang, K.; Chao, S.; Zhang, F.; Pei, Y.; Pei, Z. Recent advances in the development of rotaxanes and pseudorotaxanes based on pillar[ *n* ]arenes: from construction to application. *Chem. Commun. (Camb.),* **2019**, *55*(88), 13198-13210.
[http://dx.doi.org/10.1039/C9CC07373F] [PMID: 31631211]

[68] Li, Q.; Wu, Y.; Liu, Y.; Shangguan, L.; Shi, B.; Zhu, H. Rationally Designed Self-Immolative Rotaxane Sensor Based on Pillar[5]arene for Fluoride Sensing. *Org. Lett.,* **2020**, *22*(16), 6662-6666.
[http://dx.doi.org/10.1021/acs.orglett.0c02492] [PMID: 32806202]

[69] Roy, S.G.; Mondal, S.; Ghosh, K. Copillar[5]arene-rhodamine conjugate as a selective sensor for Hg$^{2+}$ ions. *New J. Chem.,* **2020**, *44*(15), 5921-5928.
[http://dx.doi.org/10.1039/C9NJ06264E]

[70] Cheng, H.B.; Li, Z.; Huang, Y.D.; Liu, L.; Wu, H.C. Pillararene-Based Aggregation-Induce--Emission-Active Supramolecular System for Simultaneous Detection and Removal of Mercury(II) in Water. *ACS Appl. Mater. Interfaces,* **2017**, *9*(13), 11889-11894.
[http://dx.doi.org/10.1021/acsami.7b00363] [PMID: 28317372]

[71] Shi, B.; Jie, K.; Zhou, Y.; Xia, D.; Yao, Y. Formation of fluorescent supramolecular polymeric assemblies *via* orthogonal pillar[5]arene-based molecular recognition and metal ion coordination. *Chem. Commun. (Camb.),* **2015**, *51*(21), 4503-4506.
[http://dx.doi.org/10.1039/C5CC00535C] [PMID: 25683101]

[72] Lou, X.Y.; Yang, Y.W. Manipulating Aggregation-Induced Emission with Supramolecular Macrocycles. *Adv. Opt. Mater.,* **2018**, *6*(22), 1800668.
[http://dx.doi.org/10.1002/adom.201800668]

[73] Lou, X.Y.; Song, N.; Yang, Y.W. Enhanced Solution and Solid-State Emission and Tunable White-Light Emission Harvested by Supramolecular Approaches. *Chemistry,* **2019**, *25*(51), 11975-11982.
[http://dx.doi.org/10.1002/chem.201902700] [PMID: 31334896]

[74]   Song, N.; Lou, X-Y.; Yu, H.; Weiss, P.S.; Tang, B.Z.; Yang, Y-W. Pillar[5]arene-based tunable luminescent materials *via* supramolecular assembly-induced Förster resonance energy transfer enhancement. *Mater. Chem. Front.,* **2020**, *4*(3), 950-956.
[http://dx.doi.org/10.1039/C9QM00741E]

[75]   Wang, K.; Wang, C. -Y.; Zhang, Y. Wang, K.; Wang, C.Y.; Zhang, Y.; Zhang, S.X.A.; Yang, B.; Yang, Y.W. Ditopic pillar[5]arene-based fluorescence enhancement material mediated by [c2]daisy chain formation. *Chem. Commun. (Camb.),* **2014**, *50*(67), 9458-9461.
[http://dx.doi.org/10.1039/C4CC03992K] [PMID: 25008450]

[76]   Zhang, H.; Ma, X.; Guo, J.; Nguyen, K.T.; Zhang, Q.; Wang, X.J.; Yan, H.; Zhu, L.; Zhao, Y. Thermo-responsive fluorescent vesicles assembled by fluorescein-functionalized pillar[5]arene. *RSC Advances,* **2013**, *3*(2), 368-371.
[http://dx.doi.org/10.1039/C2RA22123C]

[77]   Zhang, Y.M.; Zhu, W.; Huang, X.J.; Qu, W.J.; He, J.X.; Fang, H.; Yao, H.; Wei, T.B.; Lin, Q. Supramolecular Aggregation-Induced Emission Gels Based on Pillar[5]arene for Ultrasensitive Detection and Separation of Multianalytes. *ACS Sustain. Chem.& Eng.,* **2018**, *6*(12), 16597-16606.
[http://dx.doi.org/10.1021/acssuschemeng.8b03824]

[78]   Takashima, Y.; Hatanaka, S.; Otsubo, M.; Nakahata, M.; Kakuta, T.; Hashidzume, A.; Yamaguchi, H.; Harada, A. Expansion–contraction of photoresponsive artificial muscle regulated by host–guest interactions. *Nat. Commun.,* **2012**, *3*(1), 1270.
[http://dx.doi.org/10.1038/ncomms2280] [PMID: 23232400]

[79]   Chen, J.F.; Chen, P. Pillar[5]arene-Based Resilient Supramolecular Gel with Dual-Stimuli Responses and Self-Healing Properties. *ACS Appl. Polym. Mater.,* **2019**, *1*(8), 2224-2229.
[http://dx.doi.org/10.1021/acsapm.9b00516]

[80]   Song, N.; Chen, D.X.; Qiu, Y.C.; Yang, X.Y.; Xu, B.; Tian, W.; Yang, Y.W. Stimuli-responsive blue fluorescent supramolecular polymers based on a pillar[5]arene tetramer. *Chem. Commun. (Camb.),* **2014**, *50*(60), 8231-8234.
[http://dx.doi.org/10.1039/c4cc03105a] [PMID: 24934882]

# CHAPTER 5

# Cucurbit[n]urils Based Molecular Recognition with Fluorescence Signalling

**Liju R.**[1] and **E. Rajkumar**[1,*]

[1] *Biomimetic and Biosensor Lab, Department of Chemistry, Madras Christian College (Autonomous), Affiliated to University of Madras, Chennai-600 059, Tamilnadu, India*

**Abstract:** The development of fluorescence based supramolecules offering selectivity, sensitivity and detection in real time applications are of great interest. Cucurbit[n]urils (CB[n]), a macrocyclic synthetic host molecule consists of a varying number of glycoluril units bridged by methylene groups. CB[n]s easily forms host-guest complexes (inclusion complexes) with a wide range of analytes. The recognition of analytes in the presence of host molecules using fluorescence techniques received greater attention due to its rapid response, high sensitivity to the environment and robust adaptability. Exploiting the fluorescence properties of CB[n] based supramolecules by enhancing or quenching the fluorescence intensity in the presence of guest molecules by spectrofluorometric methods is discussed. A brief outlook on the development of fluorescence properties of CB[n] based supramolecules used for imaging and photodynamic therapy is presented and discussed.

**Keywords:** Binding affinity, Cucurbiturils, Fluorescence, Host-guest complex, Molecular Recognition, Non-covalent interactions, Supramolecules.

## INTRODUCTION

Molecular recognition is one of the fundamental supramolecular events. It is a process that involves the binding and selection of substrate (guest) by a given receptor (host)molecule with a specific function [1]. These two molecules exhibit molecular complementarity and have different combinations of non-covalent interactions such as hydrophobic interactions, electrostatic interactions, π-π interactions, intermolecular hydrogen bonding, and van der Waals interactions, which are essential for the biological processes, supramolecules, molecular biology, and supramolecular assembly [2]. The greater the affinity that exists between host and guest by the combination of forces, the greater will be the selectivity of the host molecule.

---

[*] **Corresponding author Eswaran Rajkumar**: Biomimetic and Biosensor Lab, Department of Chemistry, Madras Christian College (Autonomous), Affiliated to University of Madras, Chennai – 600 059, Tamilnadu, India; Tel: +919842303478; E-mail: rajjkumar@gmail.com

Paulpandian Muthu Mareeswaran, Palaniswamy Suresh and Seenivasan Rajagopal (Eds.)

In 1894 Emil Fischer introduced the term "lock and key" for the double complementarity principle extending both the electronic and geometrical features to explain the selectivity and specificity of the specific enzymatic reactions. Supramolecular chemistry and host-guest chemistry are limited to the intermolecular processes, whereas "recognition" can apply to both inter and intra molecular phenomena [3].

Supramolecules are molecular assemblies that are held together by intermolecular forces rather than by covalent bonds. Supramolecules easily form a highly ordered system without any direct bond formation, which facilitates the development of rapid detection of analytes such as sensors. The relatively weak nature of the intermolecular interactions that present in supramolecular assemblies can change their configurations with respect to the variety of different external stimuli, including the introduction of the target analyte. The change in configurations with respect to the external stimuli can be measured by optical signals like fluorescence and absorption changes. The system can be reversed by removing the external stimuli, due to the presence of labile intermolecular interaction sexist in the supramolecules. These inherent features of supramolecules are ideal candidates for use as chemosensors.

The detection of a variety of analytes in real time conditions is crucial for the different domains of scientific, medical and security fields [4]. There are several detection methods available for the detection of analytes varying from small organic molecules, anions, cations to whole cells and organisms. The following components are essential for the recognition/detection of analytes using chemosensors: (i) the analyte, defined as the target for recognition/detection (ii) the recognition element, defined as the part of the sensor that recognizes the analyte; (iii) the transducer, defined as the sensor component that responds to the presence of the analyte with changes in the signal.

For supramolecular chemosensors, there is a thermodynamic equilibrium established between the bound and free state, where the amount of complex formation depends on the concentration and the affinity of the analyte. The naturally occurring biological analytes such as amino acids, peptides, neurotransmitters, hormones and drugs are in the range of mM(millimolar) to nM(nanomolar) concentrations in aqueous media. The development of chemosensors with affinities $(K_a) > 10^3 M^{-1}$ is commonly preferred. The calixarene or cyclodextrin based macrocycles rarely provide $K_a$ values beyond $10^3 M^{-1}$ in aqueous media unless particularly highly charged or hydrophobic analytes are targeted [5, 6]. Cucurbit[n]urils based macrocycles exhibit much higher binding affinity towards various guest molecules and they are inert towards many reagents/chemicals and are highly biocompatible. In this chapter, we are highlighting the cucurbit[n]urils based system for the detection of various analytes using fluorescence techniques and the application of the selected supramolecules in cancer imaging and photodynamic therapy are discussed.

## Cucurbituril CB[n]

Cucurbit[n]urils (n=5,6,7,8,10,14), a family of synthetic macrocyclic host molecules are composed of varying number of glycoluril units bridged by methylene groups, possess unique guest binding properties in aqueous media. In 1905, Berhend and coworkers [7] reported the polymeric product obtained from the condensation reaction between glycoluril units and formaldehyde in acidic solution; however, a complete characterization of the product was done by Mock and coworkers [2] in 1981, the product was successfully crystalized from the reaction; a macrocycle consisting of 6 glycoluril units bound together by 12 methylene bridges. The macrocycle resembled to a pumpkin, which belongs to the family of Cucurbitaceae, hence the name "Cucurbituril" was coined. In CB[n], "n" indicates the number of glycoluril building blocks that constitute the macrocycles. Kim *et al* [8, 9]and Day *et al* [10, 11]successfully synthesised and isolated the homologues series of CB[n] (n=5,6,7,8,10,14) and the formation of a reaction mechanism of CB[n]s were reported by Isaacs *et al* [12 - 14].

## Structural Features of CB[n]s

The structural parameters of CB[n] (n =5,6,7,8,10) such as cavity volume, cavity diameter, portal diameter and outer diameter are well explored, which are determined from the X-ray crystal analysis [8, 12] (Table **1**). The portal diameter and cavity diameter of cucurbit [5]uril is 3.9Å and 5.8Å respectively. From the Table **1** and Fig. **1**, it is clear that the structural parameters are increasing with increasing "n" value, though the CB[n]s have the same height of 9.1 Å. For example the inner cavity volume of CB [5] is 82 Å, whereas the inner cavity volume of CB [10] is 870 Å$^3$, which is approximately more than ten times of CB [5] inner cavity volume. All the CB[n] exhibits good aqueous solubility in acidic conditions. The CB[n]s, with "n" as odd numbers are soluble in neutral water, whereas with "n" as even numbers are poorly soluble. The solubility of CB[n] can be increased by complexing with metal cations or CB[n]: guest complexes in aqueous solution. CB[n], with n > 10, are less explored and larger cucurbituril are converted into hemicucurbitruil or twisted/inverted CB[n] [16].

Table 1. The structural and physical parameters of CB[n] [8, 12].

|  | Mol. Weight | CB [5] [a] | CB [6] [a] | CB [7] [a] | CB [8][a] | CB [10][b] |
|---|---|---|---|---|---|---|
| Portal diameter (Å) | 830 | 2.4 | 3.9 | 5.4 | 6.9 | 9.5- 10.6 |

*(Table 1) cont.....*

| | Mol. Weight | CB [5] [a] | CB [6] [a] | CB [7] [a] | CB [8][a] | CB [10][b] |
|---|---|---|---|---|---|---|
| Cavity diameter (Å) | 996 | 4.4 | 5.8 | 7.3 | 8.8 | 11.3- 12.4 |
| Inner cavity volume (Å³) | 1163 | 82 | 164 | 279 | 479 | 870 |
| Outer diameter(Å) | 1329 | 13.1 | 14.4 | 16.0 | 17.5 | 20.0 |
| height(Å) | 1661 | 9.1 | 9.1 | 9.1 | 9.1 | 9.1 |
| Aqueous solubility $S_{H_2O}$ [mM] | | 20-30 | 0.03 | 5 | <0.01 | <0.05 |

[a]Adapted with permission from ref [8]. Copyright 2000 American Chemical Society. [b]ref [12, 17].

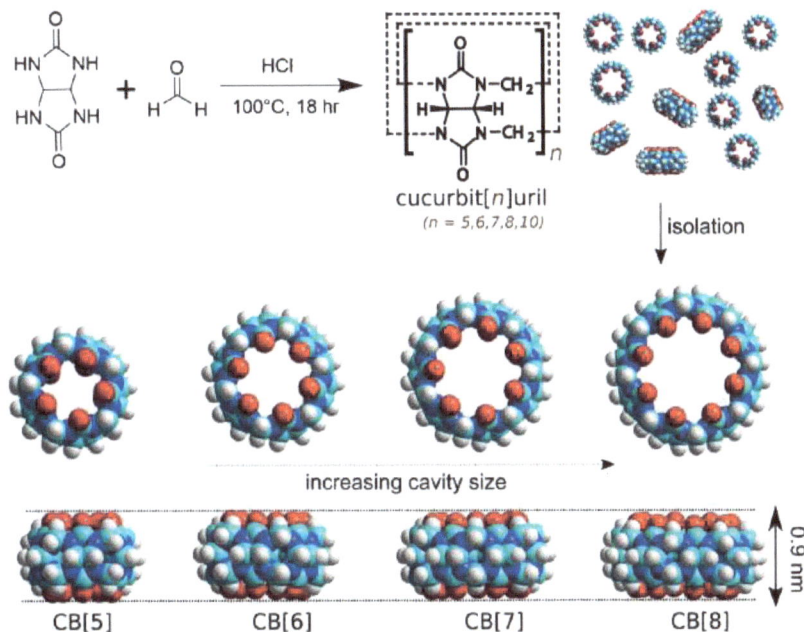

**Fig. (1).** The formation of cucurbit[n]uril (top) and Space-filling models of CB[n](bottom) [15]. Copyright 2015 from American Chemical Society.

## Physical Properties of CB[n]s

CB[n]s form different types of complexation with guest molecules (Fig. **2**). The mean diameter of the internal cavity of the CB[n]s is approximately 2Å narrower than the diameter of the cavity, which is very important for the constructive binding that produces significant steric barriers to the analyte association and dissociation [8]. Due to poor solubility in common solvents and aqueous solutions, host-guest chemistry of CB [6] was extensively studied in highly acidic conditions. In general, the CB[n]s show a poor solubility (<$10^{-5}$M) in common solvents except CB [5] and CB [7] having moderate solubility in water. One of the most interesting properties of cucurbiturils is to form a strong complexation/association with analyte molecules compared to the other types of macrocyclic host molecules such as cyclodextrins, calixarene, crown ethers, *etc*

[4, 12, 18]. Among the homologues series of CB[n] family, CB [7] based supramolecules are extensively studied due to their higher aqueous solubility and higher binding affinities with varieties of guest molecules than other CB[n] derivatives. The inner cavity of CB[n]s are largely hydrophobic which is due to fused rings of the glycoluril subunits having a macrocyclic structure, and portals of CB[n]s are highly electronegative.

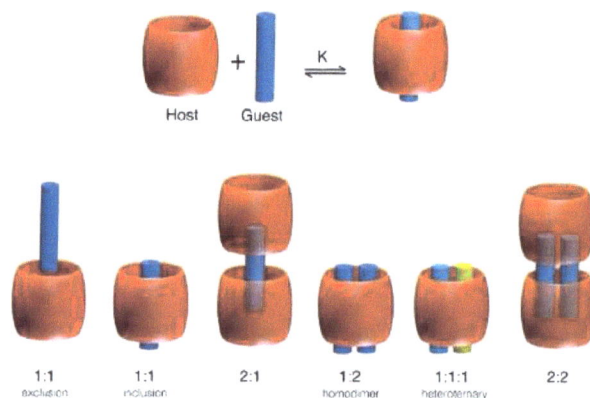

**Fig. (2).** CB[n] inclusion and exclusion complexes [15]. Copyright 2015 American Chemical Society.

The structures of CB[n]s are unique with highly symmetric, negatively charged carbonyl rims and having a hydrophobic cavity. The cavity neither has any functional groups nor electron pairs pointing towards inside. The higher hydrophobicity of the cavity arises due to non-dipolar nature of the macrocycle and it cannot engage in hydrogen bonding interactions [19]. The electrostatic potential map, shown in Fig. **3**, highlights the strength of negative charge associated with the portals of CB[n], where metal cations binding acts as a "lids" to the portal. Binding at these regions will enhance the solubility of less water-soluble CB[n]s. The low polarizability of the CB[n] cavity can be rationalized as (i) there are no bonds or lone electron pairs pointing towards inside, (ii) the hosts are non-aromatic, (iii) any residual electron density at the nitrogen is mesomerically delocalized to the carbonyl oxygens, and, potentially, (iv) the packing is sufficiently loose to avoid close guest-to wall contacts i.e packing coefficient [19].

The molecular recognition of cucurbit[n]uril based supramolecules was used by fluorescence signalling, as the read-out measurement received greater attention due to its rapid response, high sensitivity to the environment and robust adaptability. The sensitivity and selectivity will be greatly enhanced by the host-guest recognition by non-covalent interactions for the detection of various analytes such as metal ions, inorganic, biomolecules, toxins and pharmaceutical compounds.

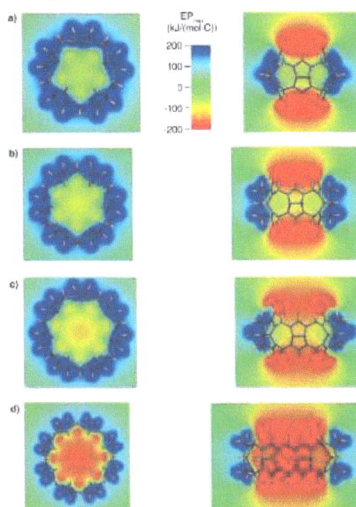

**Fig. (3).** The calculated electrostatic potential (EP) for (a) CB [5], (b) CB [6], (c) CB [7] and (d) CB [8]. Reprinted with permission from ref [19]. Copyright 2012 American Chemical Society.

## CB[n] Based Sensors

Cucurbit[n]urils are known for recognition of metal ions, biomolecules (*e.g.* proteins) and small organic molecules. CB[n] systems have been proven to be excellent chemosensor candidates because of their properties like high binding affinity as a host, size and charge, selective binding, fast complex formation *etc.* In a CB[n], the two portals, which are lined with urea carbonyl groups, are responsible for complexation with positively charged molecules through ion–dipole interaction whereas the internal hydrophobic cavity is responsible for binding of small guest molecules through hydrophobic interactions [20]. The CB[n] macrocycles are highly inert towards chemicals, which make it applicable in wide spectrum of reaction media. It also exhibits good tolerance towards extreme pH conditions. Further, CB[n] and certain acyclic CB[n] derivatives are biologically benign, i.e., non-toxic, such that *ex-vivo* and *in vivo* use of CB[n]-based materials and sensors are in use [15]. Besides CB[n]s are redox and photochemically inactive which allow them to be developed into spectrochemical and redox sensors, without disturbing the macrocyclic unit. Upon binding with CB[n], analytes can be readily recognized and quantified by exploiting their inherent spectroscopic and redox characteristics. Complexation of chromophoric or emissive molecules by CB[n] can cause significant changes in the photophysical properties of the molecules. Fluorescence spectroscopic analysis is one of the most commonly used techniques for the supramolecules containing CB[n] systems, due to their sensitivity and rapid response. For the molecular recognition of various analytes using CB[n] based supramolecules using fluorescence properties, there are three major modes of recognition of analytes such as direct sensing, associative binding assay and indicator displacement assay. The mechanisms of detections are dealt separately in this chapter.

## Detection Mechanism

In supramolecular chemistry, the sensor or molecular recognition system is generally viewed as a receptor (host) that interacts with an analyte (guest), producing a detectable change in a signal. In other words, the signal is modulated upon the binding of an analyte to a receptor. Signal can be in the form of change in pH, electrochemical characteristics, or photophysical characteristics *etc* [4]. The CB[n] macrocycles are known for their accommodative cavity sizes, that qualifies them as host/receptors in sensor applications. The size of CB[n] influences largely on the type of analyte and the detection strategies. The even numbered homologues display a low water-solubility, which is nevertheless sufficient to study their complexation with many indicator fluorescent dyes, because the low solubility is counterbalanced by the fact that all cucurbiturils larger than CB [5] can form much stronger host/guest complexes than other macrocycles like cyclodextrins with a binding constants range typically between $10^4$ to $10^{17}$ $M^{-1}$.

Fluorescence based sensors are sensitive to analyte even in sub nano-molar concentrations. The fluorescence based molecular recognition in CB[n]s takes place by various detection mechanisms; each mode of mechanism of sensing has its own advantages and disadvantages. In the presence of analyte, the responses were measured either by increase in fluorescence signal or fluorescence quenching, FRET *etc*. The most common detection mechanisms of analytes are based on Indicator displacement assays (IDA), Associative binding assay (ABA) and Direct-binding assays (DBA).

## Direct Binding Assay

Direct sensing/detection can trace its origin to the "lock and key principle", first postulated in 1894 by Emil Fischer, where a substrate shape selectively binds to the active site of an enzyme. In the paradigm of Direct-binding assays (DBA), CB[n]-derivatives (host/receptor) containing a remotely attached or framework-incorporated dye component yield a spectroscopic response upon binding of an analyte to the cavity of CB[n]. Elegant design and optimization as well as multistep synthesis are generally required for the purpose of selective sensing/detection. There are several reports of CB[n] systems-based sensors which rely on the direct mechanism. With ease of functionalization, these macrocycles were applied in a variety of sensing applications, for example, the CB [6] host has been demonstrated to recognize alkane diamines, aromatics, amino acids, and nucleobases in aqueous media [21] with affinities up to $10^6$ $M^{-1}$. CB [6] with naphthalene fluorophores have shown to sense analytes associated with over-the-counter drugs, the analyte upon binding with the probe(CB[n]). The pyridinium and amine-based analytes were inducing the quenching of naphthalene fluorescence, by photoinduced electron transfer (PET) process. For analytes without quencher moieties, the fluorescence of CB [6]-naphthalene has been increased and this enhancement of fluorescence is due to the increased rigidity of the analyte-receptor complex and restriction of vibrational/rotational modes that would otherwise cause non-radiative decay.

The CB [7] host materials have been reported widely in literature based on DBAs, as CB [7] analogous are well known for their superior solubility compared to other higher and lower CB[n]s. A Cucurbit [7]uril−Tetramethylrhodamine CB [7]-TMR conjugate has been designed to sense peptide derivative Phe-Ala-Ala and viologen. The fluorescence properties of CB [7]-TMR chemosensors showed essentially the same binding constants for the guests as unfunctionalized CB [7], which is a significant advantage(unlike in Indicator displacement assays). It was theorized that, two nearly isoenergetic conformations are adopted by the CB [7]-TMR conjugate prior to analyte binding; a strongly fluorescent one where the TMR dye is held close to the CB[n] portals, and another with a less emissive conformation. Upon binding of an analyte, the unimolecular conformation equilibrium shifts to the weaker fluorescent form, exhibiting a quenching of fluorescence [22]. In another report a supramolecular host−guest FRET pair based on a carboxyfluorescein tagged cucurbit [7]uril (CB [7] as acceptor) and the fluorescent dye 4',6-diamidino-2-phenylindole (DAPI, as donor) was developed for sensing of DNA with an excellent linear dependence of the ratiometric fluorescence intensities [23]. In another study the CB [8] analogue is used for molecular recognition of peptide derivatives in aqueous medium, CB [8] is fundamentally different from its smaller CB[n]homologues in that it can bind simultaneously and selectively to two different aromatic guests at low concentrations due to its higher cavity size. In this study CB [8] forms a 1:1:1 complex with methyl viologen(MV) and the indoles (tryptophan) guest, the tryptophan fluorescence is significantly reduced by charge transfer in presence of CB [8]: MV. It has been clearly observed that magnitude of fluorescence signal modulation depends upon the extent of binding between the analyte and the receptors [24].

**Associative Binding Assay**

Associative binding assays or ABA is an unique strategy of chemosensing wherein the higher CB[n]s like CB [8] capable of encapsulating more than one guest into their cavity, forms an associative binding with the analyte and indicator within the confined cavity. Inside the cavity the indicator and analyte molecules communicate mutually to generate distinguishable spectroscopic finger prints. The simultaneous inclusion of two aromatic molecules was reported for CB [8] with methyl viologen and naphthol molecule. The CB [8] is known to form 1:1:1 heterocomplexes with two different analytes. The associative binding of analyte with inside cavity of CB [8], induces a charge transfer (CT) between the electron deficient viologen and electron rich naphthol, giving a colour to the solution arising due to the CT. This affords an interesting way for colorimetric detection of analytes, however the sensitivity of this system was lower as the CT band are relatively weak [25]. Incorporation of fluorophore in the host molecule, as one of the guests could enhance the sensitivity of ABAs significantly. Similarly the fluorescence quenching of the indole in tryptophan unit, when it interacts with a methyl viologen, which is present in inside a cavity with pre-complexed CB [8]-MV pair have been reported. A more versatile ABA can be thought of when the first guest is a fluorophore which upon complexation to the second guest, typically the analyte, modify the first guest's fluorescence profile. With this realization numerous ABA systems have been designed and studied for sensing application. A self-assembled 1:1 complex of CB [8] with 2,7-dimethyldiazapyrenium (MDAP) or with 2,7-dimethyldiazaphenanthrenium (MDPT) have been shown to sense catechol and indole derivative, the fluorescence is quenched by interaction of the indole with MDAP inside the cavity (Fig. **4**). Other commonly reported fluorescent dyes as components for CB [8] chemosensors are perylenebisdiimide and perylenemonodiimide derivatives, and certain aryl viologens and these CB [8]:dye pair have been used as sensor for several aromatic molecules and selective peptide group recognitions [26].

**Indicator Displacement Assay (IDA)**

In an indicator displacement assay, an indicator is allowed to bind to a receptor reversibly (CB[n]). Then a competitive analyte is introduced into the receptor-indicator pair which displaces the indicator causing a change in fluorescence response in the form of change in fluorescence (FL) intensity (quenching or enhancement), shift in emission wavelength (red shift or blue shift), and/or energy and electron transfers *etc.* The common interactions between the receptors-indicator and analyte are hydrogen bonding, electrostatic interactions, and complexation. The interactions mostly depend on the geometry of the guest molecule, its hydrophobicity, and its charge and the nature of solvent system also influence the interactions to a larger extent. The main prerequisite for an IDA is that, affinity between the indicator and the CB[n]s is comparable to that between the analyte and the CB[n]s. IDA offers several advantages over traditional sensing mechanisms. In IDA the indicator is not covalently attached to the receptors, because of this non-covalent interaction it offers a chance of using several indicators to the same receptor. Further, the IDA systems with CB[n] as host works well in organic and aqueous media and are easily adaptable as quick responsive sensors for various analyte.

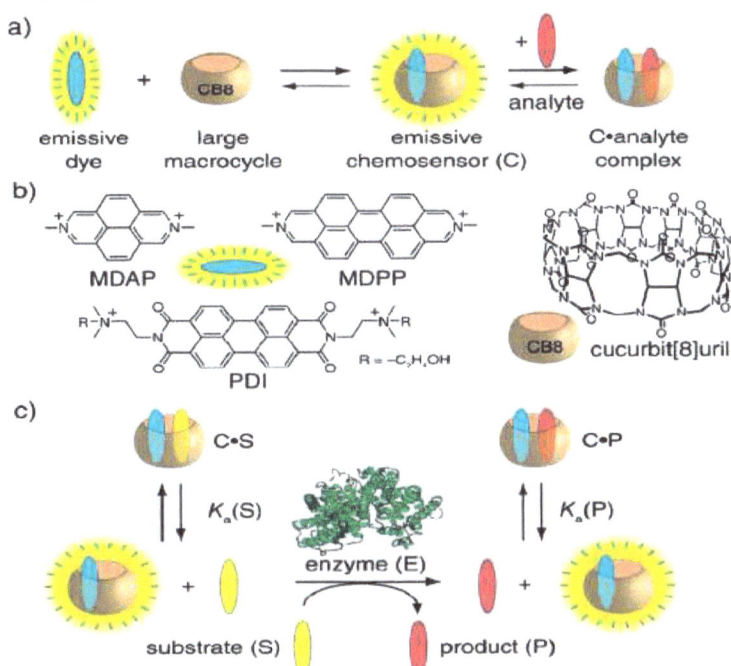

**Fig. (4).** Representations of Associative binding mechanism [27].

There are many molecular recognition systems based on CB[n]s which work on the principle of IDA. Few IDA based CB[n] molecular systems are tabulated in Table **2**. The CB [7] host is widely seen in IDA systems as it displays a favourable combination of a sufficient cavity size and high water solubility, which makes into the most promising cucurbituril host for binding of Indicator dye. IDAs are disadvantageous under continuous flow conditions (*e.g.,* flow sensing or separations) or in biological imaging applications in which the working concentrations are well below the $K_d$ value of the CB[n]-Indicator complex [28].

**Table 2. Indicator displacement assay for molecular recognition.**

| CB[n] | Indicator | Analyte | FL response | Ref |
|---|---|---|---|---|
| CB [7] | Protonated acridine orange | Cordycepin | Hypochromic shift in $\lambda_{em}$ | [30] |
| CB [7] | Luteolin and epigallocatechin gallate | Sunset Yellow | FL quenching | [31] |
| CB [7] | N-(4-(aminomethyl)benzyl)-1-(anthracen-9-yl)methanamine | Amantadine | FL quenching | [32] |
| CB [7] | Berberine chloride | Nandrolone | Quenching | [33] |
| CB [8] | Methyl-pyridinium-paracyclophane | Memantine in blood serum | Shift in $\lambda_{em}$ | [34] |
| CB [8] | Perylenebis(diimide) | Progesterone | Quenching | [33] |

Recently a new variant in competitive binding assay has been reported and called as Guest displacement assay (GDA) (Fig. **5**). The GDA is complementary to the IDA mechanism; here the guest is competitively displaced by the indicator, whereas in later the indicator is displaced by the guest (analyte), It is superior over IDA for the insoluble and/or for weakly binding analyte [29].

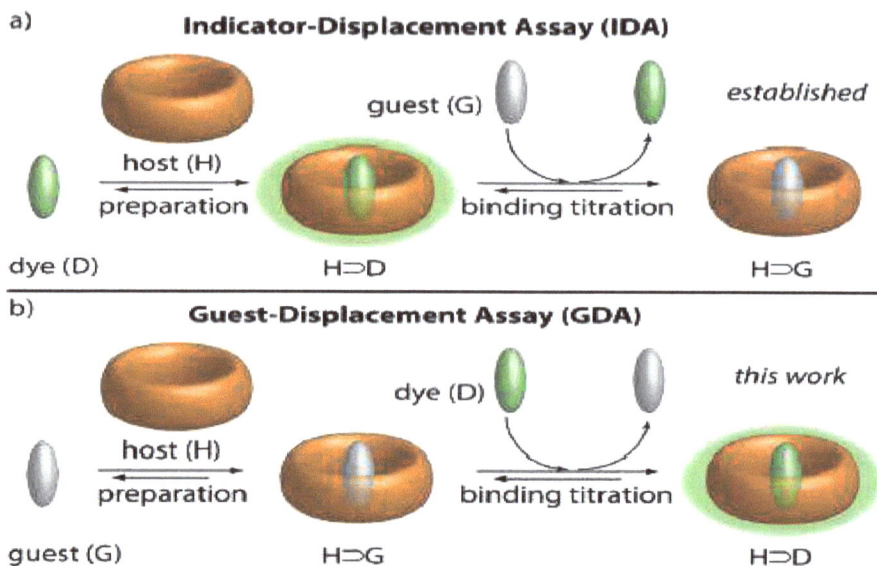

**Fig. (5).** Schematic representation of Indicator displacement assay and Guest displacement assay. Reproduced from ref [29]

## Detection of Metal Ions

Cations are important in biological systems including enzyme cofactors, signalling pathways and biologically essential electrolytes *etc.* Detection of cations is essential for their significant biological and environmental relevance. Typically, the metal ions size falls in the range of picometers ($10^{-12}$ m) to Angstroms ($10^{-10}$ m) and depends on the charge of the cations. There are several analytical tools to detect and quantify the presence of metal ions, supramolecular chemistry is one such tool based on host-guest interaction between the metal ion and supramolecule. Cucurbit[n]uril based macrocycles, especially the lower analogues (n<7), are reported for the application of sensing of metal ions. The lower CB[n], namely CB [5] and CB [6] have a diameter of 4.4 Å and 5.8 Å respectively, which is large enough to accommodate the smaller metal ions in its cavity (Fig. **6**). The smallest, CB [5] has been reported as a host for almost all the alkali earth metals ion from $Li^+$ to $Cs^+$, these metal ions have an ionic radius between 0.76 Å (Li$^+$) to 1.67 Å (Cs$^+$) which easily fit inside the CB cavity. Similarly, as size of the cation increases, higher analogues of CB[n] are chosen for the detection. For example, sensing the alkaline earth metal ions and transition metal ions, CB [6] and CB [7] systems are widely used [35].

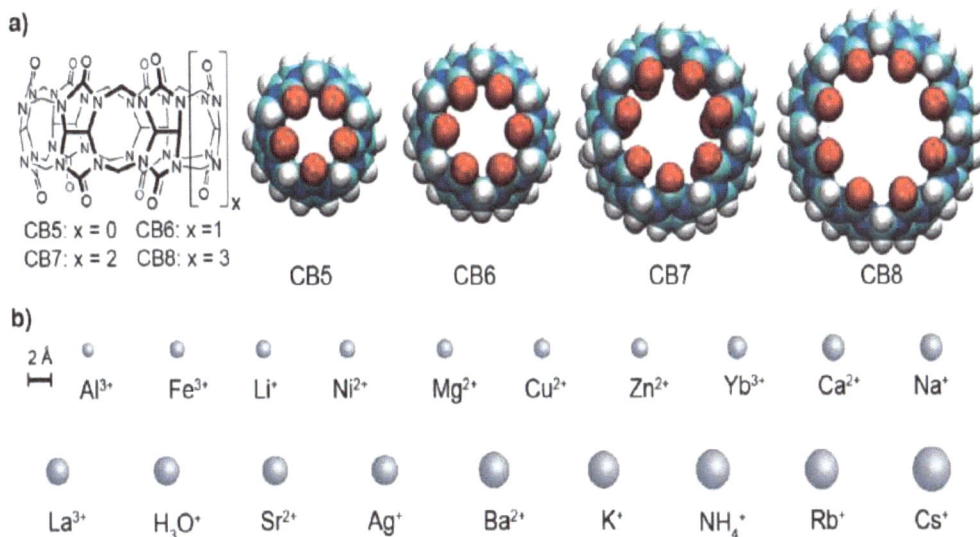

**Fig. (6).** (a) Chemical structures and their 3D representation (space filling model) of CB[n] and (b) of the metal cations [35]. Published by The Royal Society of Chemistry copyright 2019.

The competitive binding of metal ions with a 1:1 CB [7]:phthalocyanine dye were reported, where metal ions upon binding with the CB [7] displaced the phthalocyanine dye and caused a significant changes in the fluorescence characteristic of the dye [36]. Many fluorescence-based detection of metal ions by the competitive binding followed by the indicator displacement mode have been reported. Another approach of detection i.e, the associative binding assay for metal ions was also reported. CB [7] has been complexed with two coumarin derivatives as a 1:1 complex, on addition of Hg$^{2+}$, a tertiary 1:1:1 complex is formed. In the presence of Hg$^{2+}$ ion, the fluorescence of coumarin dye was drastically reduced i.e quenching of fluorescence was observed [37]. In another example, N-(2-benzimidazolylmethyl)-N,N-bis(2- pyridylmethyl)-amine (BIBPA) forming host-guest molecule with CB [7] and exhibiting two emission peaks at 299 nm and 382 nm under the excitation of 272 nm. The fluorescence intensity of host-guest molecule was increased to 30-40 fold upon addition of Zn$^{2+}$ and Cd$^{2+}$, when it was encapsulated with CB [7]. [38] The metal ion recognition is based on associative binding mechanism having the binding constant of Cd$^{2+}$ $K_b$ = (8.4 ± 0.9) × 10$^4$ L mol$^{-1}$ and Zn$^{2+}$, $K_b$ = (7.3 ± 0.7) × 10$^4$ L mol$^{-1}$ respectively.

## Detection of Amino Acids and Peptide Molecules

Despite the existence of a many amino acid and peptide selective fluorescent probes, their detection in a biological medium (aqueous solution) still remain a challenge. Cucurbit[n]uril receptors CB[n] are capable of binding protonated amines via ion–dipole interactions and hydrogen bonding to the carbonyl moieties of the ureidyl portals. There are various studies on the binding ability of amino acid with CB[n]s which provide evidences that CB[n]s are an ideal platform for amino acid or protein fragments, depending upon the binding strength, the sensitivity and selectivity of the chemosensors can be ascertained. The recognition of amino acid or peptide molecules is based on either fluorescence "Turn on" or "Turn off" of the host-guest complex.

The most commonly spotted host for the recognition of amino acids is the CB [7] molecule, because of its suitable cavity size and water solubility. Apart from CB [7], other CB[n]s were also reported. Bai et.al reported the competitive fluorescence recognition of sulphur containing amino acid, lysine and methionine, in aqueous solution by alkyl substituted cucurbit [6]uril system [39]. In the reported system, a π-conjugated bispyridinium phenylenevinylene dye formed a stable inclusion complex with CB [6] unit, giving rise to a pseudorotaxanes, which was further competitively exchanged with amino acids. A clear ratiometric fluorescence signalling was observed, in the CB [6] dye. The dye molecule present in the inner cavity exhibits emission at 480 nm upon competitive displacement of the dye by lysine and methionine. There is a decrease in host-guest interaction, in turn decrease in fluorescence intensity at 480 nm along with increase in fluorescence intensity at longer wavelength (605nm) due to the aggression emission of molecules. It has been demonstrated a naphthalene-CB [6] derivative based sensor to recognize basic amino acids in aqueous system. A size selective binding of the guest amino acid and the host have been established, this difference in binding affinities are expected to contribute to the selectivity of amino acids. Upon the binding of amino acid an enhancement of fluorescence signal "Turn on" was observed [40].

In another approach, amino acid chemosensor based on fluorescence "Turn off" was reported. The higher CB [8] analogue are known to form inclusion of more than one guest. The dicationic 2,7-dimethyldiazaphenanthrenium($DPT^{2+}$) forms a fluorescent inclusion complex with CB [8], leaving enough room inside the host. Addition of indole derivatives like tryptophan, serotonin form stable ternary complexes, this binding interaction is detected by the quenching of the fluorescence emission of $DPT^{2+}$:CB [8]. The quenching is attributed to the association of π donor tryptophan with the π acceptor 2,7-dimethylph--nanthrenium dication [24].

Gao et.al, reported the use of inverted Cucurbit[n]uril, iCB[n]s for molecular recognition of amino acids. The iCB [7] host have shown pH dependent binding towards ten essential L-α-amino acids as guests. At pH 7, the aromatic amino acid, Try and Phe showed higher binding with the iCB [7] host, whereas the non-aromatic amino acid Lys, Arg, and His lied outside the portal of the host (Fig. **7**). Amino acids such as Met, Leu, Ile with side chain alkyl moieties were accommodated within the cavity of iCB [7], while Thr and Val did not show any significant interactions with iCB [7]. At a lower pH of 3, the aromatic amino acids Try and Phe were found deeply buried in the iCB [7] cavity, under the acidic condition the alkyl side chain of the protonated Lys, Arg, and His were found inside the cavity. In neutral pH, Trp displayed an emission peak at 366 nm at an excitation wavelength of 269 nm. On successive addition of iCB [7] a decrease and a hypsochromic shift from 359 to 350 nm in the fluorescence intensity at 359 nm was observed (Fig. **8**) [41]. The host–guest inclusion complexes that comprise an inverted cucurbit [7]uril (iCB [7]) and a quinoline derivative dye, 4-(-dimethylaminostyryl) quinoline (DSQ) at different pHs were exploited as sensor for L-α-amino acids. The inclusion of DSQ inside iQ [7] at different pHs led to increased fluorescence and formation of multi-colored iCB [7]-DSQ complexes. The enhancement of DSQ fluorescence after iCB [7] encapsulation may be attributed to limited rotation ofdimethylamine and the formation of a twisted internal charge transfer (TICT) state. The DSQ:iCB [7] sensors exhibit different affinities for L-α-amino acids at different pHs.

Amino acid residue in peptide sequences are also recognized and analyzed using CB[n]s host. Werner M. Nau *et al* developed a real time analytical method for the unlabelled peptides using fluorescence spectroscopy. The assay enabled with receptor CB [7] and fluorescent dye acridine orange, a feebly fluorescent dye in aqueous solution, which becomes strongly fluorescent upon encapsulation by CB [7]. The CB [7] host selectively binds with N-terminal Phe residues due to cooperative hydrophobic and ion dipole interactions and competitively displaces the acridine orange dye consequently its fluorescence intensity drops again, leading to a "turn-off" fluorescence response [42]. The molecular recognition of the tripeptide Tyr-Leu-Ala was reported using CB [8] receptor in aqueous buffer. The tetramethylbenzobis(imidazolium) (MBBI) was used as fluorescent indicator which resulted in an unexpected discovery that MBBI can serve not only as a turn-off sensor via the simultaneous inclusion of a Trp residue but also as a turn-on sensor via the competitive displacement of MBBI upon binding of Phe- or Tyr-terminated peptides. The high stability, recyclability, and low cost of CB [8] combined with the straightforward incorporation of Tyr-Leu-Ala into recombinant proteins had made this system attractive towards the development of biological applications [43].

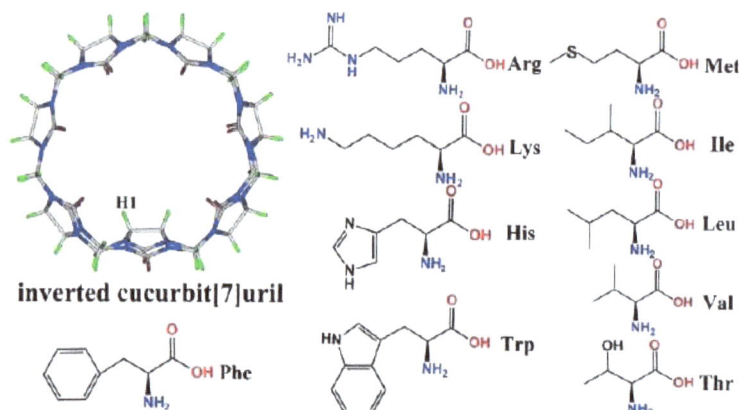

**Fig. (7).** Structure of iCB [7] and essential amino acids investigated [41]. Copyright 2019, American Chemical Society.

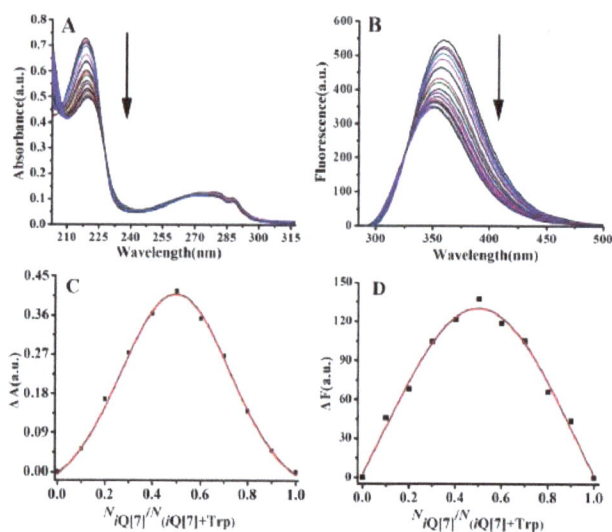

**Fig. (8).** (A) Absorption and (B) Emission spectra of tryptophan upon the addition of increasing amounts of iCB [7]. (C) ΔA and (D) ΔF vs NiCB [7]/(NiCB [7] + NTrp) plots [41]. Copyright 2019, American Chemical Society.

## Detection of Biomolecules

Molecular recognition and quantification of biological species are very important in the field of diagnosis and therapeutics. The commonly sorted biomolecules include proteins, steroids, hormones, bio-fluids, toxins, and other bioactive compounds. Supramolecular assemblies based chemosensors are widely studied for biomolecule sensing, the cucurbit[n]urils CB[n], which have become popular building blocks for chemosensors on account of their aqueous solubility, and their wide applicability in binding indicators and biomolecular analytes. The choice of CB[n] host and the mode of detection depend on the nature of analyte. For example, in the sensing of steroids, the biologically relevant steroids are mostly hydrophobic biomolecules with limited recognition motifs like hydrogen bonds, charges *etc*. Therefore, receptors with strong hydrophobic affinity are needed for the recognition of steroids. Based on the size-fitting aspects and their well-known affinities for hydrophobic guests the cucurbit[n]uril macrocycles CB [7, 8] serve as promising receptor candidates for steroidal compounds, CB [8] for: androstanes, estranes, pregnanes and CB [7] for nandrolone (Fig. **9**) have been reported [33]. These steroids have exhibited a strong binding affinity in aqueous and buffered biological media such as blood serum and gastric acid.

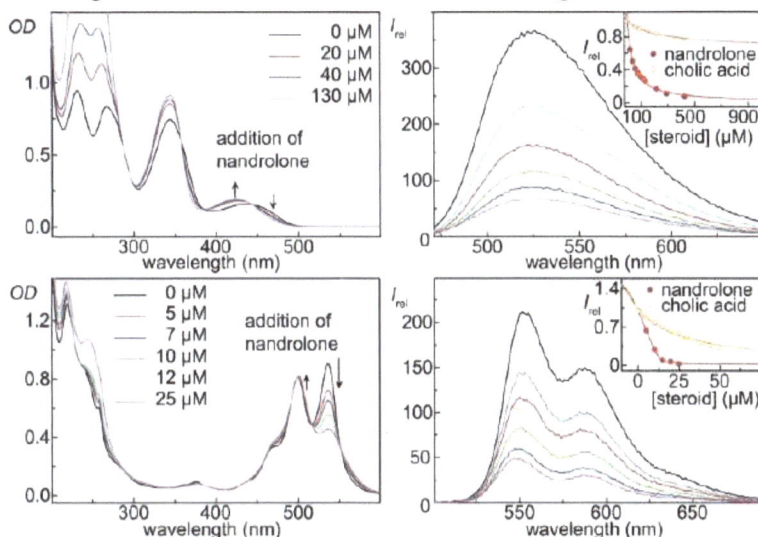

**Fig. (9).** UV−vis (left) and fluorescence (right) spectra of BC ($\lambda_{ex}$ = 445 nm) with CB8 (top) and PDI−OH ($\lambda_{ex}$=420nm) with CB8 (bottom) aqueous solutions upon addition of nandrolone or cholic acid. Insets: Fluorescence emission intensities at 520 nm (top) and 550 nm (bottom) upon addition of different concentrations of cholic acid [33]. Copyright 2016, American Chemical Society.

Steroid on binding with the CB[n] has competitively displaced reported dye which was pre-bond with the CB[n]. A covalent CB [7]:dye conjugate have been developed for selective sensing of Parkinson's drug amantadine in human urine and saliva. The host CB [7] and the reporter dye gastric acid are covalently linked using hexaethylene glycol and tetraethylene glycol as linkers, this way of covalent linking of the reporter dye enable faster response. On the addition of amantadine analytein CB[n], a $S_N2$-type guest–dye exchange mechanism operates, the chemosensors responded with emission quenching of the berberine moiety.

**Detection of Organic Compound**

Organic molecules hold an integral part in our environment, biological systems, medicines, materials *etc*. Organic compounds possess potential risk on living forms, hence sensing and quantifying organic compounds are very essential. There are several methods for the detection of organic molecules such as electroanalytical techniques, optical analysis *etc.* Fluorescence spectroscopy is one such tool which allows trace level detection of analysis. The uses of CB[n] macrocycles as platform for fluorescence-based sensing of organic molecules are widely reported. Table **3** summaries an account of organic compound detections based on CB[n], nature of analyte, and detection mechanism.

**Table 3. Common Organic Analyte detection by CB[n]s**

| Organic Analyte | CB[n] | $K_a$ | Detection mechanism[a] | Ref |
|---|---|---|---|---|
| Benzene | CB [4] | $6.9 \times 10^3$ | D | [44] |
| Aniline | CB [8] | $50.1 \times 10^3$ | ABA | [27] |
| Phenol | CB [8] | $56.0 \times 10^3$ | ABA | [27] |
| N- nitrosonornicotine | CB [6] | $2.8 \times 10^3$ | IDA | [6] |
| (-)-cotinine | CB [6] | - | IDA/D | [6] |

(Table 3) cont.....

| Organic Analyte | CB[n] | $K_a$ | Detection mechanism[a] | Ref |
|---|---|---|---|---|
| N-nitrosodimethylamine | CB [6] | - | IDA/D | [6] |
| Acetaminophen | CB [6] | $6.9 \times 10^3$ | IDA/D | [45] |
| Terpenes | CB [8] | $1.9 \times 10^8$ | IDA | [46] |
| Cyclopentanone | CB [6] | $3.2 \times 10^3$ | IDA | [47] |
| hexane-1,6-diaminium | CB [7] | $8.9 \times 10^7$ | IDA | [22] |
| 1,3,5-trinitrobenzene | CB [7] | - | D | [48] |

[a]D- Direct sensing, IDA – Indicator displacement assay, ABA – Associative binding assay.

## Cucurbit[n]urils in Imaging and in Photodynamic Therapy

Applying the principles of supramolecular chemistry to imaging, particularly with respect to molecular recognition by a designed host molecule, it is possible to generate much more sophisticated diagnostic agents [49 - 54].

Photodynamic therapies have been widely received attention to treat various types of cancers and diseases caused by the bacteria. The generation of reactive oxygen species and singlet oxygen from the molecular oxygen by light with the help of photosensitizer is an important process and it is the essential component for PDT, to destroy the cancer cells and pathogens. Özkan *et al* [55]reported the photoactive supramolecules based on porphyrin core decorated with CB [7] with suitable linkers(TPP-4CB7) (Fig. **10**). In PDT, one of the most important components is the photosensitizer, which should ideally be water-soluble with negligible dark toxicity, but it should be cytotoxic to the cancerous cell only when it is irradiated by a suitable light source. The anticancer drug, doxorubicin (DOX) loaded CB7 inactivates the cells efficiently (Fig. **11**). This multifunctional supramolecules synergistically used as a vehicle to carry anticancer drugs and exhibiting a broad-spectrum antibacterial agent with complete inhibition of bacterial growth. The porphyrin core is responsible for the PDT, and CBs can act as a host molecule for drug delivery.

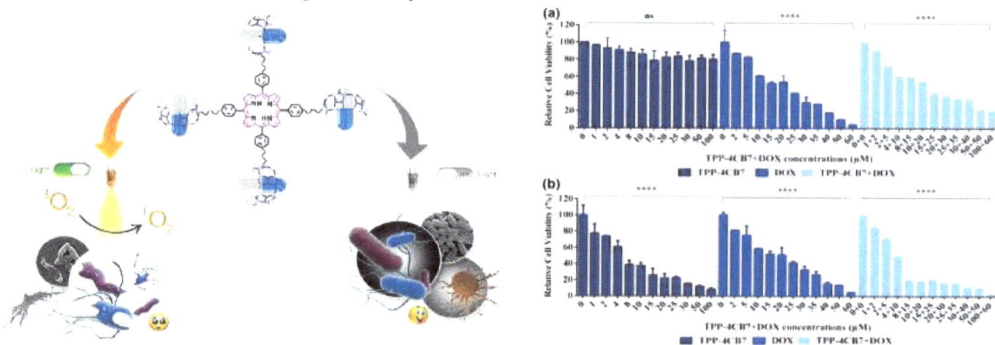

**Fig. (10).** The photoactive supramolecules based on porphyrin core decorated with CB [7] for Photodynamic therapy (a) Absence of light and (b) upon light exposure [55]. Copyright 2019, American Chemical Society.

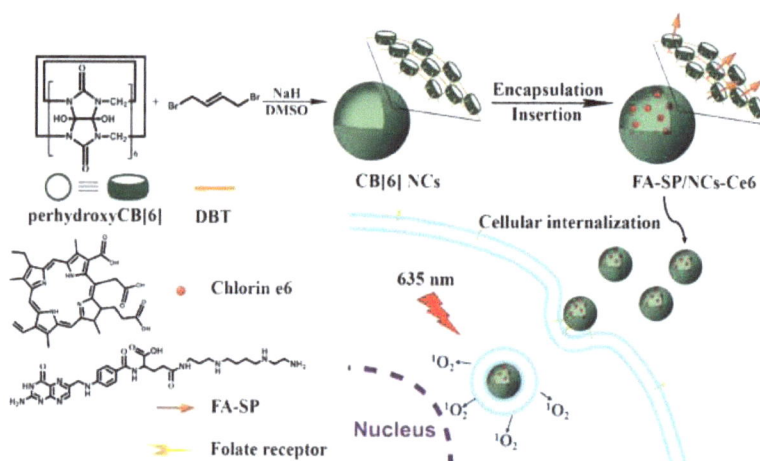

*(Fig. 11) contd.....*

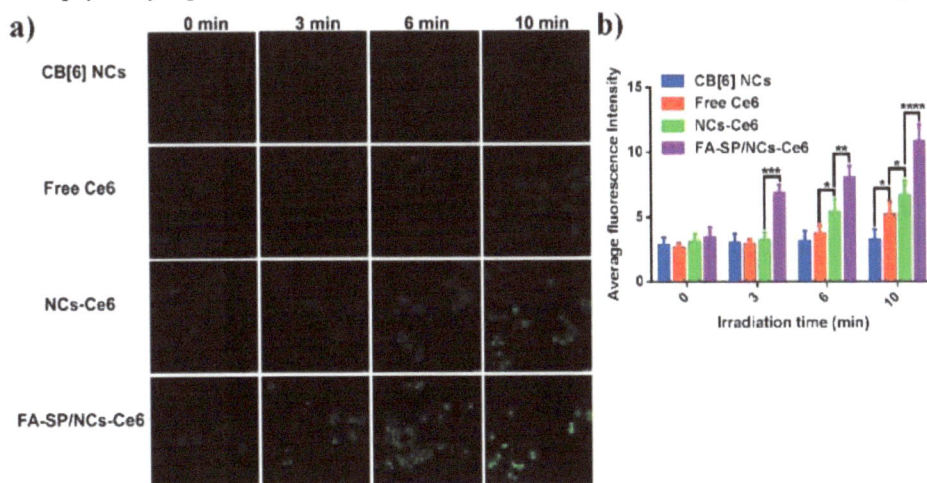

**Fig. (11).** Schematic Illustration of Polymer NanocapsulesPreparation and Construction of FA−SP/NCs-Ce6 for Targeted Delivery and PDT and their CLSM images (a) and the mean fluorescence intensity (b) incubation with the MCF-7 cell line [56]. Copyright 2019, American Chemical Society.

A covalently self-assembled polymer nanocapsules by direct alkylation of perhydroxy CB [6] with a ditopic linker, 1,4-dibromobut-2-ene(DBT) were reported. The generation of singlet oxygen by fluorescent images on the cells at each of these irradiation time points and for the longer irradiation time triggered higher intracellular green fluorescence [56, 57]. The host guest FRET pair of CB [7] forms a very stable host-guest complex with its selected guest molecules such as ferrocenemethyl (FcA), adamantyl (AdA) and didamantinyl ammonium derivatives with very high binding affinity ($K_a \approx 10^{13}$ to $10^{17}$ $M^{-1}$). In biological system, the high affinity protein based binding pairs are found in nature such as streptavidin-biotin ($K_a \approx 10^{13} M^{-1}$) and antibody-antigen ($K_a \approx 10^8$ to $10^{13} M^{-1}$). This high binding affinity molecular recognition of CB [7] and its guests molecules under the *in situ* physiological conditions without significant interference from other biological molecules selectively and specifically were exploited for the protein imaging in cells and tissues [58]. The high affinity of CB [7] based pairs were explored for imaging and they will enhance the local concentration of prodrug-conjugate adamantane (or ferrocene)- in cancer bearing mouse. This synthetic binding pair systems can be alternatives for the non-covalent anchoring because they are typically small(~1kDa), chemically robust and resistant to enzymatic degradation. The occurrence of a FRET signal inside the cells between Cy-3-CB [7] and cyanine conjugated AdA as the donor and acceptor respectively, is confirmed from florescence experiments. Control experiments demonstrate that these are highly specific host-guest recognition between CB [7] and AdA *in vivo* through the spontaneous FRET from BDP630/650-AdA in C.elegans (Figs. **12** and **13**).

Similar CB capped nanoparticle electrochemical sensors are also reported in the literature [59 - 60].

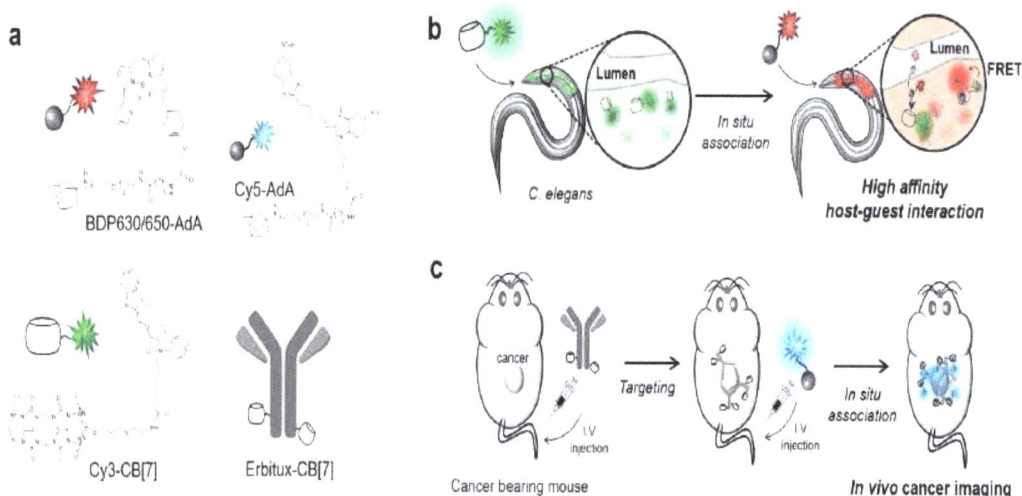

**Fig. (12).** (a) Chemical structures of Cy3-CB [7], antibody-CB [7] (Erbitux-CB [7]), BDP630/650-AdA, and Cy5-AdA, (b) Schematic illustration of locations of high-affinity host–guest interaction in live C. elegans, and (c) *in vivo* cancer imaging [58]. Copyright 2019, American Chemical Society.

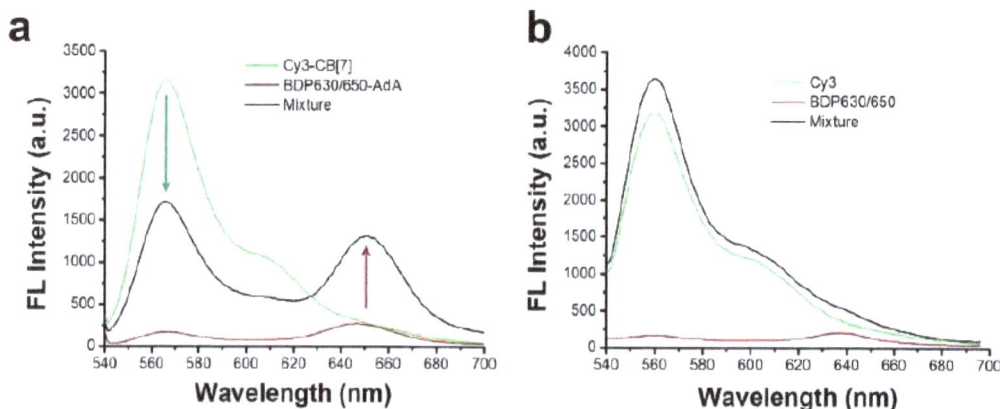

**Fig. (13).** (a) Emission spectra of Cy3-CB [7] and BDP630/650-AdA in PBS buffer containing 3% BSA. (b) Emission spectra of Cy3 and BDP630/650 (without CB [7]) in PBS buffer containing 3% BSA [58].Copyright 2019, American Chemical Society.

In addition, these antibodies integrating an active functionality to the synthetic host molecule allows selective imaging of a cancer site in the live mouse by an efficient *in situ* supramolecular latching of the fluorescent dye conjugated guest to prelocalized host molecules on a cancer site.

# CONCLUSION

In this chapter, we have highlighted CB[n]s based systems having high binding affinity host-guest interactions with various analyte molecules and their applications. A broad variety of supramolecules containing cucurbit[n]urils have been developed for the recognition of a diverse range of analytes based on host-guest complexation with improved sensitivity and specificity with rapid response and ease of usage. These systems respond in the presence of analyte with the measurable change of fluorescence signal, like change in fluorescence intensity and change in fluorescence wavelength. Furthermore, these molecules with an appropriate combination of multicomponent systems provide a suitable environment for the recognition of biomolecules, cancer imaging and therapeutics. The functionalization of CB[n] opened up the possibilities for the development of multicomponent supramolecules for applications in cellular imaging, photodynamic cancer therapy and other biological applications.

# CONSENT OF PUBLICATION

The author grants the publisher the sole and exclusive license of the full copyright of this material.

# CONFLICT OF INTEREST

The author declares no conflict of interest, financial or otherwise.

# ACKNOWLEDGEMENT

Declared none.

# REFERENCES

[1]    Cram, D.J. The design of molecular hosts, guests, and their complexes (Nobel lecture). *Angew. Chem. Int. Ed. Engl.,* **1988**, *27*(8), 1009-1020.
       [http://dx.doi.org/10.1002/anie.198810093]

[2]    Freeman, W.A.; Mock, W.L.; Shih, N.Y. Cucurbituril. *J. Am. Chem. Soc.,* **1981**, *103*(24), 7367-7368.
       [http://dx.doi.org/10.1021/ja00414a070]

[3]    Gellman, S.H. *Introduction: molecular recognition*; ACS Publications, **1997**.
       [http://dx.doi.org/10.4324/9780203392843_chapter_1]

[4]    Mako, T.L.; Racicot, J.M.; Levine, M. Supramolecular luminescent sensors. *Chem. Rev.,* **2019**, *119*(1), 322-477.
       [http://dx.doi.org/10.1021/acs.chemrev.8b00260] [PMID: 30507166]

[5]    Biedermann, F.; Schneider, H.J. Experimental binding energies in supramolecular complexes. *Chem. Rev.,* **2016**, *116*(9), 5216-5300.
       [http://dx.doi.org/10.1021/acs.chemrev.5b00583] [PMID: 27136957]

[6]    Sinn, S.; Biedermann, F. Chemical sensors based on cucurbit [n] uril macrocycles. *Isr. J. Chem.,* **2018**, *58*(3-4), 357-412.

[http://dx.doi.org/10.1002/ijch.201700118]

[7]   Behrend, R.; Meyer, E.; Rusche, F. Ueber condensationsproducte aus glycoluril und formaldehyd. *Justus Liebigs Ann. Chem.,* **1905**, *339*(1), 1-37.
[http://dx.doi.org/10.1002/jlac.19053390102]

[8]   Kim, J.; Jung, I.S.; Kim, S.Y. *et al.* New cucurbituril homologues: syntheses, isolation, characterization, and X-ray crystal structures of cucurbit [n] uril (n= 5, 7, and 8). *J. Am. Chem. Soc.,* **2000**, *122*(3), 540-541.
[http://dx.doi.org/10.1021/ja993376p]

[9]   Jon, S.Y.; Selvapalam, N.; Oh, D.H. *et al.* Facile synthesis of cucurbit[n]uril derivatives via direct functionalization: expanding utilization of cucurbit[n]uril. *J. Am. Chem. Soc.,* **2003**, *125*(34), 10186-10187.
[http://dx.doi.org/10.1021/ja036536c] [PMID: 12926937]

[10]  Day, A.; Arnold, A.P.; Blanch, R.J.; Snushall, B. Controlling factors in the synthesis of cucurbituril and its homologues. *J. Org. Chem.,* **2001**, *66*(24), 8094-8100.
[http://dx.doi.org/10.1021/jo015897c] [PMID: 11722210]

[11]  Day, A.I.; Blanch, R.J.; Arnold, A.P.; Lorenzo, S.; Lewis, G.R.; Dance, I. A cucurbituril-based gyroscane: a new supramolecular form. *Angew. Chem. Int. Ed.,* **2002**, *41*(2), 275-277.
[http://dx.doi.org/10.1002/1521-3773(20020118)41:2<275::AID-ANIE275>3.0.CO;2-M]     [PMID: 12491407]

[12]  Isaacs, L. Cucurbit[n]urils: from mechanism to structure and function. *Chem. Commun. (Camb.),* **2009**, (6), 619-629.
[http://dx.doi.org/10.1039/B814897J] [PMID: 19322405]

[13]  Huang, W.H.; Zavalij, P.Y.; Isaacs, L. Cucurbit[n]uril formation proceeds by step-growth cyclo-oligomerization. *J. Am. Chem. Soc.,* **2008**, *130*(26), 8446-8454.
[http://dx.doi.org/10.1021/ja8013693] [PMID: 18529059]

[14]  Chakraborty, A.; Wu, A.; Witt, D.; Lagona, J.; Fettinger, J.C.; Isaacs, L. Diastereoselective formation of glycoluril dimers: isomerization mechanism and implications for cucurbit[n]uril synthesis. *J. Am. Chem. Soc.,* **2002**, *124*(28), 8297-8306.
[http://dx.doi.org/10.1021/ja025876f] [PMID: 12105910]

[15]  Barrow, S.J.; Kasera, S.; Rowland, M.J.; del Barrio, J.; Scherman, O.A. Cucurbituril-based molecular recognition. *Chem. Rev.,* **2015**, *115*(22), 12320-12406.
[http://dx.doi.org/10.1021/acs.chemrev.5b00341] [PMID: 26566008]

[16]  Cheng, X.J.; Liang, L.L.; Chen, K. *et al.* Twisted Cucurbit[14]uril. *Angew. Chem. Int. Ed.,* **2013**, *52*(28), 7252-7255.
[http://dx.doi.org/10.1002/anie.201210267] [PMID: 23716359]

[17]  Assaf, K.I.; Nau, W.M. Cucurbiturils: from synthesis to high-affinity binding and catalysis. *Chem. Soc. Rev.,* **2015**, *44*(2), 394-418.
[http://dx.doi.org/10.1039/C4CS00273C] [PMID: 25317670]

[18]  Dsouza, R.N.; Pischel, U.; Nau, W.M. Fluorescent dyes and their supramolecular host/guest complexes with macrocycles in aqueous solution. *Chem. Rev.,* **2011**, *111*(12), 7941-7980.
[http://dx.doi.org/10.1021/cr200213s] [PMID: 21981343]

[19]  Biedermann, F.; Scherman, O.A. Cucurbit[8]uril mediated donor-acceptor ternary complexes: a model system for studying charge-transfer interactions. *J. Phys. Chem. B,* **2012**, *116*(9), 2842-2849.
[http://dx.doi.org/10.1021/jp2110067] [PMID: 22309573]

[20]  Kovalenko, E.A.; Pashkina, E.A.; Kanazhevskaya, L.Y.; Masliy, A.N.; Kozlov, V.A. Chemical and biological properties of a supramolecular complex of tuftsin and cucurbit[7]uril. *Int. Immunopharmacol.,* **2017**, *47*, 199-205.
[http://dx.doi.org/10.1016/j.intimp.2017.03.032] [PMID: 28427014]

[21]    Ekaterina, K.; Irina, A.; Denis, S.; Vladimir, F. Syntheses, crystal structures and physico-chemical properties of supramolecular assemblies based on cucurbit[6]uril and mono- and polynuclear bismuth(III) and mercury(II) halides. *J. Mol. Struct.,* **2019**, *1193*, 357-364.
[http://dx.doi.org/10.1016/j.molstruc.2019.05.039]

[22]    Bockus, A.T.; Smith, L.C.; Grice, A.G. *et al.* Cucurbit [7] uril–Tetramethylrhodamine conjugate for direct sensing and cellular imaging. *J. Am. Chem. Soc.,* **2016**, *138*(50), 16549-16552.
[http://dx.doi.org/10.1021/jacs.6b11140] [PMID: 27998093]

[23]    Zhang, S.; Assaf, K.I.; Huang, C.; Hennig, A.; Nau, W.M. Ratiometric DNA sensing with a host–guest FRET pair. *Chem. Commun. (Camb.),* **2019**, *55*(5), 671-674.
[http://dx.doi.org/10.1039/C8CC09126A] [PMID: 30565597]

[24]    Ling, Y.; Wang, W.; Kaifer, A.E. A new cucurbit[8]uril-based fluorescent receptor for indole derivatives. *Chem. Commun. (Camb.),* **2007**, (6), 610-612.
[http://dx.doi.org/10.1039/B611559D] [PMID: 17264907]

[25]    Biedermann, F.; Rauwald, U.; Cziferszky, M.; Williams, K.A.; Gann, L.D.; Guo, B.Y.; Urbach, A.R.; Bielawski, C.W.; Scherman, O.A. Benzobis(imidazolium)-cucurbit[8]uril complexes for binding and sensing aromatic compounds in aqueous solution. *Chemistry,* **2010**, *16*(46), 13716-13722.
[http://dx.doi.org/10.1002/chem.201002274] [PMID: 21058380]

[26]    Sindelar, V.; Cejas, M.A.; Raymo, F.M.; Chen, W.; Parker, S.E.; Kaifer, A.E. Supramolecular assembly of 2,7-dimethyldiazapyrenium and cucurbit[8]uril: a new fluorescent host for detection of catechol and dopamine. *Chemistry,* **2005**, *11*(23), 7054-7059.
[http://dx.doi.org/10.1002/chem.200500917] [PMID: 16175642]

[27]    Biedermann, F.; Hathazi, D.; Nau, W.M. Associative chemosensing by fluorescent macrocycle–dye complexes – a versatile enzyme assay platform beyond indicator displacement. *Chem. Commun. (Camb.),* **2015**, *51*(24), 4977-4980.
[http://dx.doi.org/10.1039/C4CC10227D] [PMID: 25622263]

[28]    You, L.; Zha, D.; Anslyn, E.V. Recent advances in supramolecular analytical chemistry using optical sensing. *Chem. Rev.,* **2015**, *115*(15), 7840-7892.
[http://dx.doi.org/10.1021/cr5005524] [PMID: 25719867]

[29]    Sinn, S.; Krämer, J.; Biedermann, F. Teaching old indicators even more tricks: binding affinity measurements with the guest-displacement assay (GDA). *Chem. Commun. (Camb.),* **2020**, *56*(49), 6620-6623.
[http://dx.doi.org/10.1039/D0CC01841D] [PMID: 32459225]

[30]    Dewantari, A.A.; Yongwattana, N.; Payongsri, P.; Seemakhan, S.; Borwornpinyo, S.; Ojida, A.; Wongkongkatep, J. Fluorescence Detection of Deoxyadenosine in *Cordyceps* spp. by Indicator Displacement Assay. *Molecules,* **2020**, *25*(9), 2045.
[http://dx.doi.org/10.3390/molecules25092045] [PMID: 32353945]

[31]    Wu, S.; Zhang, Y.; Yang, L.; Li, C.P. Label-Free Fluorescent Determination of Sunset Yellow in Soft Drinks Based on an Indicator-Displacement Assay. *J. Food Qual.,* **2018**, *2018*, 1-9.
[http://dx.doi.org/10.1155/2018/6302345]

[32]    Zhu, L.; Zhao, Z.; Zhang, X.; Zhang, H.; Liang, F.; Liu, S. A Highly Selective and Strong Anti-Interference Host-Guest Complex as Fluorescent Probe for Detection of Amantadine by Indicator Displacement Assay. *Molecules,* **2018**, *23*(4), 947.
[http://dx.doi.org/10.3390/molecules23040947] [PMID: 29670072]

[33]    Lazar, A.I.; Biedermann, F.; Mustafina, K.R.; Assaf, K.I.; Hennig, A.; Nau, W.M. Nanomolar binding of steroids to cucurbit [n] urils: selectivity and applications. *J. Am. Chem. Soc.,* **2016**, *138*(39), 13022-13029.
[http://dx.doi.org/10.1021/jacs.6b07655] [PMID: 27673427]

[34]    Sinn, S.; Spuling, E.; Bräse, S.; Biedermann, F. Rational design and implementation of a

cucurbit[8]uril-based indicator-displacement assay for application in blood serum. *Chem. Sci. (Camb.)*, **2019**, *10*(27), 6584-6593.
[http://dx.doi.org/10.1039/C9SC00705A] [PMID: 31367309]

[35]    Zhang, S.; Grimm, L.; Miskolczy, Z.; Biczók, L.; Biedermann, F.; Nau, W.M. Binding affinities of cucurbit[ *n* ]urils with cations. *Chem. Commun. (Camb.)*, **2019**, *55*(94), 14131-14134.
[http://dx.doi.org/10.1039/C9CC07687E] [PMID: 31696884]

[36]    Pais, V.F.; Carvalho, E.F.A.; Tomé, J.P.C.; Pischel, U. Supramolecular control of phthalocyanine dye aggregation. *Supramol. Chem.*, **2014**, *26*(9), 642-647.
[http://dx.doi.org/10.1080/10610278.2014.926011]

[37]    Aliaga, M.E.; García-Río, L.; Pessêgo, M.; Montecinos, R.; Fuentealba, D.; Uribe, I.; Martín-Pastor, M.; García-Beltrán, O. Host–guest interaction of coumarin-derivative dyes and cucurbit[7]uril: leading to the formation of supramolecular ternary complexes with mercuric ions. *New J. Chem.*, **2015**, *39*(4), 3084-3092.
[http://dx.doi.org/10.1039/C5NJ00162E]

[38]    Lindoy, L.; Cong, H.; Geng, Q-X.; Tao, Z.; Wei, G. Cucurbit [7] uril-improved recognition by a fluorescent sensor for cadmium and zinc cations. *Supramol. Chem.*, **2016**, *28*, 9-10.

[39]    Bai, Q.; Zhang, S.; Chen, H. *et al.* Alkyl substituted cucurbit [6] uril assisted competitive fluorescence recognition of lysine and methionine in aqueous solution. *ChemistrySelect*, **2017**, *2*(8), 2569-2573.
[http://dx.doi.org/10.1002/slct.201700053]

[40]    Minami, T.; Esipenko, N.A.; Zhang, B.; Isaacs, L.; Anzenbacher, P., Jr "Turn-on" fluorescent sensor array for basic amino acids in water. *Chem. Commun. (Camb.)*, **2014**, *50*(1), 61-63.
[http://dx.doi.org/10.1039/C3CC47416J] [PMID: 24196424]

[41]    Gao, Z.Z.; Kan, J.L.; Chen, L.X. *et al.* Binding and selectivity of essential amino acid guests to the inverted cucurbit [7] uril host. *ACS Omega*, **2017**, *2*(9), 5633-5640.
[http://dx.doi.org/10.1021/acsomega.7b00429] [PMID: 31457827]

[42]    Ghale, G.; Ramalingam, V.; Urbach, A.R.; Nau, W.M. Determining protease substrate selectivity and inhibition by label-free supramolecular tandem enzyme assays. *J. Am. Chem. Soc.*, **2011**, *133*(19), 7528-7535.
[http://dx.doi.org/10.1021/ja2013467] [PMID: 21513303]

[43]    Smith, L.C.; Leach, D.G.; Blaylock, B.E.; Ali, O.A.; Urbach, A.R. Sequence-specific, nanomolar peptide binding via cucurbit[8]uril-induced folding and inclusion of neighboring side chains. *J. Am. Chem. Soc.*, **2015**, *137*(10), 3663-3669.
[http://dx.doi.org/10.1021/jacs.5b00718] [PMID: 25710854]

[44]    Lagona, J.; Wagner, B.D.; Isaacs, L. Molecular-recognition properties of a water-soluble cucurbit[6]uril analogue. *J. Org. Chem.*, **2006**, *71*(3), 1181-1190.
[http://dx.doi.org/10.1021/jo052294i] [PMID: 16438536]

[45]    Minami, T.; Esipenko, N.A.; Akdeniz, A.; Zhang, B.; Isaacs, L.; Anzenbacher, P., Jr Multianalyte sensing of addictive over-the-counter (OTC) drugs. *J. Am. Chem. Soc.*, **2013**, *135*(40), 15238-15243.
[http://dx.doi.org/10.1021/ja407722a] [PMID: 24000805]

[46]    Romero, M.A.; González-Delgado, J.A.; Mendoza, J.; Arteaga, J.F.; Basílio, N.; Pischel, U. Terpenes Show Nanomolar Affinity and Selective Binding with Cucurbit[8]uril. *Isr. J. Chem.*, **2018**, *58*(3-4), 487-492.
[http://dx.doi.org/10.1002/ijch.201700119]

[47]    Florea, M.; Nau, W.M. Strong binding of hydrocarbons to cucurbituril probed by fluorescent dye displacement: a supramolecular gas-sensing ensemble. *Angew. Chem. Int. Ed.*, **2011**, *50*(40), 9338-9342.
[http://dx.doi.org/10.1002/anie.201104119] [PMID: 21948429]

[48]    Liu, W.; Ai, H.; Meng, Z.; Isaacs, L.; Xu, Z.; Xue, M.; Yan, Q. Interactions between acyclic CB[ *n* ]-

type receptors and nitrated explosive materials. *Chem. Commun. (Camb.)*, **2019**, *55*(71), 10635-10638.
[http://dx.doi.org/10.1039/C9CC05117A] [PMID: 31429448]

[49]   Park, K.M.; Hur, M.Y.; Ghosh, S.K.; Boraste, D.R.; Kim, S.; Kim, K. Cucurbit[ *n* ]uril-based amphiphiles that self-assemble into functional nanomaterials for therapeutics. *Chem. Commun. (Camb.)*, **2019**, *55*(72), 10654-10664.
[http://dx.doi.org/10.1039/C9CC05567C] [PMID: 31418758]

[50]   Li, S.; Kuok, K.I.; Ji, X.; Xu, A.; Yin, H.; Zheng, J.; Tan, H.; Wang, R. Supramolecular Modulation of Antibacterial Activity of Ambroxol by Cucurbit[7]uril. *ChemPlusChem*, **2020**, *85*(4), 679-683.
[http://dx.doi.org/10.1002/cplu.202000119] [PMID: 32253831]

[51]   Li, M.; Lee, A.; Kim, S.; Shrinidhi, A.; Park, K.M.; Kim, K. Cucurbit[7]uril-conjugated dyes as live cell imaging probes: investigation on their cellular uptake and excretion pathways. *Org. Biomol. Chem.*, **2019**, *17*(25), 6215-6220.
[http://dx.doi.org/10.1039/C9OB00356H] [PMID: 31179469]

[52]   Ding, Y.F.; Sun, T.; Li, S.; Huang, Q.; Yue, L.; Zhu, L.; Wang, R. Oral Colon-Targeted Konjac Glucomannan Hydrogel Constructed through Noncovalent Cross-Linking by Cucurbit[8]uril for Ulcerative Colitis Therapy. *ACS Appl. Bio Mater.*, **2020**, *3*(1), 10-19.
[http://dx.doi.org/10.1021/acsabm.9b00676] [PMID: 35019421]

[53]   Kim, S.; Hur, M.Y.; Kim, J.; Park, K.M.; Kim, K. Strong host-guest interaction enables facile and controllable surface modification of cucurbit[6]uril-based polymer nanocapsules for *in vivo* cancer targeting. *Supramol. Chem.*, **2019**, *31*(5), 289-295.
[http://dx.doi.org/10.1080/10610278.2019.1593413]

[54]   Wu, X.; Chen, Y.; Yu, Q.; Li, F.Q.; Liu, Y. A cucurbituril/polysaccharide/carbazole ternary supramolecular assembly for targeted cell imaging. *Chem. Commun. (Camb.)*, **2019**, *55*(30), 4343-4346.
[http://dx.doi.org/10.1039/C9CC01601E] [PMID: 30911744]

[55]   Özkan, M.; Kumar, Y.; Keser, Y.; Hadi, S.E.; Tuncel, D. Cucurbit[7]uril-Anchored Porphyrin-Based Multifunctional Molecular Platform for Photodynamic Antimicrobial and Cancer Therapy. *ACS Appl. Bio Mater.*, **2019**, *2*(11), 4693-4697.
[http://dx.doi.org/10.1021/acsabm.9b00763] [PMID: 35021467]

[56]   Sun, C.; Zhang, H.; Yue, L.; Li, S.; Cheng, Q.; Wang, R. Facile preparation of cucurbit [6] uril-based polymer nanocapsules for targeted photodynamic therapy. *ACS appl. ACS Appl. Mater. Interfaces*, **2019**, *11*(26), 22925-22931.
[http://dx.doi.org/10.1021/acsami.9b04403] [PMID: 31252492]

[57]   Özkan, M.; Hadi, S.E.; Tunç, İ.; Midilli, Y.; Ortaç, B.; Tuncel, D. Cucurbit[7]uril-Capped Hybrid Conjugated Oligomer-Gold Nanoparticles for Combined Photodynamic-Photothermal Therapy and Cellular Imaging. *ACS Appl. Polym. Mater.*, **2020**, *2*(9), 3840-3849.
[http://dx.doi.org/10.1021/acsapm.0c00540]

[58]   Li, M.; Kim, S.; Lee, A. *et al.* Bio-orthogonal supramolecular latching inside live animals and its application for *in vivo* cancer imaging. *ACS Appl. Mater. Interfaces*, **2019**, *11*(47), 43920-43927.
[http://dx.doi.org/10.1021/acsami.9b16283] [PMID: 31686496]

[59]   Song, S.; Hu, X.; Li, H.; Zhao, J.; Koh, K.; Chen, H. Guests involved CB[8] capped silver nanoparticles as a means of electrochemical signal enhancement for sensitive detection of Caspase-3. *Sens. Actuators B Chem.*, **2018**, *274*, 54-59.
[http://dx.doi.org/10.1016/j.snb.2018.07.143]

[60]   Li, H.; Hu, X.; Zhao, J.; Koh, K.; Chen, H. A label-free impedimetric sensor for the detection of an amphetamine-type derivative based on cucurbit[7]uril-mediated three-dimensional AuNPs. *Electrochem. Commun.*, **2019**, *100*, 126-133.
[http://dx.doi.org/10.1016/j.elecom.2019.02.002]

CHAPTER 6

# Rhenium(I)-Based Metallacycles for Sensing Applications

**Murugesan Velayudham**[1,*] and **Pounraj Thanasekaran**[2,*]

[1] *Department of Chemistry, Thiagarajar College of Engineering, Madurai – 625015, Tamilnadu, India*

[2] *Department of Chemistry, Pondicherry University, Kalapet, Puducherry -605 014, India*

**Abstract:** Coordination-driven self-assembly provides unique opportunities to prepare highly complex chemical systems from simple components and has led to significant progress in the construction of supramolecular materials with novel topologies and exploitable functions. During the past few decades, metallacycles have captured widespread interests due to their wide applications in catalysis, sensor, and biological relevant applications. Thus, exploring new metallacycles, studying their physical and chemical properties and applications have become one of the most attractive and exciting areas of inorganic chemistry and supramolecular chemistry. Among which, rhenium(I)-based metallacycles, constructed from the rhenium metal ions and a variety of aromatic ligands, have attracted considerable attention because of their unique potentials in light-harvesting, catalysis, sensing, biomedical, *etc*. In this chapter, we summarize the recent research progress in rhenium-based metallacycles with their synthesis, properties and potential application in host-guest chemistry.

**Keywords:** Aggregation induced emission, Aromatic molecules, Binding, Cages, Guest-Host, Hydrophobic interactions, Luminescence, Luminescent probes, Luminescent sensors, Metallacycles, Photophysics, Prisms, Quenching, Rectangle, Rectangular box, Rhenium, Self-assembly, Sensing, Weak interactions.

## INTRODUCTION

Self-assembly that is one of the important organizing principles of biological systems is a widely applied strategy in supramolecular chemistry as the driving forces are used to assemble various artificial structures from simple building blocks. The design and synthesis of self-assembled metallacycles has matured as

* **Corresponding author Murugesan Velayudham and Pounraj Thanasekaran**: Department of Chemistry, Thiagarajar College of Engineering, Madurai – 625015, Tamilnadu, India and Department of Chemistry, Pondicherry University, Kalapet, Puducherry -605 014, India; Tel: +917708446619; E-mails: velayudhamm@gmail.com, ptsekaran@gmail.com

**Paulpandian Muthu Mareeswaran, Palaniswamy Suresh and Seenivasan Rajagopal (Eds.)**

an intensely active area of research in supramolecular chemistry owing to their internal cavities with well-defined shape and size and promising potential applications (Figs. **1 - 2**) [1 - 10]. Several synthetic approaches such as directional bonding [1, 11, 12], symmetry interaction [13], and a weak-link approach [14] permit the formation of a single thermodynamic product in high yield with reducing synthetic costs. Thanks to the diversity of coordination-driven self-assembly, a large number of intriguing molecular assemblies, such as molecular triangles, squares, rectangles, cages, prisms, *etc.* have been synthesized *via* a self-assembly route using suitable organic ligands and various metal ions. For instance, the Fujita [11] and Stang [1, 15] groups used a multicomponent self-assembly method for designing large structures based on specific information set within the individual components. Raymond *et al.* significantly extended the investigations towards the dynamic behaviour, chirality and catalysis in supramolecular coordination complexes (SCCs) [16 - 18]. Nitschke *et al.* introduced the concept of 'sub-component self-assembly' [19] according to which the actual linker is formed *in situ* to allow covalent post-assembly modifications of the SCCs [20]. Therefore, metallosupramolecular architectures *via* coordination-driven self-assembly are at the forefront of supramolecular chemistry and will continue to attract focused research.

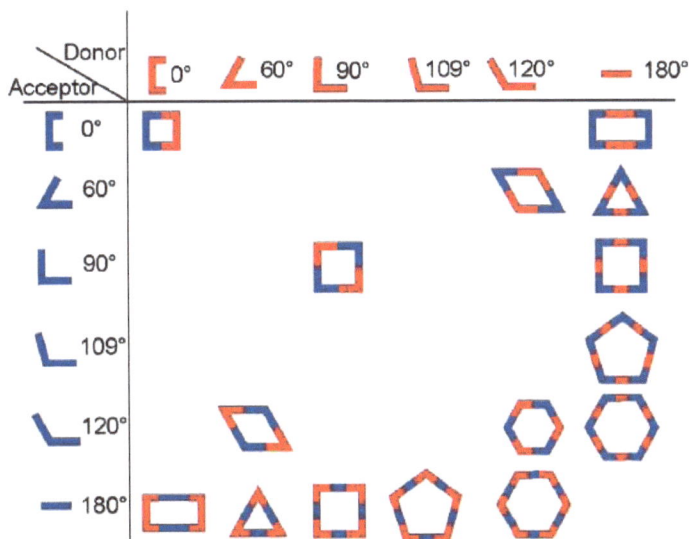

**Fig. (1).** Combination of various building units for accessing convex polygons and canonical polyhedra. Reproduced with permission from ref. 15. Copyright (2011) American Chemical Society.

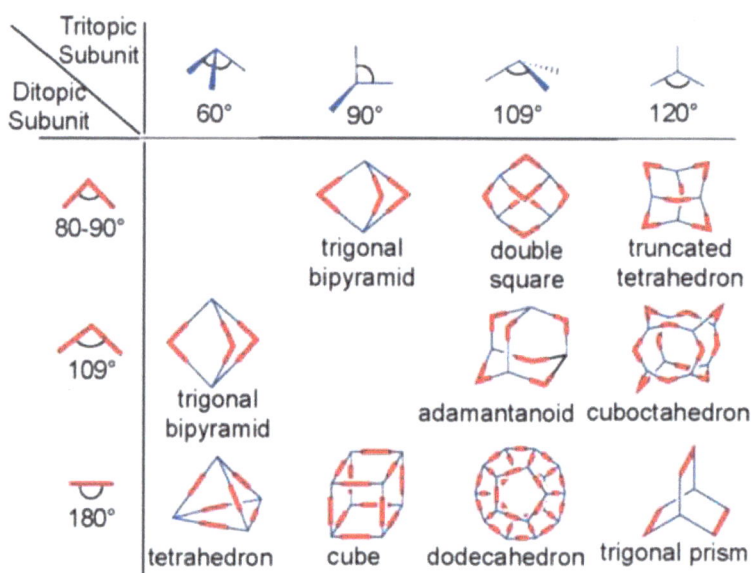

**Fig. (2).** Three-dimensional architectures formed by the combination of ditopic and tritopic subunits by the directional bonding approach. Reproduced with permission from ref. 15. Copyright (2011) American Chemical Society.

A majority of these designed metallacycles possess attractive properties such as Lewis acidity, magnetism, redox activity, absorption or luminescence properties, which allow them for use in potential applications in various areas like sensing and catalysis [21]. Especially, the inspiration for using metallacycles in host-guest applications originates from their characteristic properties, such as the ease of fine-tuning the structures, the judicious choice of metal ions and ligands with specific sizes, their coordination geometry, and the simple incorporation of essential functional group modifications [15, 22 - 27]. The photophysical properties from the UV through the visible to the near IR can be tuned by varying the structure of the ligands used, as well as the metal ions present in the metal–organic architectures, which is more favorable for biological studies. Compared with conventional compounds with covalent interactions, the host-guest system based on metallacycles and guests can provide greater flexibility [17]. These compounds offer interesting opportunities for incorporating different cavity sizes that can act as hosts for aromatic guests through π-π, hydrophobic interactions or producing a cavity with functional groups that can act as a hydrogen-bond donor or acceptor and would be expected to allow the selective uptake of hydrogen-bonding guests in addition to photochemical reactions and molecular devices [3, 15, 28]. In addition, by taking advantages of the improvement in photoluminescent properties and cavity size of these host molecules to external guests, it is possible to use such systems as effective sensors.

Among these metallacycles, rhenium(I) compounds have proven to be especially suitable for use in host-guest chemistry applications owing to their stable structures even under physiological conditions, large Stokes shifts, long lifetimes, light-switching properties, photostability, and effective singlet oxygen photosensitization *etc* [29 - 33]. By utilizing *fac*-Re(CO)$_3$ corners with suitable organic linkers, the rich possibilities for generating various Re(I)-based metallacycles such as squares, rectangles, gondolas, rotors, triangles, cages, and boxes, with predesigned geometries that are constructed in a step-wise manner or one-pot reactions in high yields are attempted, demonstrating the scalability of the size of SCCs [3, 34, 35]. The photoluminescence efficiency can also be improved by the incorporation of π-conjugated ligands or long alkyl chains *via* an aggregation process. Various rhenium-based metallacycles feature an internal cavity accessible to guest encapsulation, and thus, exploitable for various functions and applications involving host-guest chemistry. Significant progress has been made in the past decades in the rational design, properties and applications of functional Re(I)-based metallacycles, and hence, it is time to summarize the recent development in the construction, functional properties and sensing applications of such functionalized supramolecular Re(I) metallacycles. In this chapter, we highlight some recent advances on the design and preparation of Re(I)-functionalized metallacycles *via* coordination-driven self-assembly. In addition, the photophysical properties and the host-guest chemistry applications of these functionalized metallacycles are discussed as well.

### Rhenium(I)-Based Metallacycles

The advantages of using *facial*-Re(CO)$_3$ core as a metal precursor for the preparation of metallacycles are: (i) It has three vacant orthogonal coordination sites that are useful to obtain 2D and 3D structures in high yields depending on the coordinating nature of the ligands, (ii) most of the metallacycles derived from this precursor are stable in solution and external stimuli, (iii) metallacycles derived from this precursor exhibit a good thermal-, redox- and photo-stability, (iv) it permits both conventional and non-conventional methods to prepare a variety of metallacycles in a single or multi-step strategies, (v) study of *in-situ* CO stretching frequencies *via* FT-IR spectroscopy allows easily the progress of the reaction, (vi) depending on the nature of coordinating ligands, neutral and ionic metallacycles can be accomplished, (vii) tuning the redox and photophysical properties with predesigned geometries of the ligands, and (vii) the presence of cavity that are depending on the size and length of ligands, and functional units in the metallacycles allow to study the host-guest interactions. Taking advantage of these properties, several neutral and ionic Re(I)-based metallacycles have been prepared and showed their potential use as sensors *via* host-guest interaction as discussed below.

## Binuclear Re(I) Complex

A clip type structured binuclear rhenium(I) compounds [{Re(CO)$_3$(1,4-NVP)}$_2$ (μ$_2$-OR)$_2$] (**1**, R = C$_4$H$_9$; **2**, R = C$_{10}$H$_{21}$; 1,4-NVP = 4-(1-naphthylvinyl)pyridine] have been shown to act as sensors for aromatic hydrocarbons [36, 37]. Through a self-assembly process, complexes **1** and **2** were accomplished by the one-pot reaction of Re$_2$(CO)$_{10}$ with 1,4-NVP in 1-butanol or 1-decanol under solvothermal reaction condition. These complexes were characterized by various spectral techniques. In this study, environmental contaminants polyaromatic hydrocarbons (PAHs) such as pyrene, anthracene, naphthalene and phenanthrene were selected as guest molecules (Fig. **3**).

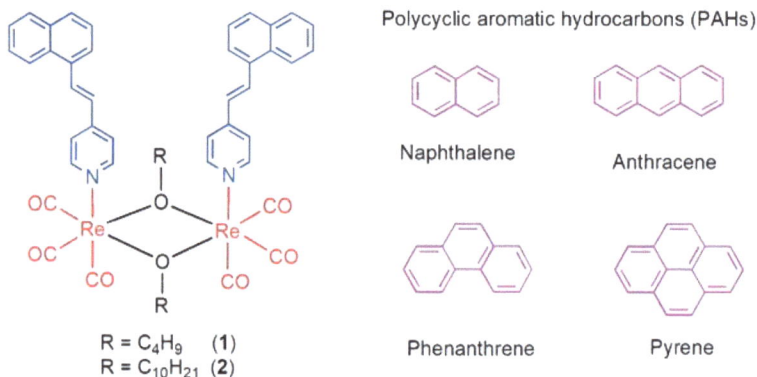

**Fig. (3).** Structure of two alkoxy bridged binuclear Re(I) complexes **1** and **2** and PAH guest molecules.

X-ray crystallographic studies of **1** and **2** showed that the distance between the two Re(I) metal centers and the face-to-face pyridyl ligand of 1,4-NVP was found to be 3.38 and 3.5 Å, respectively, indicating the presence of intramolecular Re⋯Re and π–π-stacking interactions. UV-Vis absorption spectra of **1** and **2** in dichloromethane exhibited a ligand centred intense bands at 200–300 nm and a dπ(Re) → π*(1,4-NVP) singlet metal-to-ligand charge transfer ($^1$MLCT transition) as a low energy band at 355 nm. At the excitation of 355 nm, a weak emission band at 420 nm was observed owing to the efficient quenching of triplet metal-to-ligand charge transfer ($^3$MLCT) state by triplet intraligand ($^3$IL) state of the 1,4-NVP moiety through an energy transfer pathway. In addition, the presence of low quantum yield and short excited state lifetime of **1** and **2** confirmed the assignment of low-lying π→π* transition localized on the 1,4-NVP ligand.

The peculiar feature of these complexes showed clip structures containing cavities, which might be useful to recognize aromatic planar guest molecules. UV-Vis spectral titration profiles revealed that the addition of PAHs into the solution of **1** or **2** caused an increase in the absorption intensity gradually, indicating the formation of charge transfer complex between the PAHs and the 1,4-NVP unit of **1** and **2**, which is stabilized by donor-acceptor complexation. The binding constants for the interaction between the PAHs and complex **1** or **2** were estimated using the Benesi-Hildebrand equation (eqn. 1) [38].

$$1/ \Delta A = 1/ \Delta \varepsilon \, [G] + (1 + \Delta \varepsilon K[H][G]) \tag{1}$$

where $\Delta A$ is the change in the absorption intensity of the PAHs guests upon the addition of the host **1** or **2**, $\Delta \varepsilon$ is the difference in the molar extinction coefficient of the bound and free guest molecule, [H] and [G] are the concentration of the host and guest molecules, respectively, and $K$ is the binding constant. A 1:1 host-guest complex formation was found between the PAHs and complex **1** or **2** using a double-reciprocal plot. The binding constant ($K$) values were calculated to be in the range of $10^4$ $M^{-1}$, and these values are listed in Table 1. Emission titration studies showed that effective emission quenching of PAHs took place without noticeable shift in the emission maximum upon the addition of complex **1** or **2**. The emission quenching rate constant ($k_q$) was calculated by using the Stern-Volmer equation (eqn. 2) [39]

Table 1. Ground-state binding constants ($K$), excited-state dynamic ($K_D$), static ($K_S$), Stern-Volmer constants ($K_{SV}$) and quenching rate constants ($k_q$) of compounds 1 and 2 with aromatic hydrocarbons at 298 K.

| | 1 | | | | 2 | | | |
|---|---|---|---|---|---|---|---|---|
| Guest | $K \times 10^4$ $M^{-1}$ | $K_D \times 10^4$ $M^{-1}$ | $K_S \times 10^4$ $M^{-1}$ | $k_q \times 10^{12}$ $M^{-1}s^{-1}$ | $K \times 10^4$ $M^{-1}$ | $K_D \times 10^4$ $M^{-1}$ | $K_S \times 10^4$ $M^{-1}$ | $k_q \times 10^{12}$ $M^{-1}s^{-1}$ |
| Naphthalene | 1.4 | 2.8 | 1.8 | --- | 2.3 | 3.1 | 1.9 | --- |
| Anthracene | 2.1 | 8.3 | 2.3 | --- | 2.3 | 6.2 | 2.4 | --- |
| Pyrene | 1.3 | 6.2 | 3.4 | 2.2 | 3.5 | 5.8 | 4.3 | 2.07 |
| Phenanthrene | 1.8 | 2.4 | 3.1 | --- | 2.5 | 2.7 | 2.5 | --- |

$$I_o /I = 1 + K_{SV}[Q] = 1 + k_q \tau_o[Q] \tag{2}$$

where Io and I are the emission intensities in the absence and presence of quencher and [Q] is the quencher concentration and $\tau_o$ is the lifetime of fluorophore in the absence of quencher, and $K_{SV}$ is the Stern-Volmer constant. A non-linear plot of Io/I *vs* [Q] indicated the presence of static quenching along with the dynamic quenching. The high value ($10^{12}$ $M^{-1}s^{-1}$) of the quenching rate constant, *k*q, was almost three orders higher than that of diffusion-controlled rate constant, revealing that quenching occurred along with binding of PAHs with complexes **1** and **2** [40]. The emission quenching was attributed to the intermolecular energy transfer from the π-π* emitting state of the guest to the low-lying charge transfer excited state, which returns to the ground state *via* radiationless decay. The modified Stern-Volmer equation (eqn. 3) was used to explain the non-linearity of the curve [41]

$$I_o/I = (1 + K_D[\text{host}])(1 + K_S[\text{host}]) \tag{3}$$

Where $K_D$ and $K_S$ are the dynamic and static Stern–Volmer constants, respectively. From a least-square fitting curve, the value of static ($K_S$) and dynamic ($K_D$) quenching rate was calculated and are listed in Table 1. The $K_S$ values are consistent with the values obtained from absorption study. It was found that alkoxy chains **1** and **2** did not alter the binding constants with PAHs. $^1$H NMR and density functional theory (DFT) studies supported the interaction of PAHs with these complexes *via* non-covalent interactions (Fig. 4). This works seems to be the first report on sensing aromatic hydrocarbons using clip type structured binuclear Re(I) complexes.

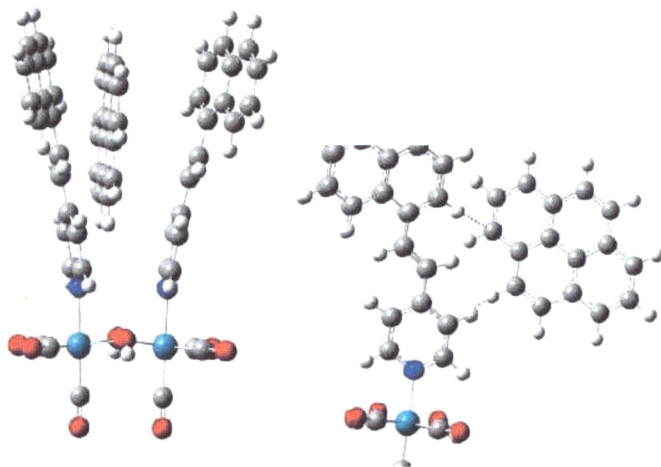

**Fig. (4).** (a) Optimized geometry of **2** with pyrene (b) C-H⋯π interaction between (host) **2** and (guest) pyrene. (need to change good figure; Reproduced with permission from ref. 37. Copyright (2019) Elsevier publishers.

In another study, the same group [42] worked on these materials to show aggregation-induced emission properties. Compounds **1** and **2** formed nano-aggregates in $CH_2Cl_2$ in the presence of poor solvent because of the presence of long alkyl chains, and showed as a luminescence sensor for the detection of nitroexplosives. Absorption spectral profile showed that upon the gradual addition of acetonitrile to a dichloromethane solution of **1** or **2**, the absorption maximum at 355 nm was blue-shifted by 15 nm with an increase in the absorbance, which can be attributed to the formation of H-aggregates through a strong $\pi$-stacking interaction of the naphthalene moiety. Emission spectral studies showed **2** had weak emission intensity at 420 nm in $CH_2Cl_2$. However, upon incremental addition of $CH_3CN$, an enormous increase in emission intensity at 420 nm by 20-fold was observed (Fig. **5a**). From these spectral studies, it was inferred that hydrophobic long alkyl chain played a major role to induce the formation of nano-aggregates upon the addition of poor solvent, acetonitrile. Compared to **1**, compound **2** that contains a long decyl group facilitated the nano-aggregate formation. The nanoscopic aggregate formation was further corroborated by transmission electron microscopic (TEM) studies (Fig.**5b**). Controlled studies showed that the addition of other organic solvents to a pure $CH_2Cl_2$ solution of **1** and **2** did not alter the formation of nano-aggregates.

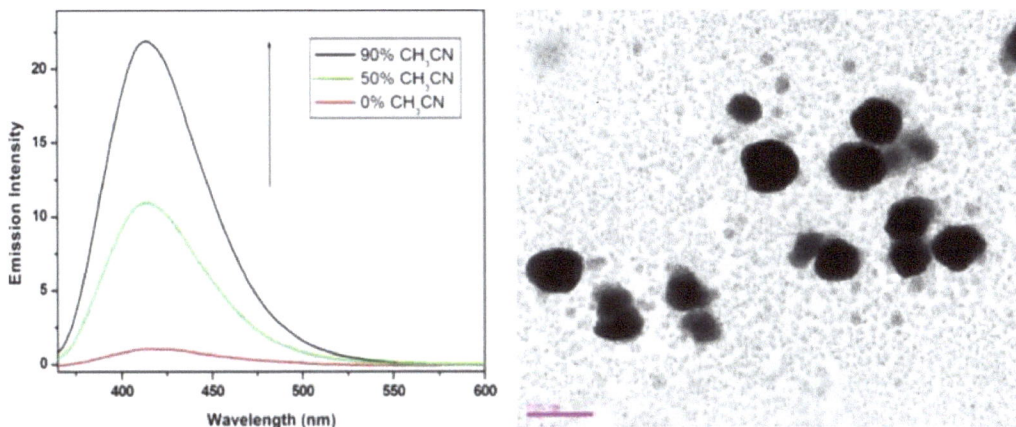

**Fig. (5).** (**a**) Emission spectra of complex **2** (20 μM) in $CH_2Cl_2$ and after the addition of $CH_3CN$ 0, 50, and 90%. (**b**) TEM image of nanoaggregates of complex **2** in $CH_2Cl_2$–$CH_3CN$ mixture (10:90 v/v). Reproduced with permission from ref. 42. Copyright (2013) American Chemical Society.

To gain insight on the potential application of these nanoaggregates, they were utilized as sensors for the detection of nitroaromatics explosive. Incremental addition of picric acid (PA) to the nanoaggregates of **2** resulted in increase in its absorption intensity along with a red-shifted absorption maximum from 326 to 366 nm to the tune of 40 nm. These spectral changes indicated a strong interaction between the PA and **2**, presumably due to the ground-state charge transfer complex formation. Similarly, luminescence spectral studies showed that addition of PA into the nanoaggregates of **2** caused the emission quenching without any considerable shift in the emission wavelength of **2** (Fig. **6**). Stern-Volmer plot analysis revealed that a plot of $I_0/I$ *vs* [PA] at the lower concentrations of [PA] was linear and the Stern-Volmer constant ($K_{sv}$) was found to be $1.0 \times 10^5$ $M^{-1}$.

However, at higher concentrations of [PA], the graph showed an upward curvature, which is attributed to superamplified quenching effect. These emission spectral changes were attributed to the formation of a charge-transfer complex between the electron rich fluorophore (donor) and electron-poor picric acid (acceptor). The quenching experiments were performed with other nitroaromatic compounds as well and the results showed that the nanoaggregates of **2** showed high selectivity toward PA, with a quenching efficiency of 94%.

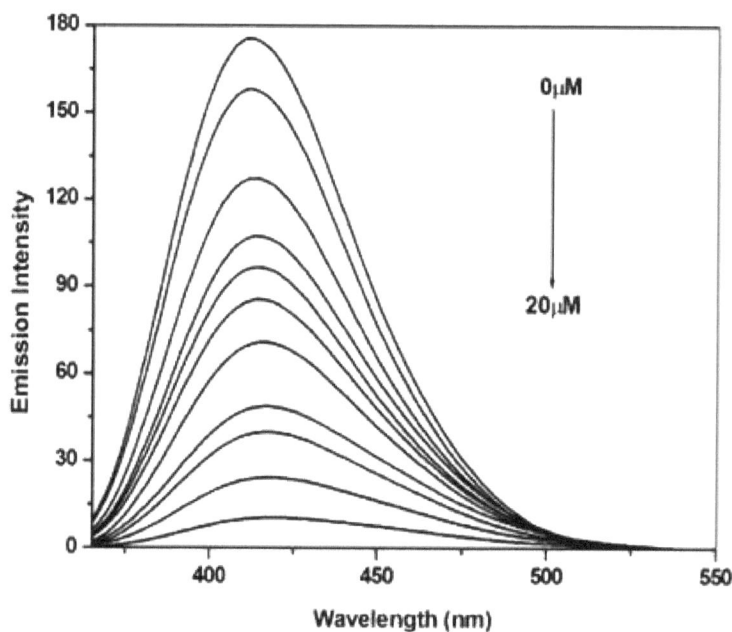

**Fig. (6).** Emission spectra of nanoaggregates of complex **2** (20 μM) in $CH_2Cl_2$ / $CH_3CN$ mixture (10:90 v/v) containing different amounts of PA. Reproduced with permission from ref. 42. Copyright (2013) American Chemical Society.

In the extension of work by Rajagopal *et al.* [43], they reported the utilization of these compounds **1** and **2** (Fig. **7**) as sensors for the detection of insulin fibrils. These compounds **1** and **2** exhibited weak fluorescence at 440 nm. In the presence of native form of insulin, compounds **1** and **2** showed a 4-fold fluorescence enhancement, which is attributed to the π–π stacking interactions between the hydrophobic nature of the naphthalene moiety in **1** and **2** and the protein.

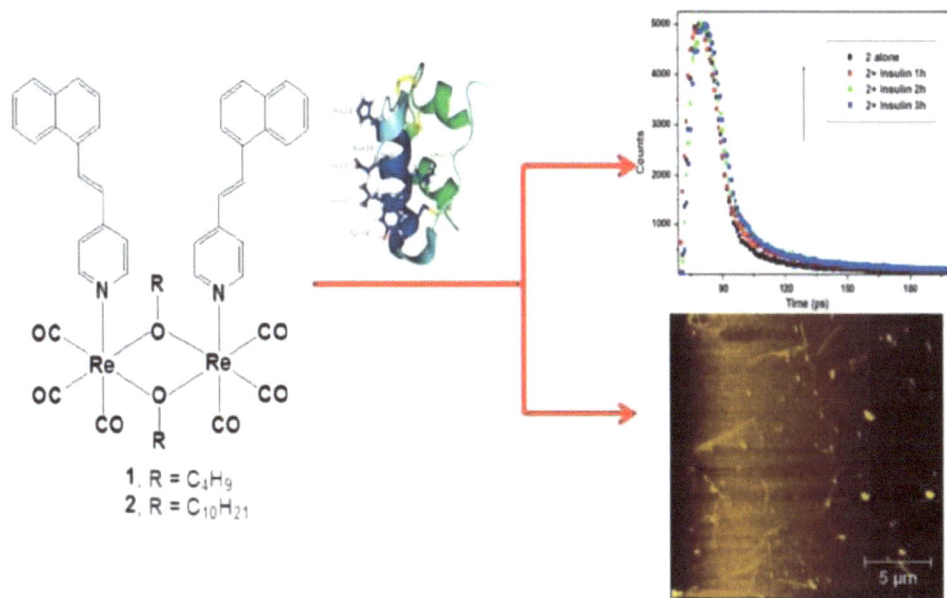

**Fig. (7).** Schematic representation of alkoxy bridged binuclear rhenium(I) complexes **1** and **2** as probes for the detection of aggregation of human insulin fibrils. Reproduced with permission from ref. 43. Copyright (2016) Elsevier.

During this time, transformation of the α-helical insulin to the β-sheet rich proto fibril structure took place. Time-resolved measurements revealed that compound **2** displayed its excited state lifetime of less than 1 ns. After the addition of insulin fibrils, it exhibited a bi-exponential decay with a lifetime of 2.1 and 1.8 ns, indicating the formation of insulin fibrils. Furthermore, AFM imaging study showed thread like structure of fibrils with a size of ~ 50 nm in the presence of **2**.

In another study, the same group [44] first reported on the selective sensing of amyloid fibril that are consorted with Alzheimer's disease (AD) using alkoxy bridged binuclear rhenium(I) complexes **1** and **2**. In the absence of amyloid-β fibrils, compounds **1** and **2** showed poor fluorescence intensity. However, upon the addition of amyloid-β fibrils, they underwent eight orders of magnitude in the fluorescence enhancement in the presence of fibrils with a blue-shift from 440 to 413 nm due to strong binding of these compounds with the hydrophobic pocket of amyloid-β fibrils *via* π-π stacking interactions Fig. (**8a**).

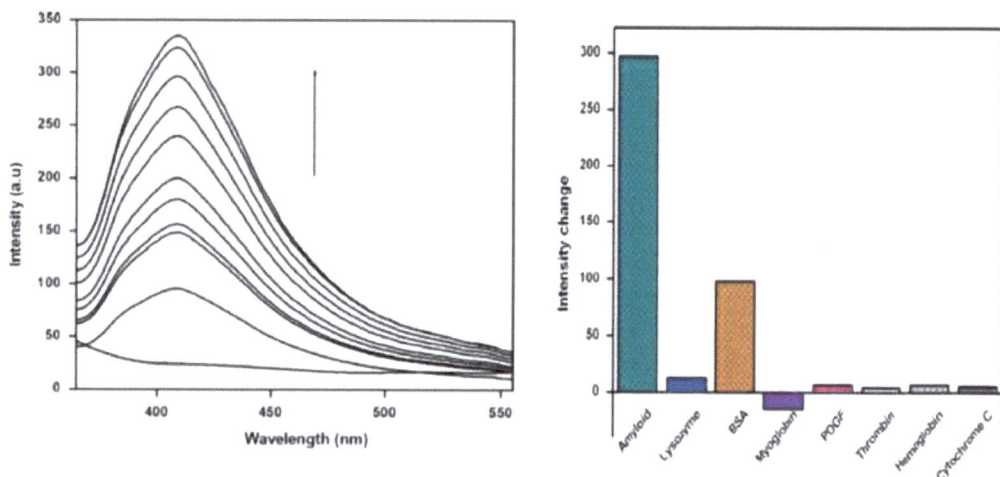

**Fig. (8).** (**a**) The luminescence titration spectra of complex **2** with various concentrations of amyloid fibrils. The concentrations of complex and amyloid-β are $1 \times 10^{-6}$ M and 0–12 μM respectively. (**b**) The selectivity of complex **2** based Aβ toward other proteins. Reproduced with permission from ref. 44. Copyright (2014) Elsevier.

The data on the luminescence intensity change of **1** and **2** in the presence of amyloid-β was analysed using Scatchard plots [45, 46] (eqn. 4).

$$\upsilon = [Re]_{free} = nK_a - \upsilon K_a \tag{4}$$

Where $\upsilon = [Re]_{bound}/[amyloid]$, n is the binding stoichiometry, [amyloid] is the concentration of the amyloid and $[Re]_{bound}$ is the concentration of the bound form of the complex.

$[Re]_{bound}$ was calculated using eqn. (5)

$$[Re]_{bound} = [Re]_{total} - [Re]_{free} \tag{5}$$

$$[Re]_{free} = [Re]_{total} [(I/I_0) - P]/(1 - P) \tag{6}$$

Where $[Re]_{total}$ and $[Re]_{free}$ are the total and free concentration of the Re(I) complex, I and $I_0$ are the emission intensity of the complex in the presence and absence of amyloid, respectively, and P is max($I/I_0$), which can be obtained as 1/(y-intercept) from the linear plot of $I_0/I$ *vs* [amyloid]$^{-1}$.

Scatchard plot analysis showed that compounds **1** and **2** were able to bind with amyloid-β fibrils with the binding constants of $2.2 \times 10^5$ and $2.0 \times 10^5$ M$^{-1}$, respectively. The detection limit of the sensing β-amyloid was found to be 2.0 μM. These compounds showed the selectivity toward amyloid over other proteins such as lysozyme, BSA, PDGF, thrombin, myoglobin, hemoglobin and cytochrome C (Fig. **8b**). AFM and TEM techniques supported the evidence for the formation of fibrils with a size of 30–40 nm upon the addition of amyloid with compounds **1** and **2**.

In 2018, metallacavitands **3−6** ([{$(CO)_3$Re($\mu$−dhaq)Re$(CO)_3$}($\mu$-L$^1$)] (**3**), [{$(CO)_3$Re($\mu$−dhaq)Re$(CO)_3$}($\mu$-L$^2$)] (**4**), [{$(CO)_3$Re($\mu$−dhaq)Re$(CO)_3$}($\mu$-L$^3$)] (**5**), and [{$(CO)_3$Re($\mu$−CA)Re$(CO)_3$}($\mu$-L$^1$)] (**6**)) were prepared by self-assembly reaction of Re$_2$(CO)$_{10}$ with rigid donor linkers (1,4-dihydroxy-9,10-anthraquinone (H$_2$-dhaq)/chloranilic acid (H$_2$-CA)) and flexible N-donor ligands (bis(4-(naphtho[2,3-*d*]imidazol-1-ylmethyl)phenyl)methane (L$^1$) or bis(4-(benzimidazo--1-ylmethyl)phenyl)methane (L$^2$) or bis(4-(2-nonylbenzimidazo--1-ylmethyl)phenyl)methane (L$^3$)) in toluene under solvothermal reaction condition (Fig. **9**) [47].

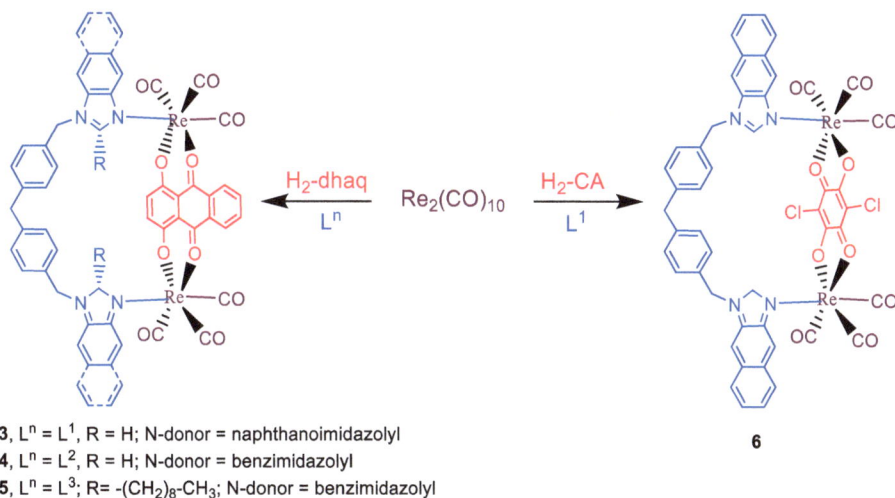

3, L$^n$ = L$^1$, R = H; N-donor = naphthanoimidazolyl
4, L$^n$ = L$^2$, R = H; N-donor = benzimidazolyl
5, L$^n$ = L$^3$; R= -(CH$_2$)$_8$-CH$_3$; N-donor = benzimidazolyl

6

**Fig. (9).** Synthesis of metallacavitands, **3−6**.

A single crystal X-ray diffraction analysis of **3** showed that compound **3** had a distorted pentagonal-shaped structure that consisted of [2+1+1] assembly of two *fac*-Re(CO)$_3$ cores, one dhaq$^{2-}$, and one nitrogen donor ligand L$^1$ with a Re$\cdots$Re of 8.390 Å (Fig. **10**). It possessed a hydrophobic cavity where a toluene guest molecule is resided inside. Compounds **4−6** showed a similar type of structure as revealed by X-ray studies. UV-Vis absorption spectra of metallacavitands, **3−6** showed that high energy bands appeared at less than 350 nm and low energy bands above 350 nm were assigned to ligand-based $\pi$–$\pi$* and MLCT-based d$\pi$–$\pi$* transitions, respectively. The dhaq$^{2-}$ moiety in **3−5** displayed its intraligand transition as a weak band at 590−650 nm. Upon excitation at their MLCT bands, compounds **3**−**6** exhibited only ligand-based emission. Fluorescence titration analysis showed that addition of nitroderivatives such as nitrobenzene (NB), 2-nitrotoluene (2-NT), 4-nitrotoluene (4-NT), and 2,4-dinitrotoluene (DNT) into a DMSO solution of compound 3 ($\lambda_{em}$ = 375 nm) resulted in the fluorescence quenching through electron transfer from **3** to nitroderivatives. B-H analysis (eqn. 1) revealed that the binding constants of **3** with 2,4-DNT, NB, 4-NT and 2-NT were found to be $2.5 \times 10^3$, $4.7 \times 10^2$, $4.5 \times 10^2$ and $1.1 \times 10^3$ $M^{-1}$, respectively. Among planar aromatic guests, anthracene ($K = 1.3 \times 10^4$ $M^{-1}$) exhibited higher binding ability toward **3** compared to naphthalene ($K = 1.7 \times 10^3$ $M^{-1}$). These results indicated that the cavity size of **3** was accessible to accommodate these guest molecules through $\pi \cdots \pi$ and C-H$\cdots \pi$ interactions.

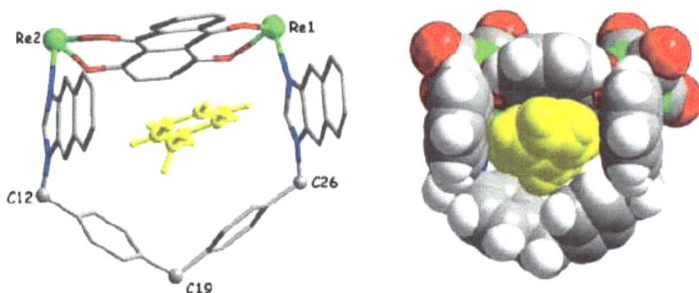

**Fig. (10).** Molecular structure of **3** (left: H atoms and six CO units were removed to show cyclic framework; right: space-filling view of **3**. C = gray, H = white, N = blue, O = red, Re = green, toluene = yellow. Reproduced with permission from ref. 47. Copyright (2018) The Royal Society of Chemistry.

## Trinuclear Re(I) Complex

A cylindrical trimer based on rhenium bisimine moiety that showed ³MLCT emission characteristic was reported through a complex-as-ligand approach.[48] Treatment of bispyridylpyridone with Re(CO)$_5$X (X = Cl, Br) resulted in the formation of [Re(4′-oxo-η²-terpy)(CO)$_3$]X (terpy = 2,2′:6′,2″-terpyridyl, X = Cl, Br), which experienced halide abstraction in the presence of AgBF$_4$ to generate a trimer, [{(η−Re-4′-oxo-1,1′-η$^{Re}$-1″-η$^{Ag}$-terpy)Re(CO)$_3$}$_3$Ag]⁺ (**7**) (Fig. 11).

In this reaction, the rhenium terpyridine moiety underwent cyclotrimerization *via* a 4′-oxygen group to generate a pseudocylindrical species in which the pyridine groups acted as the walls of a triangular cavity and unbound pendant pyridine of each unit that coordinated to silver ion not only formed caps of the cylinder but also triggered the emission behavior of trimer, **7**. The molecular dimensions of the cavity size of this equilateral geometry were 8.0 and 4.6 Å with respect to sides and inscribed circle, respectively. Upon photolysis using sunlight or 405 nm laser, compound **7** underwent photolytic loss of silver stopper to generate a Re(I)-trimer [{(η−4′-oxo-η-2,2′-terpy)Re(CO)$_3$}$_3$] (**8**) (Fig. **12**). The removal of silver stopper was reversible process upon treatment of **8** with silver salt. Compound **8** was analogous to **7** but silver ion was missing at the end of the channel while the uncoordinated pyridine nitrogen atoms pointed toward the macrocyclic cavity of **8**, which was available for sensing cations.

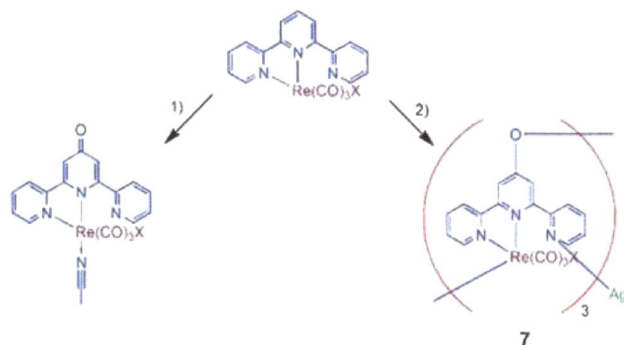

**Fig. (11).** Abstraction of halides from complex, [Re(4′-oxo-η²-terpy)(CO)₃]X. Conditions: 1) 1 equiv. AgBF₄, MeCN; 2) 3 equiv AgBF₄, MeCN. The oxygen atom in **7** coordinates to a neighboring rhenium atom.

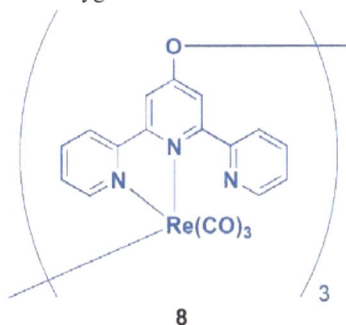

**Fig. (12).** Structure of rhenium(I) trimer **8**.

Compounds **7** and **8** exhibited broad absorption bands at 348 and 352 nm, respectively which were attributed to the $d\pi-\pi^*$ transition of $^1$MLCT-based absorbance. Upon excitation, they displayed a broad $^3$MLCT based emission at 550–570 nm with a lifetime of 20 ns region, which were short values compared to usual rhenium tricarbonyl complexes. Compound **8** was stable under light irradiation compared to compound **7**. As evidenced from the overlay spectra of excitation and emission of **7** and **8**, silver ion in **7** participated significant photophysical properties through its binding. The authors suggested that their unique structures and photophysical properties might be useful for potential applications in host–guest chemistry and sensing.

**Tetranuclear Re(I) Complex**

In 2006, alkoxy- or thiolato- ligand was used as a one of the ligands to prepare

Re(I) molecular rectangles for the sensing of pyrene molecule and silver ion [49]. Alkoxy- or thiolato-bridged molecular rectangles [{(CO)$_3$Re($\mu$-ER)$_2$Re(CO)$_3$}$_2$($\mu$-bpy)$_2$] ((ER = SC$_4$H$_9$, **9a**; SC$_8$H$_{17}$, **9b**; OC$_4$H$_9$, **10a**; OC$_{12}$H$_{25}$, **10b**; bpy = 4,4'-bipyridine) were obtained in high yields by the self-assembly reaction of Re$_2$(CO)$_{10}$ with 4,4'-bipyridine (bpy) in the presence of aliphatic alcohols or thiols under solvothermal as well as refluxing conditions (Fig. **13**).

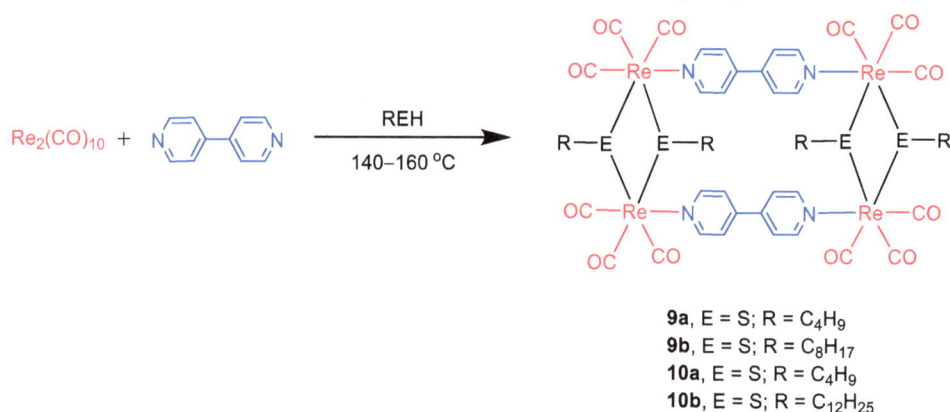

**9a**, E = S; R = C$_4$H$_9$
**9b**, E = S; R = C$_8$H$_{17}$
**10a**, E = S; R = C$_4$H$_9$
**10b**, E = S; R = C$_{12}$H$_{25}$

**Fig. (13).** Re(I)-based molecular rectangles **9** and **10**.

B-H analysis from UV-Vis absorption study (eqn. (1)) revealed that the host-guest interaction between the rectangles and pyrene took place and their binding constant ($K$) values are given in Table 2. This can be attributed to the formation of donor-acceptor complexation as a result of charge transfer complex between the electron-rich pyrene and the 4,4'-bpy of rectangles. Fluorescence titration data showed that addition of pyrene to a solution of rectangles caused the fluorescence quenching due to an intermolecular energy transfer from the emitting $\pi$–$\pi$* state of the pyrene to the low-lying CT excited state of rectangles, which returns to the ground-state *via* radiationless decay. The quenching rate constants along with dynamic and static Stern-Volmer constants were derived from Stern-Volmer analyses (eqns. (2) and (3)) and the results are listed in Table 2. The $K_S$ values obtained from fluorescence study are corroborative to the results obtained from absorption study, and are supportive of a 1:1 complexation model.

**Table 2. Ground-state binding constants ($K$), excited-state dynamic ($K_D$), static ($K_S$), Stern-Volmer constants, and quenching rate constants ($k_q$) of hosts 9 and 10 with pyrene at 298 K.**

| Host | $K$, M$^{-1}$ | $K_D$, M$^{-1}$ | $K_S$, M$^{-1}$ | $k_q$, M$^{-1}$ s$^{-1}$ |
|---|---|---|---|---|
| **9a** | $1.9 \times 10^4$ | $7.4 \times 10^4$ | $3.6 \times 10^4$ | $2.3 \times 10^{12}$ |
| **9b** | $2.3 \times 10^4$ | $1.2 \times 10^5$ | $5.1 \times 10^4$ | $3.2 \times 10^{12}$ |

(Table 2) cont.....

| Host | $K$, M$^{-1}$ | $K_D$, M$^{-1}$ | $K_S$, M$^{-1}$ | $k_q$, M$^{-1}$ s$^{-1}$ |
|------|------|------|------|------|
| 10a | $9.7 \times 10^3$ | $3.5 \times 10^4$ | $1.7 \times 10^4$ | $1.1 \times 10^{12}$ |
| 10b | $1.1 \times 10^4$ | $4.8 \times 10^4$ | $1.9 \times 10^4$ | $1.5 \times 10^{12}$ |

The solid-state structure of a grown **9a**·pyrene complex showed that the H atoms of pyridyl group in **9a** were interacted with the π-cloud of the pyrene with a dihedral angle of ~95° and H(pyridyl)···C(pyrene) distances of 2.769−3.295 Å in a 1:1 ratio (Fig. **14**). A packing diagram study of **9a** revealed that pyrene guest molecule was not resided within the cavity of the rectangles because of insufficient accessible sites but the spaces between the bpy linkers (6.40 Å) of **9a** were able to accommodate pyrene molecules through CH···π interactions. This work was one of the rarely designed studies for the host-guest interaction.

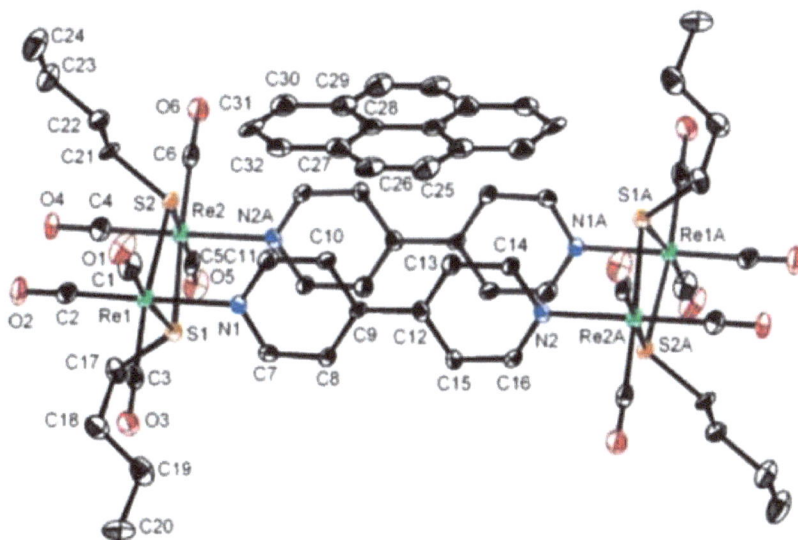

**Fig. (14).** Crystallographic drawing of [**9a**·pyrene] showing CH···π interactions in the solid state. Reproduced with permission from ref. 49. Copyright (2006) American Chemical Society.

Thiolate rectangle **9a** was able to interact with metal ions, especially silver salt. A single-crystal X-ray analysis of $[\{9a\cdot(Ag^+)_2(NO_3^-)_2(C_3H_6O)_2\}(C_3H_6O)]$ that was grown by slow evaporation of **9a** with AgNO$_3$ in acetone revealed that the thiolato linkers of **9a** were linked with Ag atom with a distance of 2.4405(17) Å, which was further coordinated with two nitrate ions with the Ag-O distances of 2.308(5)-2.542(5) Å and linked by one acetone molecule with a distance of 2.432(5) Å (Fig. **15**). This kind of arrangement led to the formation of a one-dimensional supramolecular array *via* S···Ag···O interactions.

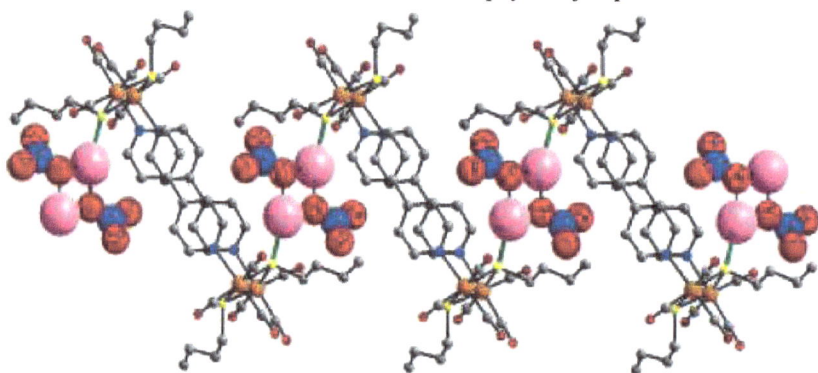

**Fig. (15).** Crystallographic drawing indicating the inclusion of AgNO$_3$ moieties by host **9a** through silver-thiolate side-arm interactions and formation of a linear supramolecular array. The coordinated acetone molecules are omitted for the sake of clarity. The AgNO$_3$ moieties are enlarged for clarity. Reproduced with permission from ref 49. Copyright (2006) American Chemical Society.

In the above study, the OFF-ON phenomenon of aggregation-induced emission enhancement (AIEE) in Re(I) molecular rectangle, **10b** (Fig. **16**) and its characteristic host-guest reaction with electron donor and acceptor molecules are reported [50]. UV-Vis absorption spectrum of **10b** in CH$_3$CN displayed a ligand-centered transition at 241 with a shoulder at 266 nm, and a metal-to-ligand charge transfer transition at 385 nm. Interestingly, the addition of water (90%, v/v) into the CH$_3$CN solution of **10b** resulted in the change of the absorption spectra along with a red-shifted band from 385 to 390 nm. In addition, a scattered light pattern with an exponential frequency function that tails down at 600 nm was observed, indicating the formation of nanoparticle suspensions. Compound **10b** displayed a weak emission at 666 nm. Surprisingly, a dramatic emission enhancement along with a blue-shifted emission wavelength from 666 to 602 nm was noticed upon increasing the water content (90%) in acetonitrile solution of **10b**. In addition, the quantum yield and lifetime of **10b** were also increased with gradual addition of water. Hence, molecular aggregation induced by long alkoxy chain of **10b** was responsible for this appropriate explanation for these experimental observations. This result demonstrated that aggregation reduced the solvent exposure to block vibrational/torsional energy relaxation, thus enhancing emission. Light scattering studies confirmed that aggregates of Re(I)-rectangle **10b** exhibited an average mass of $1.43 \times 10^6$ with a diameter in the range of 262−265 nm.

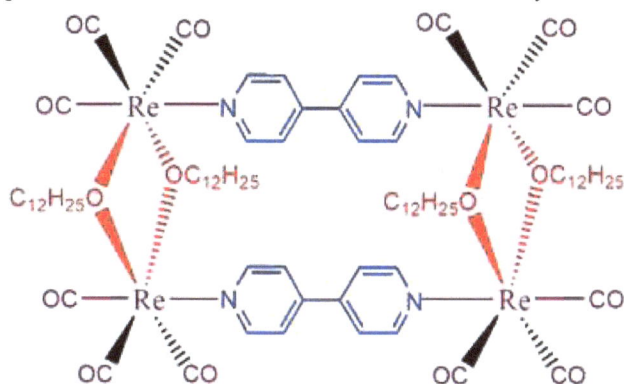

**Fig. (16).** Structure of rhenium(I)-rectangle, **10b**.

Aromatic molecules such as quinones and amines were selected as guest molecules and were tested to study the host-guest interaction with the Re(I)-aggregates. UV-Vis spectral studies did not show any noticeable changes in the aggregates. However, the emission intensity of the excited Re(I)-aggregates decreased when these guest molecules were added (Fig. **17**). The quenching rate constants ($k_q$), static and dynamic Stern-Volmer constants for the excited- state reaction of Re(I)-aggregates with quinones and amines were calculated from eqns. (2) and (3) and these values are summarized in Tables 3 and 4, respectively. The close in proximity resulting from aggregation promoted by the hydrophobicity of the Re(I) and quenchers caused the effective emission quenching. From the above experimental results, excited Re(I)-rectangles served

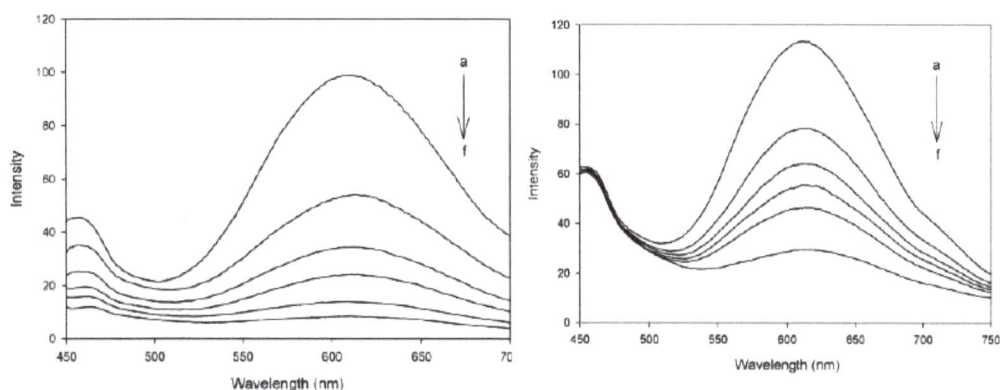

**Fig. (17).** Luminescence quenching of the aggregated form of Re(I)-rectangle **10b** with a) 2,5-dichloro-1-4-benzoquinone and b) *N*-ethylaniline: (a) 0, (b) $4 \times 10^{-3}$, (c) $8 \times 10^{-3}$, (d) $12 \times 10^{-3}$, (e) $16 \times 10^{-3}$, and (f) $20 \times 10^{-3}$ M quencher in 50% aqueous acetonitrile at 298 K. Reproduced with permission from ref. 50. Copyright (2007) American Chemical Society.

**Table 3.** Dynamic ($K_D$) and static Stern-Volmer ($K_S$) constants and quenching rate constants for the aggregated form of Re(I)-rectangle 10b with electron acceptors in 50% acetonitrile-water mixture at 298 K.

| quenchers | $K_D$, M$^{-1}$ | $K_S$, M$^{-1}$ | $k_q$, M$^{-1}$s$^{-1}$ |
|---|---|---|---|
| benzoquinone | $2.8 \times 10^1$ | $1.4 \times 10^1$ | $2.0 \times 10^8$ |
| methyl-*p*-benzoquinone | $2.4 \times 10^3$ | $4.1 \times 10^2$ | $7.6 \times 10^9$ |
| 2,5-dichloro-1,4-benzoquinone | $2.6 \times 10^2$ | $7.6 \times 10^1$ | $1.9 \times 10^9$ |
| 1,4-naphthaquinone | $1.2 \times 10^2$ | $1.7 \times 10^1$ | $8.9 \times 10^8$ |
| 1,4-chloranil | $6.2 \times 10^2$ | $9.4 \times 10^1$ | $4.5 \times 10^9$ |
| 1,2-chloranil | $1.2 \times 10^3$ | $2.1 \times 10^2$ | $8.6 \times 10^9$ |
| duroquinone | $2.1 \times 10^2$ | $8.4 \times 10^1$ | $1.5 \times 10^9$ |
| tetracyanoquinodimethane (TCNQ) | $5.3 \times 10^4$ | $3.2 \times 10^4$ | $3.8 \times 10^{11}$ |

**Table 4.** Dynamic and static Stern-Volmer constants and quenching rate constants for the aggregated form of Re(I)-rectangle 10b with electron donors in 50% acetonitrile-water mixture at 298 K.

| quenchers | $K_D$, M$^{-1}$ | $K_S$, M$^{-1}$ | $k_q$, M$^{-1}$s$^{-1}$ |
|---|---|---|---|
| aniline | 1.4 | 0.7 | $1.0 \times 10^7$ |
| *N*-methylaniline | $4.1 \times 10^1$ | $4.5 \times 10^1$ | $3.0 \times 10^8$ |
| *N*-ethylaniline | $8.8 \times 10^1$ | $3.7 \times 10^1$ | $6.4 \times 10^8$ |
| *N,N*-dimethylaniline | $8.2 \times 10^1$ | $3.9 \times 10^1$ | $6.0 \times 10^8$ |
| *N,N*-diethylaniline | $3.8 \times 10^2$ | $1.3 \times 10^2$ | $2.8 \times 10^9$ |
| *p*-anisidine | $9.3 \times 10_1$ | $1.1 \times 10^2$ | $6.8 \times 10^8$ |
| *N,N*-dimethyl-*p*-toluidine | $3.8 \times 10^2$ | $8.5 \times 10^2$ | $2.8 \times 10^9$ |
| diphenylamine | $1.3 \times 10^3$ | $1.3 \times 10^3$ | $9.2 \times 10^9$ |
| *p*-phenylenediamine | 6.0 | 0.9 | $4.4 \times 10^7$ |
| *N,N,N',N'*-tetramethyl-phenylenediamine | $4.0 \times 10^2$ | $4.9 \times 10^2$ | $2.9 \times 10^9$ |
| benzidine | $2.3 \times 10^3$ | $2.2 \times 10^3$ | $1.7 \times 10^{10}$ |
| *N,N,N',N'*-tetramethyl-benzidine | $2.6 \times 10^3$ | $1.6 \times 10^3$ | $1.9 \times 10^{10}$ |

as electron donors or acceptors under different conditions. Surprisingly, bulky quenchers showed higher quenching rate constants than those of other quenchers. In addition, the $k_q$ values were gradually increased, which depended on the oxidation/reduction potentials of quinones and amines. In this reaction, the individual rectangle, monomer **10b** did not have accessible cavity size to accommodate guest molecules. These guest molecules were able to bind with the hydrophobic domain consisting of the aliphatic chains of aggregated Re(I) species. Therefore, aggregates of **10b** was considered as a new class of rectangle to be a *"surfactant with a host-guest type recognition site."*

In 2014, Manimaran *et al.* [51] reported the preparation of selenium bridged Re(I) rectangles, $[\{(CO)_3Re(\mu-SeR)_2Re(CO)_3\}_2(\mu-L)_2]$ (**11–13**) (L = 4,4'-bipyridine (bpy), R = $C_6H_5$ (**11a**), $CH_2C_6H_5$ (**11b**); *trans*-1,2-bis(4-pyridyl)ethylene (bpe), R = $CH_2C_6H_5$ (**12**), and 1,4- bis[2-(4-pyridyl)ethenyl]benzene (bped), R = $CH_2C_6H_5$ (**13**)) by the oxidative addition of diselenide derivatives with $Re_2(CO)_{10}$ in the presence of pyridyl derivatives at 140 °C through a self-assembly synthetic route (Fig. **18**).

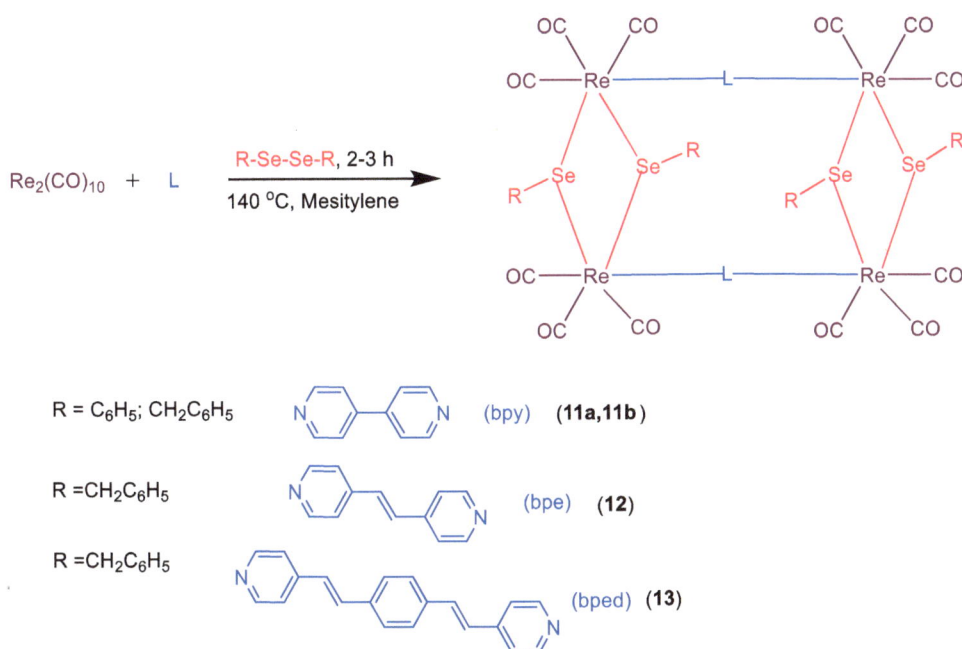

**Fig. (18).** Preparation of Re(I)-based rectangles **11–13** *via* a self-assembly route.

Single-crystal X-ray diffraction studies of **11a** and **12** revealed that pyridyl ligands were interacted through a weak π–π stacking interaction (3.63–3.711 Å) with a dimension in the range of 3.90 × 13.84–3.99 × 11.50 Å. Electronic absorption and emission spectral studies were carried out to understand the binding ability of **11** and **12** with pyrene and triphenylene guest molecules. An enhanced absorption intensity of guest molecules was observed upon the addition of complexes **11** and **12**, indicating the formation of donor-acceptor complexation. The binding constants were deduced from the Benesi–Hildebrand equation (1) and the binding value was found in the range of $10^3$–$10^5$ $M^{-1}$ (Table

5). Emission profiles showed that effective emission quenching of guest molecules by complexes **11** and **12** took place owing to the energy transfer from the $\pi-\pi^*$ state of the guest molecules to the low-lying CT excited state, which leads to returning the ground state through non-radiation pathway. From the quenching measurements, the Stern−Volmer quenching constants ($K_{sv}$) of complexes **11** and **12** with pyrene and triphenylene guests were obtained and the values were in the range from $6.0 \times 10^4$ to $1.4 \times 10^5$ $M^{-1}$ (Table 5). In addition, X-ray crystallographic studies supported the host-guest interaction of complexes **11** and **12** with pyrene and triphenylene through CH$\cdots\pi$ interactions (Fig. 19) while the packing diagram supported a 1:1 binding mode, as revealed by B-H analysis.

**Table 5. Binding constants ($K$) and Stern−Volmer constants ($K_{SV}$) for the host−guest systems of molecular rectangles 11a, 11b and 12 with pyrene and triphenylene.**

| Compound | pyrene | | triphenylene | |
|:---:|:---:|:---:|:---:|:---:|
| | $K$, $M^{-1}$ | $K_{SV}$, $M^{-1}$ | $K$, $M^{-1}$ | $K_{SV}$, $M^{-1}$ |
| **11a** | $6.7 \times 10^5$ | $6.0 \times 10^4$ | $2.2 \times 10^5$ | $8.4 \times 10^4$ |
| **11b** | $1.6 \times 10^5$ | $6.2 \times 10^4$ | $2.4 \times 10^4$ | $1.4 \times 10^5$ |
| **12** | $7.2 \times 10^5$ | $7.2 \times 10^4$ | $1.6 \times 10^5$ | $1.3 \times 10^5$ |

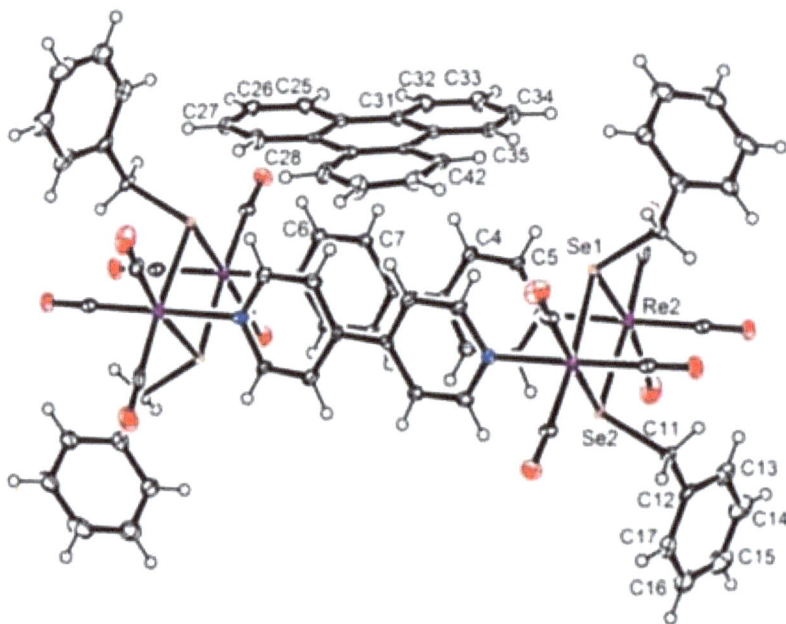

**Fig. (19).** ORTEP diagram of **11b** with a triphenylene guest with thermal ellipsoids at the 40% probability level. Reproduced with permission from ref. 51. Copyright (2014) American Chemical Society.

In 2014, Manimaran and co-workers reported the preparation and host-guest study of supramolecular rectangles [{(CO)$_3$Re(μ-η$^4$-L)Re(CO)$_3$}$_2$(μ-pbin)$_2$] (**14−17**) by the reaction of Re$_2$(CO)$_{10}$ with oxamide ligands (**14**, H$_2$L = N,N'-dibutyloxamide; **15**, H$_2$L = N,N'-dioctyloxamide; **16**, H$_2$L = N,N'-didodecyloxamide; **17**, H$_2$L = N,N'-dibenzyloxamide) and the ester-containing pyridyl ligand, phenyl-1,--bis(isonicotinate) (pbin) in mesitylene *via* an orthogonal bonding approach under solvothermal condition (Fig. **20**) [52]. The ORTEP diagrams revealed that compounds **14** and **17** featured a rectangular architecture with a dimension of ~5.67 × 19.61 Å (Fig. **21**). In these rectangles, Re atoms were occupied at the four corners, while the longer and shorter edges were occupied by the pbin and dibutyloxamidato bridges, respectively. One of the butyl groups on each oxamidato edge was directed toward the cavity of the rectangle, and the other butyl group was directed away from the cavity.

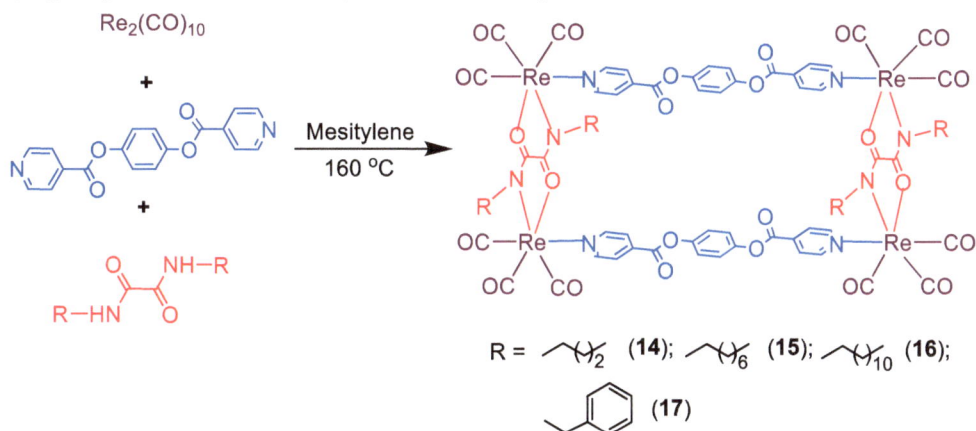

**Fig. (20).** Self-assembly of metallarectangles, **14−17**.

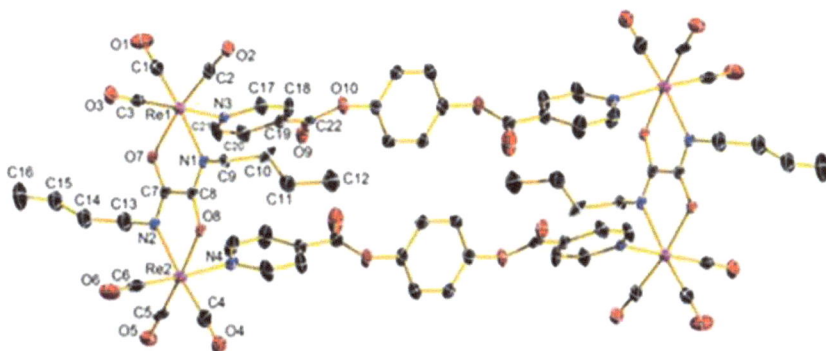

**Fig. (21).** ORTEP diagram of [{(CO)$_3$Re(μ-η$^4$-dibutyloxamidato)Re(CO)$_3$}$_2$(μ-pbin)$_2$] (**14**) with thermal ellipsoids drawn at the 30% probability level. Reproduced with permission from ref. 52. Copyright (2014) American Chemical Society.

UV-Vis absorption spectra of **14−17** showed the ligand centered absorption bands at < 350 nm and $^1$MLCT absorption band at > 350 nm. They showed a $^3$MLCT emission band at 609−611 nm. Because of the available cavity size of these rectangles, they were utilized as molecular receptors for the binding of aromatic compounds such as 3,3′-methylenedianiline, 4,4′-methylenedianiline and 1,5-diaminoanthraquinone using absorption and emission techniques (Fig. **22**). Absorption spectral studies revealed that binding constants for the reaction of rectangle **15** with these guest molecules were calculated to be in the range of $1.5–2.5 \times 10^5$ M$^{-1}$ using the Benesi-Hildebrand eqn. (1). Fluorescence studies showed that emission intensity of guest molecules got quenched by host **2**, showing the $K_{SV}$ values of $1.3–8.3 \times 10^5$ M$^{-1}$ (using the eqn. 2), which were in good agreement with the absorption spectral studies (Table 6). These studies confirmed the host-guest interaction of host **15** and these guest molecules. This interaction was presumably attributed to the N-H$\cdots$O hydrogen bonding interactions between -NH$_2$ groups of guest molecule and O atoms of ester groups present in rectangle **15**, along with C−H$\cdots\pi$ and $\pi\cdots\pi$ interactions. Besides the host-guest interaction, rectangles **15** and **16** underwent molecular aggregation process upon increasing the water content, and the results were discussed in detail. In another study, the same group developed a series of aminoquinonato-bridged rhenium(I)-rectangles, $[\{(CO)_3Re(\mu-\eta^4-L)Re(CO)_3\}_2(\mu-N-L'-N)_2]$ (**18−21**) by the self-assembly reaction of Re$_2$(CO)$_{10}$ with aminoquinone ligands (H$_2$L = 2,5-bis(*n*-butylamino)-1,4-benzoquinone (bbbq), 2,5-bis-(phenethylamino)--,4-benzoquinone (bpbq), and dinucleating linear pyridyl ligands (N-L′-N = 4,4′-bipyridine (bpy) and *trans*-1,2-bis(4-pyridyl)ethylene (bpe)) (Fig. **23**) [53].

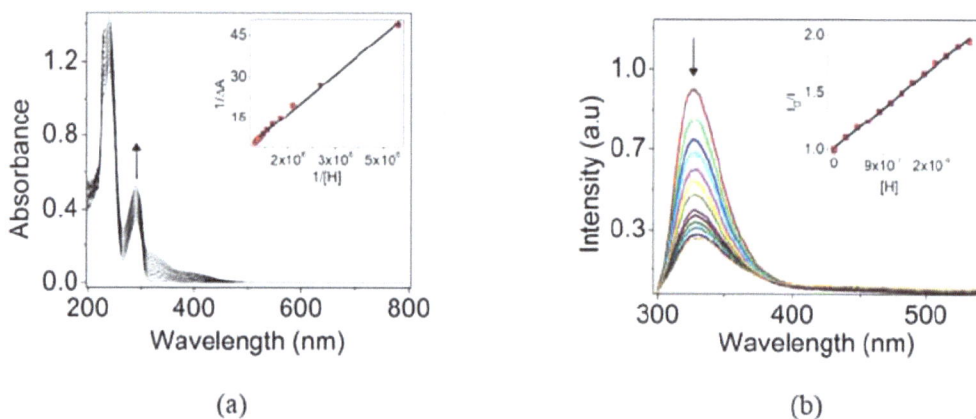

(a)            (b)

**Fig. (22).** Electronic absorption spectra of 3,3′-methylenedianiline ($2 \times 10^{-5}$ M) with an incremental addition of host **15** ($2–24 \times 10^{-7}$ M) in dichloromethane and inset shows the corresponding Benesi-Hildebrand plot. (b) Emission intensity of 3,3′-methylenedianiline ($2 \times 10^{-5}$ M) decreasing with incremental addition of host **15** ($2–24 \times 10^{-7}$ M) and inset shows the corresponding Stern-Volmer plot. Reproduced with permission from ref. 52. Copyright (2014) American Chemical Society.

**Table 6. Binding constants ($K$) and Stern-Volmer constants ($K_{SV}$) for the reaction of host 15 with guest species.**

| Guest | $K$ (M$^{-1}$) | $K_{SV}$ (M$^{-1}$) |
|---|---|---|
| 3,3′-methylenedianiline | $2.2 \times 10^5$ | $3.9 \times 10^5$ |
| 4,4′-methylenedianiline | $1.5 \times 10^5$ | $1.3 \times 10^5$ |
| 1,5-diaminoanthraquinone | $2.5 \times 10^5$ | $8.3 \times 10^5$ |

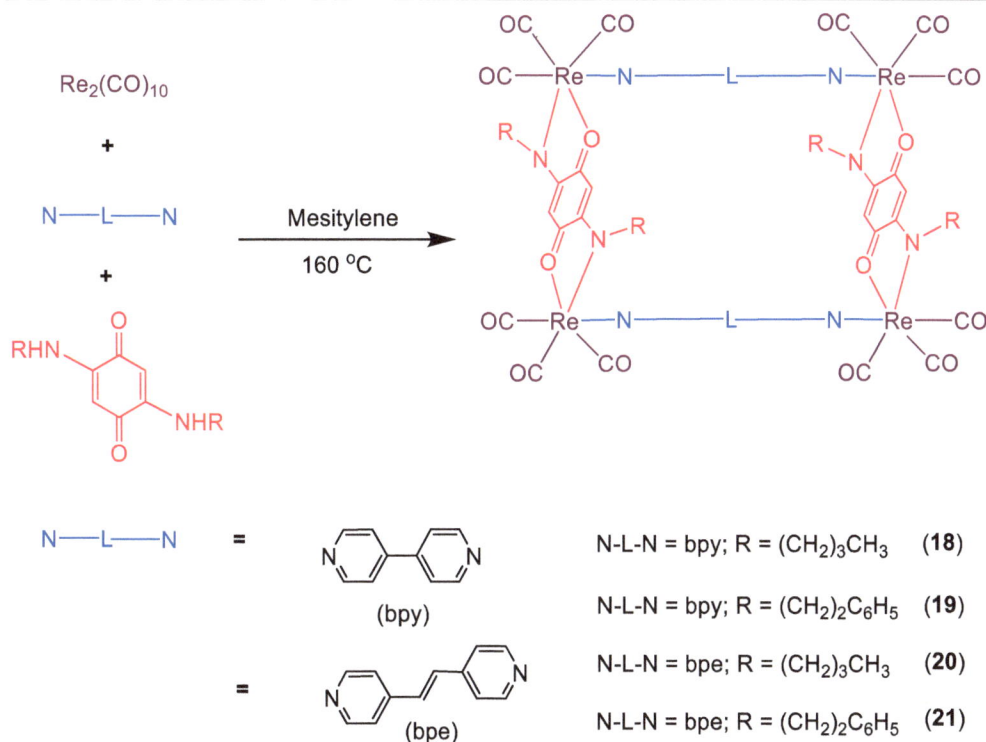

**Fig. (23).** Self-assembly of aminoquinonato-bridged Re(I)-rectangles, **18–21**.

X-ray crystallographic study revealed that compound **18** displayed a rectangular architecture in which four *fac*-Re(CO)$_3$ moieties occupied the corners, and aminoquinonato- and pyridyl linkers acted as two short and two longer edges, respectively, with a dimension of ~8.14 × 11.51 Å. The presence of porous structure in rectangles allowed them to function as sensor for sensing aromatic compounds. To understand the sensing behaviour, the interaction of metallarectangles **18** and **20** toward planar aromatic guests such as pyrene and triphenylene was studied using absorption and emission techniques. UV−vis absorption spectral titration experiments showed that absorbance of pyrene was enhanced upon the gradual addition of rectangle **18**, indicating the formation of host –guest complex between **18** and pyrene through donor-acceptor interaction. The binding constant was calculated to be $1.3 \times 10^5$ M$^{-1}$ with a 1:1 ratio, based on the linear Benesi–Hildebrand plot (eqn. 1). In case of triphenylene guest, the binding constant was found to be $3.1 \times 10^5$ M$^{-1}$. In addition, fluorescence spectral studies also supported the host-guest interaction between the host **18** and pyrene or triphenylene guest molecules, showing the Stern–Volmer constants ($K_{sv}$) of $1.9 \times 10^5$ M$^{-1}$ and $1.3 \times 10^5$ M$^{-1}$, respectively. These values are corroborated with the results obtained from absorption studies (Table 7).

**Table 7. Binding constants ($K$) and Stern–Volmer constants ($K_{SV}$) for the host–guest systems of molecular rectangles 18 and 20 with pyrene and triphenylene.**

| Compound | pyrene | | triphenylene | |
|---|---|---|---|---|
| | $K$, M$^{-1}$ | $K_{SV}$, M$^{-1}$ | $K$, M$^{-1}$ | $K_{SV}$, M$^{-1}$ |
| **18** | $1.3 \times 10^5$ | $1.9 \times 10^5$ | $3.1 \times 10^5$ | $1.3 \times 10^5$ |
| **20** | $2.6 \times 10^5$ | $1.5 \times 10^5$ | $1.1 \times 10^5$ | $1.3 \times 10^5$ |

To gain more insights on host-guest interaction, single crystals were grown with metallarectangle **20** and aromatic guest, pyrene. A single-crystal X-ray analysis of (**20**⊐pyrene)·(DMF) confirmed the formation of 1:1 donor–acceptor complex between **20** and pyrene with a dimension of ~8.15 × 13.74 Å (Fig. **24**). In the porous cavity of rhenium rectangle **20**, a pyrene guest molecule was stacked inside the cavity and a solvent DMF molecule was present outside its cavity.

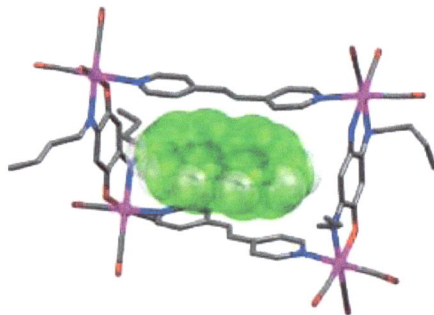

**Fig. (24).** (a). Molecular structure of (**20**⊐pyrene)·(DMF) showing the presence of pyrene guest in rectangular cavity (DMF molecule is not shown for clarity). Reproduced with permission from ref. 53. Copyright (2015) American Chemical Society.

Compared to the guest-free rectangle **20**, the overall structure of **20**⊐pyrene·DMF differed considerably, implying that accommodation of large planar guest pyrene imposed significant conformational strain on the host **20**. Previously, an orthogonal-bonding approach i.e. simultaneous incorporation of *fac*-(CO)$_3$Re-cores, neutral ditopic N-donors, and dianionic bis(chelating) ligands, was utilized to prepare rectangles rather than two molecular squares. Rectangular box $[\{(CO)_3Re(\mu\text{-}CA)Re(CO)_3\}_2(\mu\text{-}bpy)_2]\cdot$mesitylene (**22**·mesitylene) was assembled by reacting equimolar amounts of $Re_2(CO)_{10}$, 4,4′-bipyridine (bpy) and chloranilic acid ($H_2CA$) in a one-pot method [54]. By following a similar procedure, rectangular box $[\{(CO)_3Re(\mu\text{-}CA)Re(CO)_3\}_2(\mu\text{-}bpe)_2]\cdot$2toluene (**23**·2toluene) was synthesized using *trans*-1,2-bis(4-pyridyl)ethylene (bpe) instead of bpy. Upon changing the assembly unit from $H_2CA$ to 5,8-dihydroxy-1,4-naphthoquinone ($H_2$dhnq) with a slightly larger bridging length, compounds $[\{(CO)_3Re(\mu\text{-}dhnq)Re\text{-}(CO)_3\}_2(\mu\text{-}L)_2]$ G (**24** G, L = bpy, G = 2benzene; **25** G, L = bpe, G = toluene) were also synthesized Fig. (**25**).

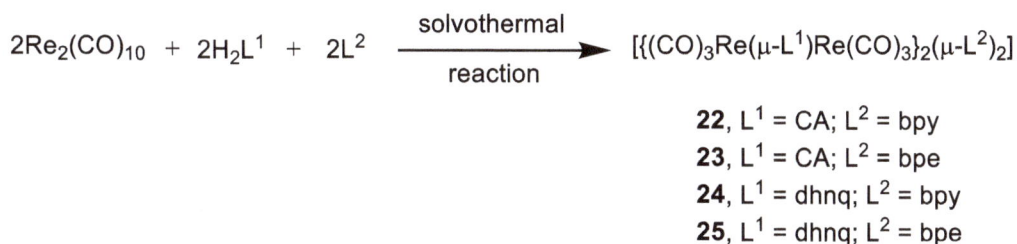

$$2Re_2(CO)_{10} \;+\; 2H_2L^1 \;+\; 2L^2 \;\xrightarrow[\text{reaction}]{\text{solvothermal}}\; [\{(CO)_3Re(\mu\text{-}L^1)Re(CO)_3\}_2(\mu\text{-}L^2)_2]$$

**22**, $L^1$ = CA; $L^2$ = bpy
**23**, $L^1$ = CA; $L^2$ = bpe
**24**, $L^1$ = dhnq; $L^2$ = bpy
**25**, $L^1$ = dhnq; $L^2$ = bpe

**Fig. (25).** Synthesis of rectangular boxes **22–25**.

A single-crystal X-ray diffraction analysis showed that compound **22** adopted a tetranuclear rectangular architecture with a dimension of $11.4 \times 8.1 \times 6.3$ Å. In this structure, each octahedral fashioned Re center was coordinated with three CO groups, one bpy and one dianionic $CA^{2-}$ ligand. The dianionic $CA^{2-}$ acted as a tetradentate ligand through coordinating its four oxygen atoms with two Re(I) atoms. Interestingly, one mesitylene guest was resided inside the cavity of the rectangular box **22** through weak $\pi\cdots\pi$ (3.6–4.6 Å) and C–H$\cdots\pi$ interactions (Fig. **26**).

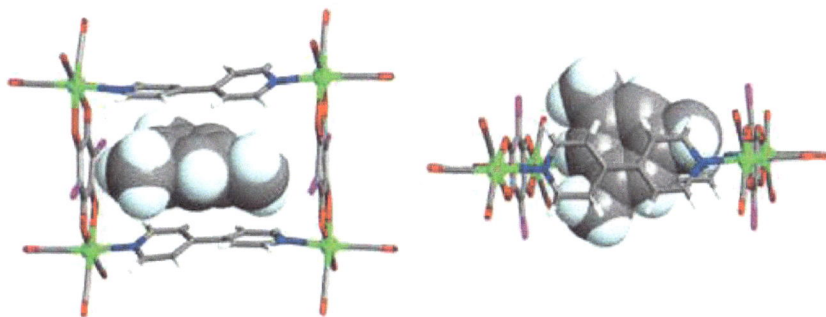

**Fig. (26).** Top and side views of the crystal structure for the **22**·mesitylene (stick and space-filling representation). C gray, Cl pink, H lime, N blue, Re green. Reproduced with permission from ref. 54. Copyright (2008) The Royal Society of Chemistry.

UV-Vis absorption spectra of **22**−**25** in THF displayed a ligand-centered π−π* transition at 230, 262 and 276 nm and MLCT transitions at 325−444 nm. In addition, intraligand transitions of the quinone moiety were also seen as weak bands at 497–677 nm. Absorption titration analysis revealed that upon the addition of host **22**, the absorption intensity of aromatic guest molecules such as benzene and mesitylene, was increased, showing the host-guest interaction between them. B-H plot analysis (eqn. 1) showed that benzene exhibited higher affinity ($K$ = 1.7 × 10$^5$ M$^{-1}$ for **22** and $K$ = 5.0 × 10$^4$ M$^{-1}$ for **23**) than that of mesitylene ($K$ = 4.2 × 10$^4$ M$^{-1}$ for **22** and $K$ = 4.0 × 10$^3$ M$^{-1}$ for **23**). The high $K$ value of **22** toward benzene was attributed to the effective π−π interaction of benzene with the bpy linkers of **22**. These studies demonstrated that the complementary size and shape of the rectangular box and the guest molecules are critical for the effective binding.

The combination of suitable building blocks using an orthogonal bonding approach allowed to prepare unique gondola-shaped structures with crown-ethe--like recognition sites. Under reflux condition, compounds **26** and **27** were achieved by reacting equimolar amounts of Re$_2$(CO)$_{10}$, 2,5-bis(5-*tert*-butyl-2-benzoxazolyl)-thiophene (tpbb), and 1,4-dihydroxy-9,10-anthraquinone (H$_2$-dhaq) or 1,2,4-trihydroxy-9,10-anthraquinone (H$_2$-thaq) (Fig. **27**) in quantitative yields [55].

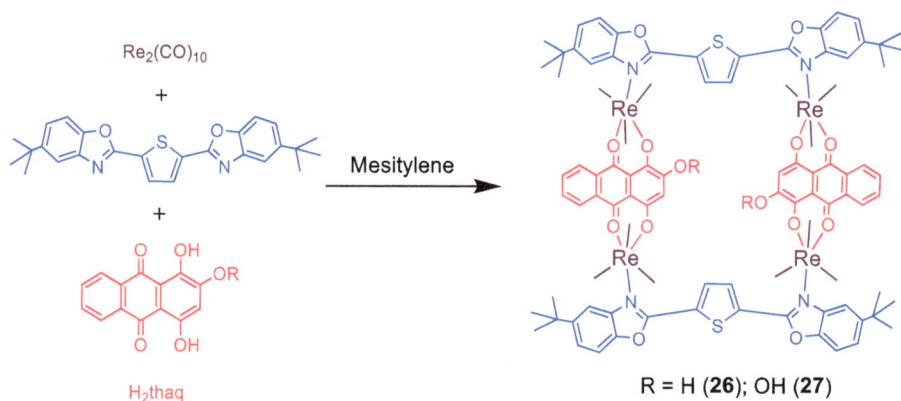

**Fig. (27).** Self-assembly of metallacycles, **26** and **27**.

A single-crystal X-ray diffraction analysis showed that compound **26** consisted of four Re atoms at the corners, two tpbb ligands connected with Re atoms through the benzoxazoline N atoms as molecular clips, and hydroxyquinones bridged through adjacent phenolate and quinone oxygens to dirhenium units, thus featuring an unusual gondola-shaped structure (size: 5.6 Å × 7.0 Å × 17.8 Å) (Fig.

**28**). Surprisingly, four methanol guest molecules were resided inside the hydrophobic cavity of the metallacycle, **26**.

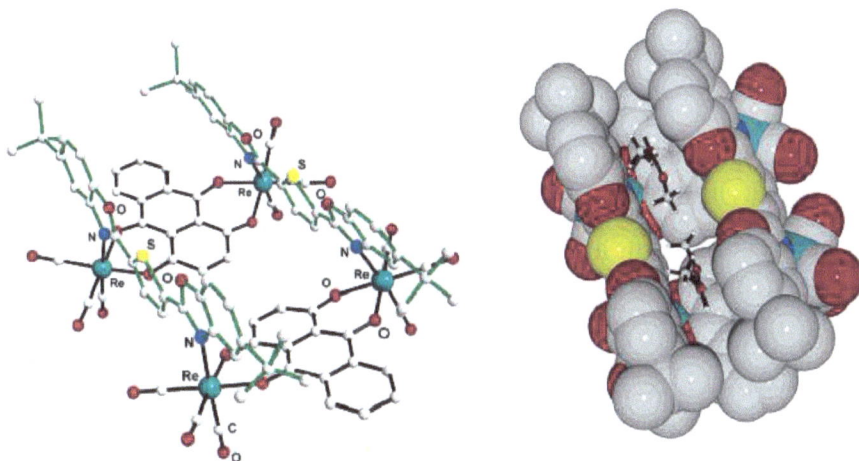

**Fig. (28).** Crystal structure of the metallacycle **26**; ball and stick representation (left); space-filling representation (right) with four methanol guests (shown in ball and stick model) occupied in the intramolecular cavity of **26**. Reproduced with permission from ref. 55. Copyright (2006) American Chemical Society.

Compounds **26** and **27** in $CH_2Cl_2$ displayed intense $\pi-\pi^*$ transitions of the dhaq and tpbb ligands at 357, 378, 397 nm, Re $\rightarrow$ tpbb MLCT transition at 420 nm and intraligand transition of the dhaq unit as weak absorption bands at 585-632 nm. Upon excitation at MLCT band, compound **26** displayed structured emission bands centered at 438 nm with a quantum yield of 0.179 and a lifetime of 1.4 ns. These fluorescent data suggested the origin of emission from the singlet $\pi-\pi^*$ excited state. The crown-ether site feature of these metallacycles made them as sensors for cation. For this study, various metal ions had been selected. Especially, addition of $Hg^{II}$ ion to a solution of **26** at 438 nm resulted in the effective emission quenching while the emission intensity at 490 nm gradually was increased. B-H analysis (eqn. 1) from emission studies showed a binding constant of $1.3 \times 10^3$ $M^{-1}$ with the formation of a 1:1 complex. However, other metal ions such as $Li^I$, $Sr^{II}$, $Co^{II}$, $Ni^{II}$, $Cu^{II}$, $Zn^{II}$, $Pb^{II}$, and $Ag^I$ did not exhibit any significant changes in the absorption and emission spectra of **26**. This binding ability was attributed to the coordination of $Hg^{II}$ ion with the uncoordinated sulfur atoms of the tpbb ligands and the adjacent hard (O atoms) Lewis base sites of **26** for the selective metal ion recognition. Thus, the emission enhancement of **26** was obtained by the chelation of metal ions thereby increasing the overall rigidity of the complex, making non-radiative decay from the excited state less probable. In addition, fluorescence assay showed that compound **26** was able to recognize anthracene selectively with a binding constant of $3.8 \times 10^3$ $M^{-1}$ among other guest molecules such as pyrene, naphthalene and benzene. The preferential binding of anthracene with this metallacycle **26** was achieved by shape complementarity with the anthraquinone moiety. These studies demonstrated the selective binding ability of **26** toward mercury cations and anthracene molecule *via* multiple functional sites.

Inorganic-based vase-shaped calixarenes are very scarcely reported in the study of host–guest inclusions due to their interior cavity. A series of organometallic calixarenes, [{(CO)₃Re}₄(L)₂(4,7-phen)₂]·2C₇H₈ (**28**, L = 1,2,4-trishydroxy-9,-0-anthraquinone (thaq), 4,7-phen = 4,7-phenanthroline; **29**, L = tetrahydroxy-1,--quinone (thq); **30**, L = 5,8-dihydroxy-1,4-naphthaquinone (dhnq)), were constructed by the self-assembly reaction of Re₂(CO)₁₀ with dianionic bischelators including H₂thaq, H₂thq, or H₂dhnq, and a neutral, 60° angular building block, 4,7-phenanthroline (4,7-phen) under solvothermal reaction (Fig. **29**)[56]. In these structures, a specific bending angle of 120° possessed by 4,7-phen as well as bis-chelating unit of quinone moieties with Re metal centers resulted in the formation of metal calix [4]arenes.

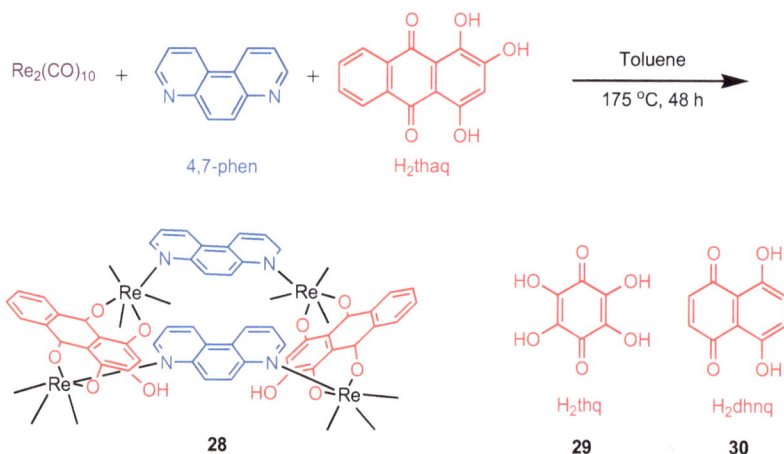

**Fig. (29).** One-pot self-assembly of compounds **28–30**.

X-ray analysis revealed that compound **27** consisted of four *fac*-Re(CO)₃ corners, two thaq ligands, and two 4,7-phen ligands to generate bowl-shaped structure in which one toluene molecule was trapped within the cavity, and the other toluene molecule was located outside the cavity. The distance between the two hydroxyl groups in the upper and lower rim of **29** was found to be 12.71 and 3.38 Å, respectively. One toluene guest molecule was residing inside the bowl cavity of **29** (Fig. **30**). These compounds were considered as metallacalix [4]arenes because four Re corners were similar to methylene linkers, the two thaq ligands were analogous to the 1,3-arranged phenol units, and the 4,7-phen ligands mimicked the 2,4-arene blocks.

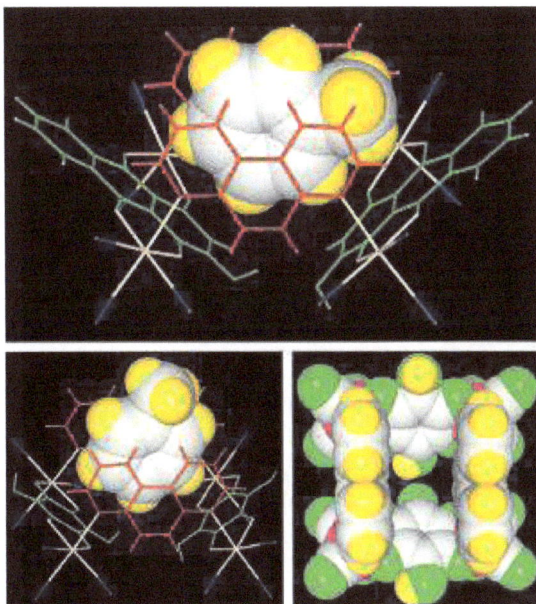

**Fig. (30).** Molecular structures of **28**·$C_7H_8$ (top), **29**·$C_7H_8$ (bottom left, ball-and-stick and space-filling mixed models), and a space-filling model of **29** (bottom right). Reproduced with permission from ref. 56. Copyright (2011) Wiley VCH.

UV-Vis absorption spectrum of **28–30** in THF showed intense ligand-centered π–π* transition of both 4,7-phen and the quinone ligands at 230–350 nm, moderate MLCT absorption bands at 420-485 nm and a weak absorption at 570−685 nm owing to intraligand transitions of the quinone moiety. Upon increasing the conjugation of the quinone ligands, MLCT bands of **28–30** were found to be blue-shifted, indicating the non-planarity of quinone ligands. The electron withdrawing quinone ligand was expected to enhance the non-radiative decay that may lead to quench the emission behavior of **28–30**. Fluorescence titration analysis revealed that addition of **29** to a $CH_3CN$ solution of aromatic guest molecules such as naphthalene, anthracene, phenanthrene, and pyrene caused the emission quenching along with a red-shifted emission wavelength from 337 to 357 nm through a complex formation (Fig. **31**). Modified B-H plot analyses (eqns. (2) and (3)) showed their binding was calculated to be $8.0 \times 10^4$, $1.7 \times 10^4$, $6.7 \times 10^4$, and $8.4 \times 10^4$ $M^{-1}$, respectively, for the interaction of **29** with naphthalene, anthracene, phenanthrene, and pyrene guest molecules.

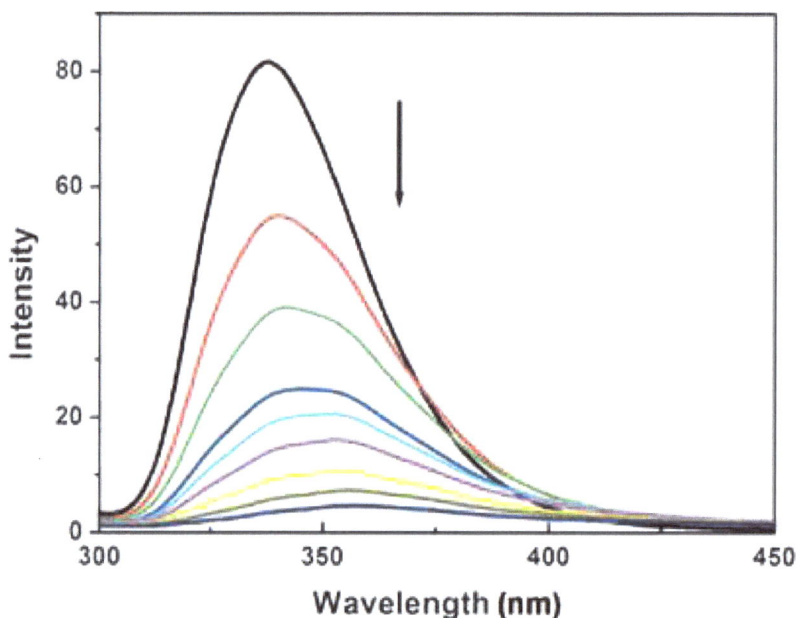

**Fig. (31).** Changes in the emission spectra of naphthalene (0.14 μm) with the addition of compound **29** in CH$_3$CN. The arrow indicates the changes that result from progressively increasing concentration of **29**.

This binding interaction was attributed to the π−π and CH···π interactions between **29** and the naphthalene guest molecule, as revealed by $^1$H NMR analysis. These observations suggested the recognition ability of these metallacycles toward aromatic guest molecules in a manner similar to inserting a coin into a slot machine.

Amide-functionalized metallacycles are rarely reported to act as hosts for the sensing of aromatic molecules. Manimaran and co-workers [57] reported a series of amide functionalized chalcogen bridged tetrarhenium metallacyclophanes, **31**−**38** ([{(CO)$_3$Re(μ-ER)$_2$Re(CO)$_3$}$_2$(μ-bpce)$_2$] (**31**, ER = SC$_4$H$_9$; **32**, ER = SCH$_2$C$_6$H$_5$; **33**, ER = SC$_5$H$_4$FeC$_5$H$_5$; and **34**, ER = SeC$_6$H$_5$) and [{(CO)$_3$Re(μ-ER)$_2$Re(CO)$_3$}$_2$(μ-bpcpm)$_2$] (**35**, ER = SC$_4$H$_9$; **36**, ER = SCH$_2$C$_6$H$_5$; **37**, ER = SC$_6$H$_5$CH$_3$; and **38**, ER = SeC$_6$H$_5$)) by the reaction of Re$_2$(CO)$_{10}$ with amide functionalized flexible ditopic ligands N,N′-bis(4-pyridylcarboxamid-)-1,2-ethane (bpce) or semi-flexible ligand N,N′-bis(4-(4-pyridylcar-ox-amide)phenyl)methane (bpcpm) in the presence of alkyl/aryl thiols or selenols in quantitative yields (Fig. **32**). The authors got good quality of crystals for **31**−**33** and **37** for X-ray studies.

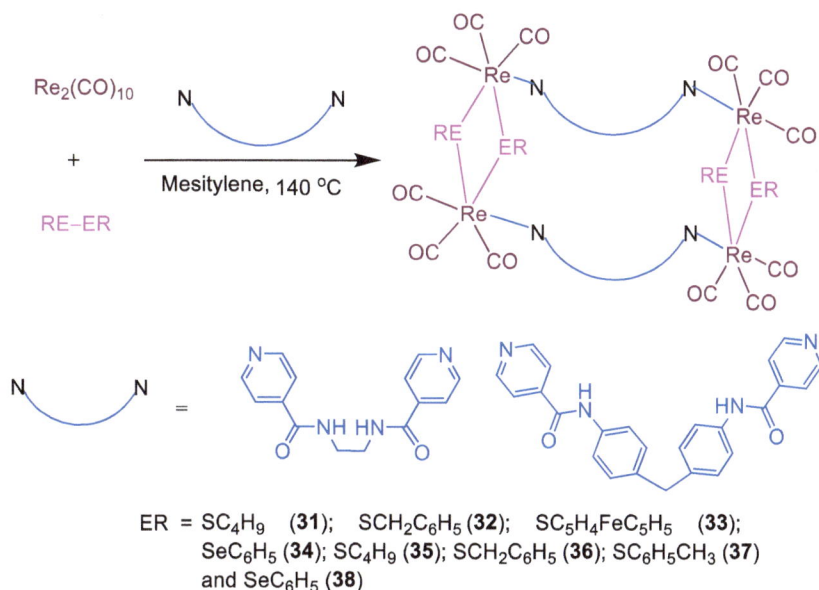

ER = SC$_4$H$_9$ **(31)**; SCH$_2$C$_6$H$_5$ **(32)**; SC$_5$H$_4$FeC$_5$H$_5$ **(33)**;
SeC$_6$H$_5$ **(34)**; SC$_4$H$_9$ **(35)**; SCH$_2$C$_6$H$_5$ **(36)**; SC$_6$H$_5$CH$_3$ **(37)**
and SeC$_6$H$_5$ **(38)**

**Fig. (32).** Synthesis of amide-functionalized flexible tetranuclear metallacyclophanes, **31–38**.

X-ray crystallographic analysis of **37**·DMF revealed that it exhibited a tetranuclear hammock-shaped architecture in which a distorted octahedral geometry of Re metal center in the *fac*-Re(CO)$_3$ core was coordinated with a N-atom of pyridyl group of bpcpm ligand and two sulfur atoms of *p*-tolylsulfide moiety (Fig. **33**). The shorter and longer Re···Re distances were found to be 3.80 and ~22.68 Å, respectively. A number of intramolecular CH···π interaction between CH of the phenyl group and π cloud of the phenyl group of the bpcpm was observed.

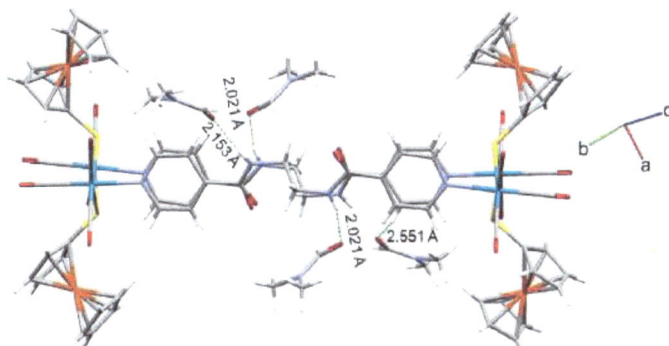

**Fig. (33).** ORTEP diagram of **37**•DMF with thermal ellipsoids drawn at 30% probability level. Reproduced with permission from ref. 57. Copyright (2018) American Chemical Society.

For the host–guest interaction studies, aromatic molecules such as 1,3,5-trimethoxybenzene, 2,4,5-trimethoxybenzaldehyde, and 3,4,5-trimethoxybenza--dehyde were selected as guest molecules with respect to metallacyclophanes **32** and **37**. Absorption spectral studies showed that the absorption intensity of guest molecules was gradually increased upon the gradual addition of hosts **32** and **37**, indicating the host-guest complex formation in the ground state (Fig. **34**). The binding constant was estimated using the B-H analysis (eqn. 1) from the absorption analysis, and their values were found in the range of $1.96 \times 10^4$–$1.83 \times 10^5$ M$^{-1}$. Fluorescence titration studies showed that hosts **32** and **37** caused the emission quenching of guest molecules, leading to the formation of host-guest interaction. The Stern-Volmer constants for these reactions were calculated from the Stern-Volmer eqn. (2) and their values were found in the range of $7.38 \times 10^4$–$1.28 \times 10^5$ M$^{-1}$ (Table 8). It was proposed that NH$\cdots$O hydrogen-bonding, CH$\cdots\pi$ and $\pi\cdots\pi$ interactions were responsible for the host–guest interaction.

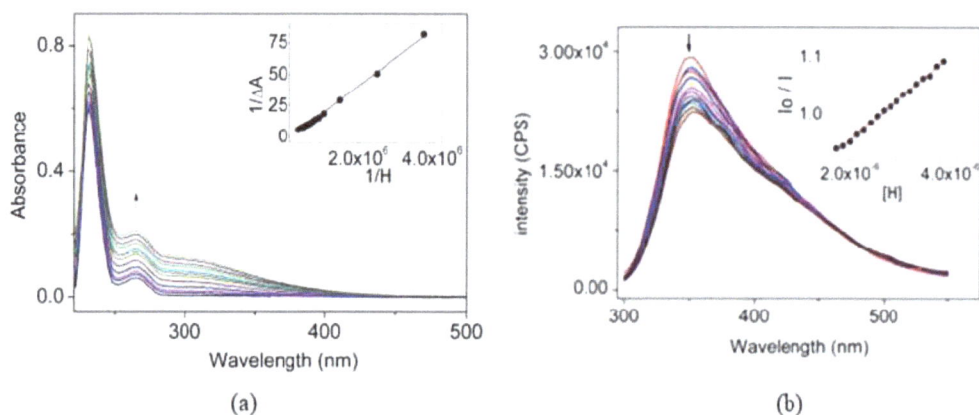

(a)                                    (b)

**Fig. (34).** (**a**) Absorption spectra of 1,3,5-trimethoxy benzene ($3.3 \times 10^{-5}$ M) upon incremental addition of host **32** ($1.3$–$26 \times 10^{-7}$ M) in tetrahydrofuran and inset shows the corresponding Benesi- Hildebrand plot ($K = 1.83 \times 10^5$ M$^{-1}$). (**b**) Emission spectra of 1,3,5-trimethoxy benzene ($3.3 \times 10^{-5}$ M) with incremental addition of host **32** ($1.3$–$26 \times 10^{-7}$ M) and inset shows the corresponding Stern-Volmer plot. Reproduced with permission from ref 57. Copyright (2018) American Chemical Society.

**Table 8. Binding constants ($K$) and Stern−Volmer constants ($K_{SV}$) for host−guest systems of 32 and 37 with few guests.**

| Guest | 32 | | 37 | |
|---|---|---|---|---|
| | $K$, M$^{-1}$ | $K_{SV}$, M$^{-1}$ | $K$, M$^{-1}$ | $K_{SV}$, M$^{-1}$ |
| 1,3,5-trimethoxy benzene | $1.83 \times 10^5$ | $7.38 \times 10^4$ | $9.80 \times 10^4$ | $1.27 \times 10^5$ |
| 2,3,5-trimethoxy benzaldehyde | $3.24 \times 10^4$ | $1.17 \times 10^5$ | $1.96 \times 10^4$ | $1.15 \times 10^5$ |
| 3,4,5-trimethoxy benzaldehyde | $1.35 \times 10^5$ | $1.31 \times 10^5$ | $9.54 \times 10^4$ | $1.28 \times 10^5$ |

## Hexanuclear Re(I) Complex

Organic-pillared prismatic cages usually have large enough cavities to accommodate large size of aromatic molecules. Through an effective orthogonal-bonding approach, a neutral triangular metalloprism, $[\{(CO)_3Re(\mu\text{-ind})Re(CO)_3\}_3$ $(\mu_3\text{-tpt})_2]$ (**39**) was self-assembled by the reaction of $Re_2(CO)_{10}$, 2,4,6-tris(-pyridyl)-1,3,5-triazine (tpt), and indigo (ind) in a mixture of toluene–acetone at 160 °C (Fig. **35**) [58]. The crystal structure of **39** showed that it adopted a $M_6L_3L'_2$ cage structure with 11.56–13.56 Å trigonal edges and 6.38–6.48 Å cage heights, in which the tpt ligands bowed slightly inward to minimize the centroid–centroid distance (3.5 Å) between the two triazine groups, thereby maximizing face-to-face π–π stacking interactions (Fig. **36**).

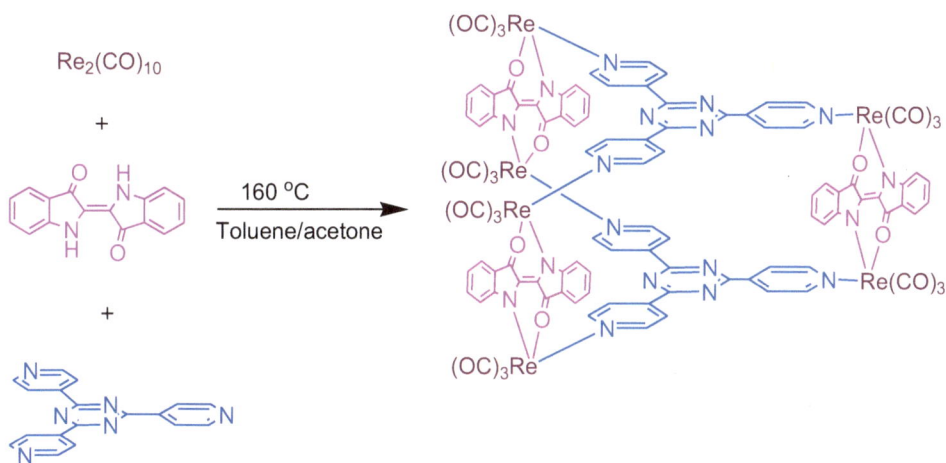

**Fig. (35).** Preparation of Re(I)-based metalloprism **39**.

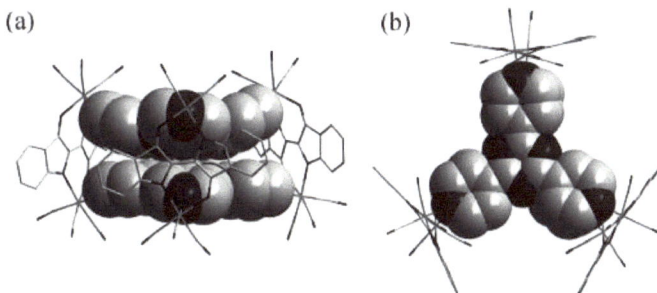

**Fig. (36).** Side (**a**) and top (**b**) views of the molecular structure of **39**; tpt ligands are drawn as space-filling models. Reproduced with permission from ref 58. Copyright (2008) The Royal Society of Chemistry.

Metalloprism **39** exhibited intense π–π* transition of both the tpt and indigo ligands at 200–330 nm, and low intensity Re → indigo and Re → tpt MLCT transitions at 335 (sh) and 410 nm, respectively, in THF. Upon excitation at 335 nm, it showed an emission band at 392 nm (with a quantum yield and lifetime of 0.01 and 5.5 ns, respectively), which was assigned to an indigo ligand localized π–π* excited state. An emission titration analysis revealed that the emission intensity of **39** was rapidly quenched upon the addition of nitroaromatic compounds with $k_q$ and $K_{SV}$ values of $8.0 \times 10^{10}$ to $3.2 \times 10^{11}$ M$^{-1}$ s$^{-1}$ and $2.0 \times 10^2$–$4.2 \times 10^3$ M$^{-1}$, respectively. During the sensing studies, the emission wavelength of **39** was red-shifted from 392 to 397 nm. These findings suggested that ground-state CT complex formation between the electron rich **39** and the electron deficient nitroaromatic compounds was responsible for the observed quenching process.

In another study, Manimaran *et al.* reported the preparation of Re(I)-based prisms and studied better selectivity in guest binding. Multicomponent self-assembly of Re$_2$(CO)$_{10}$ with oxamide ligands (H$_2$L = N,N′-dibutyloxamide (**40**), N,N′-dioctyloxamide (**41**), N,N′-didodecyloxamide (**42**) and N,N′-dibenzyloxamide (**43**)) and phenyl-1,3,5-tris(isonicotinate) (ptin) in mesitylene under solvothermal conditions has resulted in the formation of supramolecular hexarhenium prisms, [{(CO)$_3$Re(μ-η$^4$-L)Re(CO)$_3$}$_3$(μ$_3$-ptin)$_2$] (**40−43**) [59] (Fig. **37**).

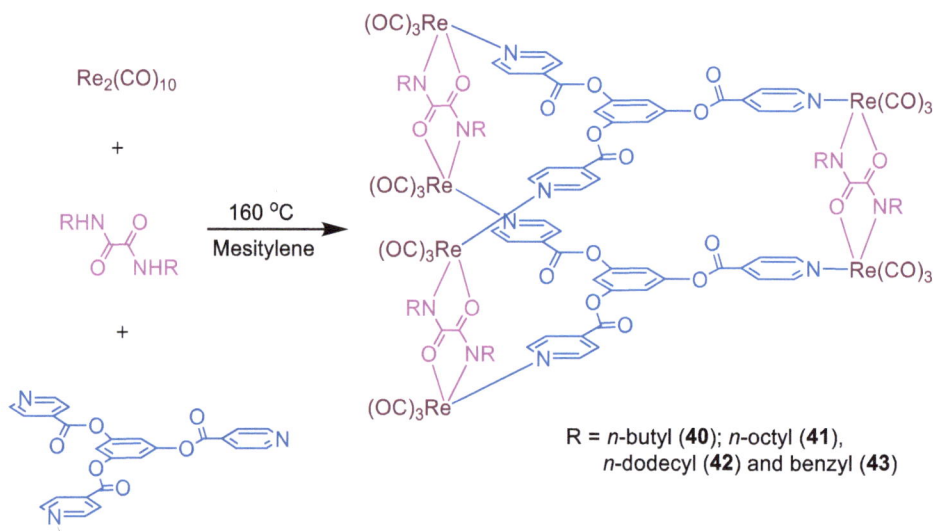

**Fig. (37).** Self-assembly of Re(I)-metallaprisms **40–43**.

An ORTEP diagram of **43** revealed a distorted trigonal prismatic architecture of $M_6L_2L'_3$ type with ~16.29−17.93 Å trigonal edges and ~5.68 Å pillar heights, in which two ptin ligands occupying the opposite trigonal faces were connected with three {(CO)$_3$Re(μ-η$^4$-dibenzyloxamidato)Re(CO)$_3$} pillars. In this structure, two ptin ligands were staggered to form a double-rosette helicity with *P* configuration while pyridyl rings created a three-bladed propeller chirality with Λ configuration (Fig. **38**).

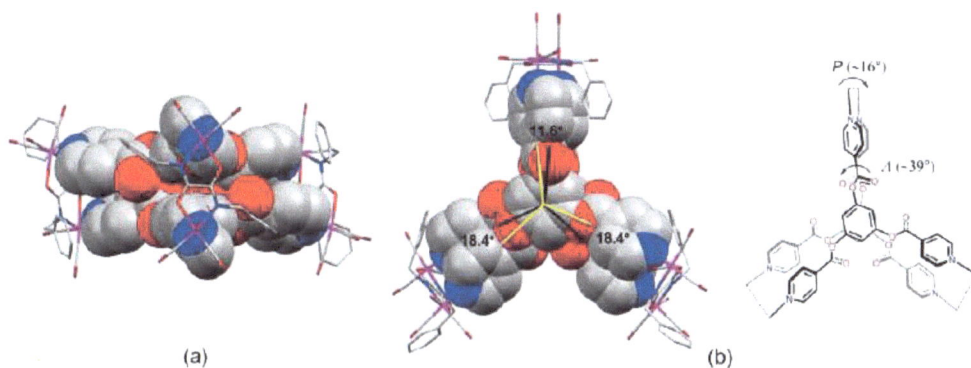

**Fig. (38).** (**a**) van der Waals representation of **43** depicting the twisting of pyridyl moieties (Λ isomer). (**b**) Bailar twist angles and schematic representation of double *P* (double rosette) and Λ (propeller) helicity. Reproduced with permission from ref. 59. Copyright (2015) American Chemical Society.

UV-Vis absorption spectra of **40−43** showed intense bands at 220−280 nm and two weak bands at 310−400 nm, corresponding to the intraligand and MLCT transitions, respectively. Upon exciting at 388−391 nm, compounds **40−43** exhibited broad band emissions at 610−612 nm with quantum yields of $3.12 \times 10^{-4} - 2.66 \times 10^{-4}$. The authors believed that ester functionality presented in **40−43** might facilitate the study of host-guest interactions with suitable aromatics. For this study, aromatic alcohols such as catechol, *p*-chloro-*m*-cresol, resorcinol, phloroglucinol and an amino acid, L-tryptophan were chosen as guest molecules.

Absorption spectral studies showed that addition of **41** to a solution of guest molecules caused an increase in the absorption intensity due to the formation of host-guest complexation. B-H analysis (eqn. 1) was used to estimate the binding constants (*K*) for these studies and their *K* values are listed in Table 9. Emission profile assay revealed that an effective quenching of guest molecules took place upon the addition of prism **41**, indicating the formation of host-guest complex, which was further confirmed by $^1$H NMR studies. The Stern-Volmer constant values ($K_{SV}$) were obtained by using the eqn. (2) and these values are given in Table 9. These studies pointed out that donor-acceptor complexation play a predominant role in the host-guest study of these molecules. In addition, an emission enhancement of **41** and **42** was observed through the formation of aggregates upon increasing the water content to the acetonitrile solution of these compounds.

**Table 9. Binding constants ($K$) and Stern-Volmer constants ($K_{SV}$) for the interaction of 41 with various guest species.**

| Guest | $K$, M$^{-1}$ | $K_{SV}$, M$^{-1}$ |
|---|---|---|
| Catechol | $1.91 \times 10^5$ | $7.03 \times 10^5$ |
| Resorcinol | $3.10 \times 10^5$ | $5.70 \times 10^5$ |
| *p*-chloro-*m*-cresol | $9.61 \times 10^5$ | $3.13 \times 10^5$ |
| Phloroglucinol | $3.20 \times 10^5$ | $2.50 \times 10^5$ |
| L-tryptophan | $4.20 \times 10^5$ | $2.60 \times 10^5$ |

## Octanuclear Re(I) Complex

Though many successful examples of bi-, tri-, tetra- and hexa-nuclear Re(I) based metallacycles are now found in the literature, the structures of octanuclear Re(I)-based metallacycles are still rare since more sophisticated preparation procedure is required. In 2003, Lu *et al.* [60] reported the preparation of first crystallographically characterized Re-based octametallic prismatic boxes, $[\{(CO)_3Re(\mu_2\text{-}OR)_2\text{-}Re(CO)_3\}_4(\mu_4\text{-}tpeb)_2]$ (**44**, R = $C_8H_{17}$; **45**, R = $C_{12}H_{25}$; **46**, R = $C_7H_7$) which were obtained by the self-assembly reaction of $Re_2(CO)_{10}$ with the tetradentate ligand, 1,2,4,5-tetraethynyl(4-pyridyl)benzene (tpeb) in the presence of respective alcohols in excellent yields (Fig. 39).

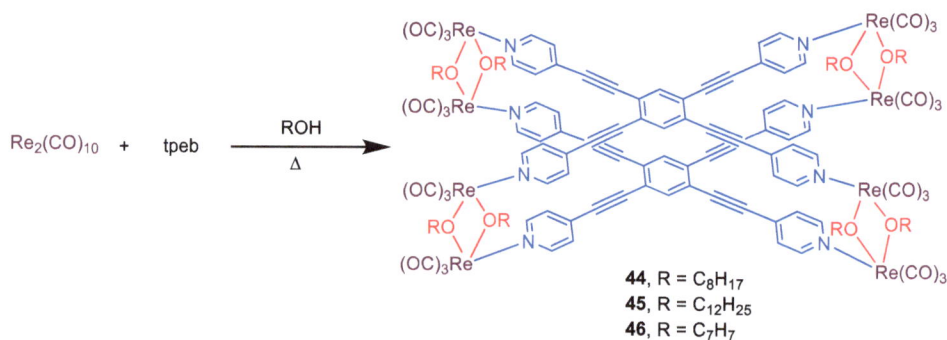

**Fig. (39).** Synthesis of octarhenium(I) rectangular prisms, **44**–**46**.

A single-crystal X-ray crystallographic study of **46** revealed a rectangular prismatic architecture in which two tpeb moieties were connected with four

$\{(CO)_3Re(\mu_2\text{-}OCH_2C_6H_5)_2Re(CO)_3\}$ edge moieties with a dimension of 18.070(6) × 10.150(7) × 3.390(2) Å (Fig. 40). The two tpeb ligands in **46** were arranged in a planar array with an effective $\pi$–$\pi$ stacking. UV-Vis absorption spectra of **44**–**46** showed a tpeb-based $\pi$–$\pi^*$ transition at 318 nm and a MLCT (Re → tpeb) transition at 363 (sh) nm in THF. They did not show any luminescence at room temperature however, they were emissive at 624-628 nm at 77 K. The luminescence property of **44**–**46** in organic solution was significantly improved in the presence of water owing to the formation of aggregates.

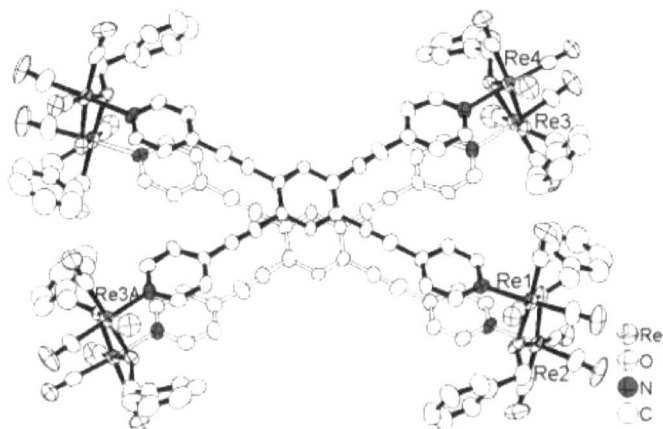

**Fig. (40).** The crystal structure of **46** at 40% ellipsoids. Solvent molecules and hydrogen atoms are omitted for clarity. Reproduced with permission from ref. 60. Copyright (2003) American Chemical Society.

B-H analysis from absorption/emission spectral studies showed that compounds **44**–**46** interacted with pyrene guest molecules in a 1:1 complex formation with the binding constants of 2.2–9.2 × 10$^4$ M$^{-1}$. Modified B-H plot analyses (eqns. 2 and 3) from emission titration studies showed that static quenching was responsible for the efficient fluorescence quenching of pyrene by **44**–**46** with quenching rate constants of 2.1–2.6 × 10$^{13}$ M$^{-1}$s$^{-1}$. $^1$H NMR study also supported that these compounds were able to recognize pyrene presumably *via* the surface of the tpeb ligand of these compounds because of insufficient cavity size (4 Å). The significance of these compounds is: (i) air and moisture stable, (ii) obtained in excellent yields, (iii) have better luminescent properties in pyridine, (iv) able to recognize aromatic guest molecules eventhough the size of the cavity is small.

## CONCLUSIONS

This chapter reports on an ongoing effort to develop molecular architectures by fusing the organic framework to the *fac*-Re(CO)$_3$ corners through self-assembly processes and study their interaction with various guest molecules. We have discussed here that different geometric metallacyclic structures ranging from dimeric metallacyclophanes to unprecedentedly large metallocycles can be prepared, in a predictable way, by assembling suitable polypyridyl bridging ligands with *fac*-Re(CO)$_3$ corners. The use of rigid, conjugated and flexible ligands in these systems provides access to interesting molecular architectures without the complete loss of control of the self-assembly process. In fact, these metallacyclic structures have been designed to fine-tune the dimensions of the cavity and their hydrophobicity and hydrogen bonding properties, as well as to adjust to the influence of the solvent in driving molecular recognition. Although such a variety of metal–organic macrocyclic architectures has been reported, those involving the use of non-covalent interactions as well as luminescence changes that depend on the nature of the guests, which would be attractive for chemo- and biosensing, have been rarely explored.

Concerning other possible developments, more efficient synthetic and assembling strategies should be developed to construct emissive higher order metallacycles based on *fac*-Re(CO)$_3$ cores materials effectively and facilely with higher efficiency and broader applications. The emergence of novel advanced optical devices and technologies may facilitate luminescence-tunable metallacyclic structures to combine their superiority with many important research frontiers, such as 3D printing, anti-counterfeiting, and even forensic investigation. Re(I) tricarbonyl complexes have unlocked a new avenue for the treatment of cancer *via* photodynamic therapy to improve the overall survival rate of cancer patients. Furthermore, the introduction of various functionalities to the chemical composition of the metallacycles can provide a platform for the imaging of specific biomolecules and targeted drug delivery for biochemical applications. Notably, multicavity metallacycles have the potential to permit the binding of multiple guests within a single assembly and could open up new applications to achieve the combination therapy of different cytotoxic agents. In these cases, when the increased stability affects the biological action of the metallacycles, guest-encapsulation can be exploited to turn the toxic species into non-toxic or vice versa, and thus, transform them from toxic anticancer agents to safe drug-delivery vehicles. Overall, we hope that the myriad of possible metallacyclic structures and their limitless modularity and tunability will encourage further developments in the exciting fields of materials science and medicinal inorganic chemistry communities.

## CONSENT FOR PUBLICATION

The author grants the publisher the sole and exclusive license of the full copyright of this material.

## CONFLICT OF INTEREST

None Declare

## ACKNOWLEDGEMENTS

M.V. thanks the Principal and the management of Thiagarajar College of Engineering, Madurai for their support and encouragement. P.T gratefully acknowledges the financial assistance supported by the Ministry of Science and Technology, Taiwan under Award Number MOST 109-2811-M-030-500.

## REFERENCES

[1]     Leininger, S.; Olenyuk, B.; Stang, P.J. Self-assembly of discrete cyclic nanostructures mediated by transition metals. *Chem. Rev.,* **2000**, *100*(3), 853-908.
[http://dx.doi.org/10.1021/cr9601324] [PMID: 11749254]

[2]     Smulders, M.M.J.; Riddell, I.A.; Browne, C.; Nitschke, J.R. Building on architectural principles for three-dimensional metallosupramolecular construction. *Chem. Soc. Rev.,* **2013**, *42*(4), 1728-1754.
[http://dx.doi.org/10.1039/C2CS35254K] [PMID: 23032789]

[3]     Thanasekaran, P.; Lee, C-H.; Lu, K-L. Neutral discrete metal–organic cyclic architectures: Opportunities for structural features and properties in confined spaces. *Coord. Chem. Rev.,* **2014**, *280*, 96-175.
[http://dx.doi.org/10.1016/j.ccr.2014.07.012]

[4]     Dumele, O.; Chen, J.; Passarelli, J.V.; Stupp, S.I. Supramolecular energy materials. *Adv. Mater.,* **2020**, *32*(17), e1907247.
[http://dx.doi.org/10.1002/adma.201907247] [PMID: 32162428]

[5]     Sepehrpour, H.; Fu, W.; Sun, Y.; Stang, P.J. Biomedically relevant self-assembled metallacycles and metallacages. *J. Am. Chem. Soc.,* **2019**, *141*(36), 14005-14020.
[http://dx.doi.org/10.1021/jacs.9b06222] [PMID: 31419112]

[6]     Hong, T.; Zhang, Z.; Sun, Y.; Tao, J-J.; Tang, J-D.; Xie, C.; Wang, M.; Chen, F.; Xie, S-S.; Li, S.; Stang, P.J. Chiral metallacycles as catalysts for asymmetric conjugate addition of styrylboronic acids to α,β-enones. *J. Am. Chem. Soc.,* **2020**, *142*(23), 10244-10249.
[http://dx.doi.org/10.1021/jacs.0c01563] [PMID: 32433874]

[7]     Goswami, A.; Saha, S.; Biswas, P.K.; Schmittel, M. (Nano)mechanical motion triggered by metal coordination: from functional devices to networked multicomponent catalytic machinery. *Chem. Rev.,* **2020**, *120*(1), 125-199.
[http://dx.doi.org/10.1021/acs.chemrev.9b00159] [PMID: 31651154]

[8]     Sinawang, G.; Osaki, M.; Takashima, Y.; Yamaguchi, H.; Harada, A. Supramolecular self-healing materials from non-covalent cross-linking host-guest interactions. *Chem. Commun. (Camb.),* **2020**, *56*(32), 4381-4395.
[http://dx.doi.org/10.1039/D0CC00672F] [PMID: 32249859]

[9]     Pöthig, A.; Casini, A. Recent developments of supramolecular metal-based structures for applications in cancer therapy and imaging. *Theranostics,* **2019**, *9*(11), 3150-3169.
[http://dx.doi.org/10.7150/thno.31828] [PMID: 31244947]

[10]    Song, B.; Kandapal, S.; Gu, J.; Zhang, K.; Reese, A.; Ying, Y.; Wang, L.; Wang, H.; Li, Y.; Wang, M.; Lu, S.; Hao, X-Q.; Li, X.; Xu, B.; Li, X. Self-assembly of polycyclic supramolecules using linear metal-organic ligands. *Nat. Commun.,* **2018**, *9*(1), 4575.
[http://dx.doi.org/10.1038/s41467-018-07045-9] [PMID: 30385754]

[11]   Fujita, M.; Umemoto, K.; Yoshizawa, M.; Fujita, N.; Kusukawa, T.; Biradha, K. Molecular paneling *via* coordination. *Chem. Commun. (Camb.),* **2001**, (6), 509-518.
[http://dx.doi.org/10.1039/b008684n]

[12]   Sauvage, J-P. Transition metal-containing rotaxanes and catenanes in motion: Toward molecular machines and motors. *Acc. Chem. Res.,* **1998**, *31*(10), 611-619.
[http://dx.doi.org/10.1021/ar960263r]

[13]   Caulder, D.L.; Raymond, K.N. Supermolecules by design. *Acc. Chem. Res.,* **1999**, *32*(11), 975-982.
[http://dx.doi.org/10.1021/ar970224v]

[14]   Gianneschi, N.C.; Masar, M.S., III; Mirkin, C.A. Development of a coordination chemistry-based approach for functional supramolecular structures. *Acc. Chem. Res.,* **2005**, *38*(11), 825-837.
[http://dx.doi.org/10.1021/ar980101q] [PMID: 16285706]

[15]   Chakrabarty, R.; Mukherjee, P.S.; Stang, P.J. Supramolecular coordination: self-assembly of finite two- and three-dimensional ensembles. *Chem. Rev.,* **2011**, *111*(11), 6810-6918.
[http://dx.doi.org/10.1021/cr200077m] [PMID: 21863792]

[16]   Davis, A.V.; Yeh, R.M.; Raymond, K.N. Supramolecular assembly dynamics. *Proc. Natl. Acad. Sci. USA,* **2002**, *99*(8), 4793-4796.
[http://dx.doi.org/10.1073/pnas.052018299] [PMID: 11880606]

[17]   Seeber, G.; Tiedemann, B.E.F.; Raymond, K.N. Supramolecular chirality in coordination chemistry. *Top. Curr. Chem.,* **2006**, *265*, 147-183.
[http://dx.doi.org/10.1007/128_033]

[18]   Brown, C.J.; Toste, F.D.; Bergman, R.G.; Raymond, K.N. Supramolecular catalysis in metal-ligand cluster hosts. *Chem. Rev.,* **2015**, *115*(9), 3012-3035.
[http://dx.doi.org/10.1021/cr4001226] [PMID: 25898212]

[19]   Zhang, D.; Ronson, T.K.; Nitschke, J.R. Functional capsules *via* sub-component self-assembly. *Acc. Chem. Res.,* **2018**, *51*(10), 2423-2436.
[http://dx.doi.org/10.1021/acs.accounts.8b00303] [PMID: 30207688]

[20]   Roberts, D.A.; Pilgrim, B.S.; Nitschke, J.R. Covalent post-assembly modification in metallosupramolecular chemistry. *Chem. Soc. Rev.,* **2018**, *47*(2), 626-644.
[http://dx.doi.org/10.1039/C6CS00907G] [PMID: 29142998]

[21]   Cook, T.R.; Stang, P.J. Recent Developments in the Preparation and Chemistry of Metallacycles and Metallacages via Coordination. *Chem. Rev.,* **2015**, *115*(15), 7001-7045.
[http://dx.doi.org/10.1021/cr5005666] [PMID: 25813093]

[22]   Korendovych, I.V.; Roesner, R.A.; Rybak-Akimova, E.V. Molecular recognition of neutral and charged guests using metallomacrocyclic hosts. *Adv. Inorg. Chem.,* **2006**, *59*, 109-173.
[http://dx.doi.org/10.1016/S0898-8838(06)59004-X]

[23]   Schneider, H.J.; Yatsimirsky, A.K. Selectivity in supramolecular host-guest complexes. *Chem. Soc. Rev.,* **2008**, *37*(2), 263-277.
[http://dx.doi.org/10.1039/B612543N] [PMID: 18197343]

[24]   Pluth, M.D.; Raymond, K.N. Reversible guest exchange mechanisms in supramolecular host-guest assemblies. *Chem. Soc. Rev.,* **2007**, *36*(2), 161-171.
[http://dx.doi.org/10.1039/B603168B] [PMID: 17264920]

[25]   Yang, X.; Yuan, D.; Hou, J.; Sedgwick, A.C.; Xu, S.; James, T.D.; Wang, L. Organic/inorganic supramolecular nano-systems based on host/guest interactions. *Coord. Chem. Rev.,* **2021**, *428*, 213609.
[http://dx.doi.org/10.1016/j.ccr.2020.213609]

[26]   Sun, Y.; Chen, C.; Liu, J.; Stang, P.J. Recent developments in the construction and applications of platinum-based metallacycles and metallacages *via* coordination. *Chem. Soc. Rev.,* **2020**, *49*(12), 3889-3919.

[http://dx.doi.org/10.1039/D0CS00038H] [PMID: 32412574]

[27]   Zhang, M.; Yin, S.; Zhang, J.; Zhou, Z.; Saha, M.L.; Lu, C.; Stang, P.J. Metallacycle-cored supramolecular assemblies with tunable fluorescence including white-light emission. *Proc. Natl. Acad. Sci. USA,* **2017**, *114*(12), 3044-3049.
[http://dx.doi.org/10.1073/pnas.1702510114] [PMID: 28265080]

[28]   Ousaka, N.; Yamamoto, S.; Iida, H.; Iwata, T.; Ito, S.; Hijikata, Y.; Irle, S.; Yashima, E. Water-mediated deracemization of a bisporphyrin helicate assisted by diastereoselective encapsulation of chiral guests. *Nat. Commun.,* **2019**, *10*(1), 1457.
[http://dx.doi.org/10.1038/s41467-019-09443-z] [PMID: 30926811]

[29]   Ramos, L.D.; da Cruz, H.M.; Morelli Frin, K.P. Photophysical properties of rhenium(i) complexes and photosensitized generation of singlet oxygen. *Photochem. Photobiol. Sci.,* **2017**, *16*(4), 459-466.
[http://dx.doi.org/10.1039/C6PP00364H] [PMID: 28054064]

[30]   Manav, N.; Kesavan, P.E.; Ishida, M.; Mori, S.; Yasutake, Y.; Fukatsu, S.; Furuta, H.; Gupta, I. Phosphorescent rhenium-dipyrrinates: efficient photosensitizers for singlet oxygen generation. *Dalton Trans.,* **2019**, *48*(7), 2467-2478.
[http://dx.doi.org/10.1039/C8DT04540B] [PMID: 30694280]

[31]   Wolcan, E. Photosensitized generation of singlet oxygen from rhenium(I) complexes: A review. *Inorg. Chim. Acta,* **2020**, *509*, 119650.
[http://dx.doi.org/10.1016/j.ica.2020.119650]

[32]   Ko, C-C.; Yam, V.W-W. Transition metal complexes with photochromic ligands—photosensitization and photoswitchable properties. *J. Mater. Chem.,* **2010**, *20*(11), 2063-2070.
[http://dx.doi.org/10.1039/B919418E]

[33]   Ko, C-C.; Yam, V.W-W. Coordination compounds with photochromic ligands: Ready tunability and visible light-sensitized photochromism. *Acc. Chem. Res.,* **2018**, *51*(1), 149-159.
[http://dx.doi.org/10.1021/acs.accounts.7b00426] [PMID: 29265804]

[34]   Thanasekaran, P.; Liao, R.T.; Liu, Y.H.; Rajendran, T.; Rajagopal, S.; Lu, K.L. Metal-containing molecular rectangles: Synthesis and photophysical properties. *Coord. Chem. Rev.,* **2005**, *249*(9-10), 1085-1110.
[http://dx.doi.org/10.1016/j.ccr.2004.11.006]

[35]   Thanasekaran, P.; Lee, C.C.; Lu, K.L. One-step orthogonal-bonding approach to the self-assembly of neutral rhenium-based metallacycles: synthesis, structures, photophysics, and sensing applications. *Acc. Chem. Res.,* **2012**, *45*(9), 1403-1418.
[http://dx.doi.org/10.1021/ar200243w] [PMID: 22721174]

[36]   Sathish, V.; Babu, E.; Ramdass, A.; Lu, Z-Z.; Chang, T-T.; Velayudham, M.; Thanasekaran, P.; Lu, K-L.; Li, W-S.; Rajagopal, S. Photoswitchable alkoxy-bridged binuclear rhenium(i) complexes – A potential probe for biomolecules and optical cell imaging. *RSC Advances,* **2013**, *3*(40), 18557-18566.
[http://dx.doi.org/10.1039/c3ra42627k]

[37]   Sathish, V.; Krishnan, M.M.; Velayudham, M.; Thanasekaran, P.; Lu, K-L.; Rajagopal, S. Host-guest interaction studies of polycyclic aromatic hydrocarbons (PAHs) in alkoxy bridged binuclear rhenium (I) complexes. *Spectrochim. Acta A Mol. Biomol. Spectrosc.,* **2019**, *222*, 117160.
[http://dx.doi.org/10.1016/j.saa.2019.117160] [PMID: 31176159]

[38]   Murakami, Y.; Kikuchi, J.I.; Suzuki, M.; Matsuura, T. Syntheses of macrocyclic enzyme models. Part 6. Preparation and guest-binding behaviour of octopus cyclophanes. *J. Chem. Soc., Perkin Trans. 1,* **1988**, (6), 1289-1299.
[http://dx.doi.org/10.1039/p19880001289]

[39]   Lakowicz, J.R. *Principles of Fluorescence Spectroscopy,* 3rd ed; Springer Press: New York, **2006**.
[http://dx.doi.org/10.1007/978-0-387-46312-4]

[40]   Sun, S.S.; Anspach, J.A.; Lees, A.J.; Zavalij, P.Y. Synthesis and electrochemical, photophysical, and

anion binding properties of self-assembly heterometallic cyclophanes. *Organometallics,* **2002**, *21*(4), 685-693.
[http://dx.doi.org/10.1021/om0109096]

[41]   Harrison, B.S.; Ramey, M.B.; Reynolds, J.R.; Schanze, K.S. Amplified fluorescence quenching in a poly(*p*-phenylene)-based cationic polyelectrolyte. *J. Am. Chem. Soc.,* **2000**, *122*(35), 8561-8562.
[http://dx.doi.org/10.1021/ja000819c]

[42]   Sathish, V.; Ramdass, A.; Lu, Z-Z.; Velayudham, M.; Thanasekaran, P.; Lu, K-L.; Rajagopal, S. Aggregation-induced emission enhancement in alkoxy-bridged binuclear rhenium(I) complexes: application as sensor for explosives and interaction with microheterogeneous media. *J. Phys. Chem. B,* **2013**, *117*(46), 14358-14366.
[http://dx.doi.org/10.1021/jp407939j] [PMID: 24175920]

[43]   Sathish, V.; Ramdass, A.; Lu, Z-Z.; Velayudham, M.; Thanasekaran, P.; Lu, K.L.; Rajagopal, S. Sensing of insulin fibrillation using alkoxy-bridged binuclear rhenium(I) complexes. *Inorg. Chem. Commun.,* **2016**, *73*, 49-51.
[http://dx.doi.org/10.1016/j.inoche.2016.09.015]

[44]   Sathish, V.; Babu, E.; Ramdass, A.; Lu, Z-Z.; Velayudham, M.; Thanasekaran, P.; Lu, K-L.; Rajagopal, S. Alkoxy bridged binuclear rhenium (I) complexes as a potential sensor for β-amyloid aggregation. *Talanta,* **2014**, *130*, 274-279.
[http://dx.doi.org/10.1016/j.talanta.2014.06.070] [PMID: 25159409]

[45]   Lo, K.K.; Tsang, K.H.K.; Hui, W.K.; Zhu, N. Synthesis, characterization, crystal structure, and electrochemical, photophysical, and protein-binding properties of luminescent rhenium(I) diimine indole complexes. *Inorg. Chem.,* **2005**, *44*(17), 6100-6110.
[http://dx.doi.org/10.1021/ic050531u] [PMID: 16097831]

[46]   Dam, T.K.; Roy, R.; Pagé, D.; Brewer, C.F. Negative cooperativity associated with binding of multivalent carbohydrates to lectins. Thermodynamic analysis of the "multivalency effect". *Biochemistry,* **2002**, *41*(4), 1351-1358.
[http://dx.doi.org/10.1021/bi015830j] [PMID: 11802737]

[47]   Bhol, M.; Shankar, B.; Sathiyendiran, M. Rhenium(i) based irregular pentagonal-shaped metallacavitands. *Dalton Trans.,* **2018**, *47*(13), 4494-4500.
[http://dx.doi.org/10.1039/C8DT00574E] [PMID: 29505049]

[48]   Coogan, M.P.; Fernández-Moreira, V.; Kariuki, B.M.; Pope, S.J.A.; Thorp-Greenwood, F.L. A rhenium tricarbonyl 4′-oxo-terpy trimer as a luminescent molecular vessel with a removable silver stopper. *Angew. Chem. Int. Ed. Engl.,* **2009**, *48*(27), 4965-4968.
[http://dx.doi.org/10.1002/anie.200900981] [PMID: 19472241]

[49]   Manimaran, B.; Lai, L.J.; Thanasekaran, P.; Wu, J.Y.; Liao, R.T.; Tseng, T.W.; Liu, Y.H.; Lee, G.H.; Peng, S.M.; Lu, K.L.C.H. CH...π interaction for rhenium-based rectangles: an interaction that is rarely designed into a host-guest pair. *Inorg. Chem.,* **2006**, *45*(20), 8070-8077.
[http://dx.doi.org/10.1021/ic0604720] [PMID: 16999404]

[50]   Thanasekaran, P.; Wu, J.Y.; Manimaran, B.; Rajendran, T.; Chang, I.J.; Rajagopal, S.; Lee, G.H.; Peng, S.M.; Lu, K-L. Aggregate of alkoxy-bridged Re(I)-rectangles as a probe for photoluminescence quenching. *J. Phys. Chem. A,* **2007**, *111*(43), 10953-10960.
[http://dx.doi.org/10.1021/jp0742315] [PMID: 17918811]

[51]   Manimaran, B.; Vanitha, A.; Karthikeyan, M.; Ramakrishna, B.; Mobin, S.M. Self-assembly of selenium-bridged rhenium(I)-based metallarectangles: Synthesis, characterization, and molecular recognition studies. *Organometallics,* **2014**, *33*(2), 465-472.
[http://dx.doi.org/10.1021/om400673f]

[52]   Nagarajaprakash, R.; Divya, D.; Ramakrishna, B.; Manimaran, B. Synthesis and spectroscopic and structural characterization of oxamidato-bridged rhenium(I) supramolecular rectangles with ester functionalization. *Organometallics,* **2014**, *33*(6), 1367-1373.

[http://dx.doi.org/10.1021/om400776m]

[53]    Govindarajan, R.; Nagarajaprakash, R.; Manimaran, B. Synthesis, structural characterization, and host
        – guest studies of aminoquinonato-bridged Re(I) supramolecular rectangles. *Inorg. Chem.,* **2015**,
        *54*(22), 10686-10694.
        [http://dx.doi.org/10.1021/acs.inorgchem.5b01543] [PMID: 26528890]

[54]    Liao, R-T.; Yang, W-C.; Thanasekaran, P.; Tsai, C-C.; Sathiyendiran, M.; Liu, Y-H.; Rajendran, T.;
        Lin, H-M.; Tseng, T-W.; Lu, K-L. Rhenium-based molecular rectangular boxes with large inner cavity
        and high shape selectivity towards benzene molecule. *Chem. Commun. (Camb.),* **2008**, (27), 3175-
        3177.
        [http://dx.doi.org/10.1039/b802777c] [PMID: 18594733]

[55]    Sathiyendiran, M.; Liao, R-T.; Thanasekaran, P.; Luo, T-T.; Venkataramanan, N.S.; Lee, G-H.; Peng,
        S-M.; Lu, K-L. Gondola-shaped luminescent tetrarhenium metallacycles with crown-ether-like
        multiple recognition sites. *Inorg. Chem.,* **2006**, *45*(25), 10052-10054.
        [http://dx.doi.org/10.1021/ic061886w] [PMID: 17140209]

[56]    Sathiyendiran, M.; Tsai, C-C.; Thanasekaran, P.; Luo, T-T.; Yang, C-I.; Lee, G-H.; Peng, S-M.; Lu,
        K-L. Organometallic calixarenes: syceelike tetrarhenium(I) cavitands with tunable size, color,
        functionality, and coin-slot complexation. *Chemistry,* **2011**, *17*(12), 3343-3346.
        [http://dx.doi.org/10.1002/chem.201003181] [PMID: 21322075]

[57]    Karthikeyan, M.; Ramakrishna, B.; Vellaiyadevan, S.; Divya, D.; Manimaran, B. Amide-
        functionalized chalcogen-bridged flexible tetranuclear rhenacycles: Synthesis, characterization,
        solvent effect on the structure, and guest binding. *ACS Omega,* **2018**, *3*(3), 3257-3266.
        [http://dx.doi.org/10.1021/acsomega.7b02075] [PMID: 31458582]

[58]    Wu, J-Y.; Chang, C-H.; Thanasekaran, P.; Tsai, C-C.; Tseng, T-W.; Lee, G-H.; Peng, S-M.; Lu, K-L.
        Unusual face-to-face $\pi$-$\pi$ stacking interactions within an indigo-pillared $M_3$(tpt)-based triangular
        metalloprism. *Dalton Trans.,* **2008**, (44), 6110-6112.
        [http://dx.doi.org/10.1039/b809489f] [PMID: 18985240]

[59]    Nagarajaprakash, R.; Govindarajan, R.; Manimaran, B. One-pot synthesis of oxamidato-bridged
        hexarhenium trigonal prisms adorned with ester functionality. *Dalton Trans.,* **2015**, *44*(26), 11732-
        11740.
        [http://dx.doi.org/10.1039/C5DT01102G] [PMID: 26050748]

[60]    Manimaran, B.; Thanasekaran, P.; Rajendran, T.; Liao, R.T.; Liu, Y.H.; Lee, G.H.; Peng, S.M.;
        Rajagopal, S.; Lu, K.L. Self-assembly of octarhenium-based neutral luminescent rectangular prisms.
        *Inorg. Chem.,* **2003**, *42*(16), 4795-4797.
        [http://dx.doi.org/10.1021/ic034172j] [PMID: 12895097]

**CHAPTER 7**

# Recent Developments in the Dynamics of Fluorescently Labelled Macromolecules

**Kandhasamy Durai Murugan**[1,*], **Pandi Muthirulan**[2] and **Vijayanand Chandrasekaran**[3]

[1] *Department of Bioelectronics and Biosensors, Alagappa University, Karaikudi-630003, Tamilnadu, India*

[2] *Department of Chemistry, Lekshmipuram College of Arts and Science, (Affiliated to MS University, Tirunelveli) Neyyoor-629802, Tamilnadu, India*

[3] *Department of Chemistry, School of Advanced Sciences, Vellore Institute of Technology, Vellore-632014, Tamilnadu, India*

**Abstract:** There is considerable interest in the photophysics and photochemistry of water-soluble macromolecules functionalized as pendant or copolymerized on the macromolecular backbone itself. A promising feature of functionalized macromolecules is that a large variety of chemical modifications based on molecular design is possible as compared to conventional organized assemblies such as micelles and vesicles. Photoactive macromolecules have important applications in photoresists, xerography, photocuring of paints and resins, and solar energy conversion systems. These macromolecular systems are broadly classified into two categories: (1) in which chromophores are directly attached to the backbone of the macromolecule as a pendant and (2) in which the macromolecule acts as a host to the photosensitizing molecules. Various aspects of photochemical and photophysical processes in polymers are discussed earlier in detail. Time resolved fluorescence techniques have been extensively used to study the dynamics of natural and synthetic macromolecules. This book chapter covers the investigations on the dynamics polymers in solution using a variety of time resolved techniques ranging from a few femtoseconds to several seconds.

**Keywords:** Dynamics, Macromolecule, Photochemistry, Time Resolved, Timescale of motion.

## INTRODUCTION

Enumerable applications of macromolecular self-assemblies in several fields force us to understand their properties not only in solution but also in various forms of matter [1 - 7]. The properties of macromolecular systems, in general, includes

---

\* **Corresponding author Kandhasamy Durai Murugan:** Department of Bioelectronics and Biosensors, Alagappa University, Karaikudi-630003, Tamilnadu, India; Tel: +917708446619; E-mail: kdmurugan@gmail.com

**Paulpandian Muthu Mareeswaran, Palaniswamy Suresh and Seenivasan Rajagopal (Eds.)**

their conformations and dynamics which are governed by various factors like pH of the solution, concentration, molecular weight, temperature and the presence of smaller molecular weight additives [8]. Polymers, especially polyelectrolytes, undergo several conformations in solution and films. Two postulates concerning the mechanism of the conformational transition of polymers in solution are proposed as shown in Fig. (**1**) [9].

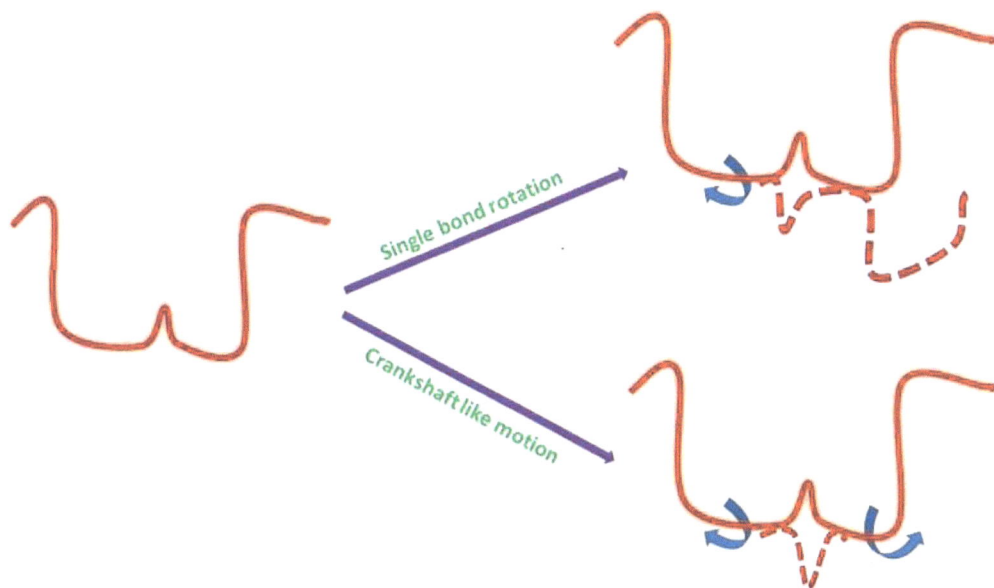

**Fig. (1).** Schematic representation of the conformational transition in a polymer chain.

Rotation around one bond, with the rest of the molecules remaining conformationally unchanged, is one way that requires a large portion of the chain to swing through the viscous medium with a prohibitive expenditure of energy. To avoid this difficulty, Kuhn and Kuhn proposed a theory more than 60 years ago. Accordingly, two temporally correlated rotations would take place so that only a small loop would move which is later described as a "crankshaft–like motion [10]. However, this mechanism, in which two energy barriers have to be surmounted simultaneously implies that the activation energy for the conformational transition in the center of a long–chain molecule is substantially higher (and the transition much slower) than in an analogous small molecule [11]. Later it is shown that a single hindered rotation is required for a conformational transition of a polymer [12]. It is conjectured that the stress introduced into the chain molecule by a hindered rotation is relieved by small distortions of the internal angle of rotation, which requires a very small expenditure of energy [13]. Later, Winnik and co–workers studied the kinetics of conformational transition of poly(styrene) in toluene and estimated the activation energy of the cyclization [14]. The calculated activation energy coincides with that required for a single bond rotation.

Naturally, the dynamics are related to the structure of the polyelectrolytes and the studies on the structural properties of the polyelectrolytes to some extent provide information on the dynamical properties of the polyelectrolytes at different length scales [15]. However, detailed investigations at different time regions are required to understand the dynamics of polyelectrolytes. The dynamics of macromolecules span several time domains depending upon the nature of the motion of the molecular systems as shown in Table 1. In the shorter time region, the vibrational motion and the bending of the bonds takes place on the atomic scale while at a longer time scale the entire molecular motions and the folding are the prominent processes occurring in the polymers [16].

**Table 1. Time scale and molecular motion.**

| Time Scale | Amplitude | Description |
|---|---|---|
| $10^{-15} - 10^{-12}$s | $0.001 - 0.1$ Å | – bond stretching, angle bending<br>– constraint dihedral motion |
| $10^{-12} - 10^{-9}$s | $0.1 - 10$ Å | – unhindered surface side chain motion<br>– loop motion |
| $10^{-9} - 10^{-6}$s | $1 - 100$ Å | – folding in small peptides<br>– helix coil transition |
| $10^{-6} - 10^{-1}$s | $10 - 100$ Å | – protein folding<br>– domain motion |

In the case of macromolecules such as polypeptides and synthetic polymers, the dynamical processes involving the motions of the side chains and loops take place in the sub–picosecond to nanosecond time scale. The coil transitions occur in a few nanoseconds to microseconds in solution [17].

## POLY(ACRYLIC ACIDS) – STRUCTURAL TRANSITION AND DYNAMICS

The conformations and dynamics of poly(acrylic acids) are subjects of intense investigation over several decades. The ionization of these polymers in solution is a function of the pH of the solution. Hence, these polymers show different conformations regulated by electrostatic interactions, hydrogen bonding, hydrophobic and other weak interactions, which are dependent upon the pH of the solution [18]. Poly(acrylic acid), PAA, is gradually expanded to a linear chain by increasing the pH of the solution [19]. On the other hand, poly(methacrylic acid), PMAA, which has a methyl substituent at the α–carbon, shows a different behaviour from PAA. The polyelectrolyte, PMAA, maintains its compact coil conformation upto certain pH and the coil opens up to a linear structure in an aqueous solution when the pH of the solution is increased above a critical pH value(≈ 5.0) [20 - 22]. The conformational transitions of poly(acrylic acids) are investigated by using several techniques including viscosity measurements, scattering methods, calorimetric methods and spectroscopic methods [23 - 26].

The dynamics of PMAA in its various states and at different time domains are studied by various techniques. The coil collapse of PMAA in an aqueous solution and solid state as a factor of hydrogen bonding is studied by $^1$H–NMR spectroscopy [27]. The swelling behavior of PMAA in an aqueous solution is studied by using atomic force microscopy in the slow time regime (i.e., in seconds) [28]. The hydration behavior of PMAA films in an aqueous solution is studied by time–resolved ATR–FTIR spectroscopy and the proposed mechanism of hydration is shown in Fig. (**2**) [29]. During the hydration process, when intra– and inter– side–chain hydrogen bonds in the PMAA are dissociated and the film is swelled by the water storming, then non–hydrogen–bond carboxyl groups instantly hydrate with water molecules and equilibrate to the side–on form, which is the most stable structure for the carboxyl groups. Though the above studies provide potential information on the structure and dynamics of PMAA, the time resolution is limited to seconds only.

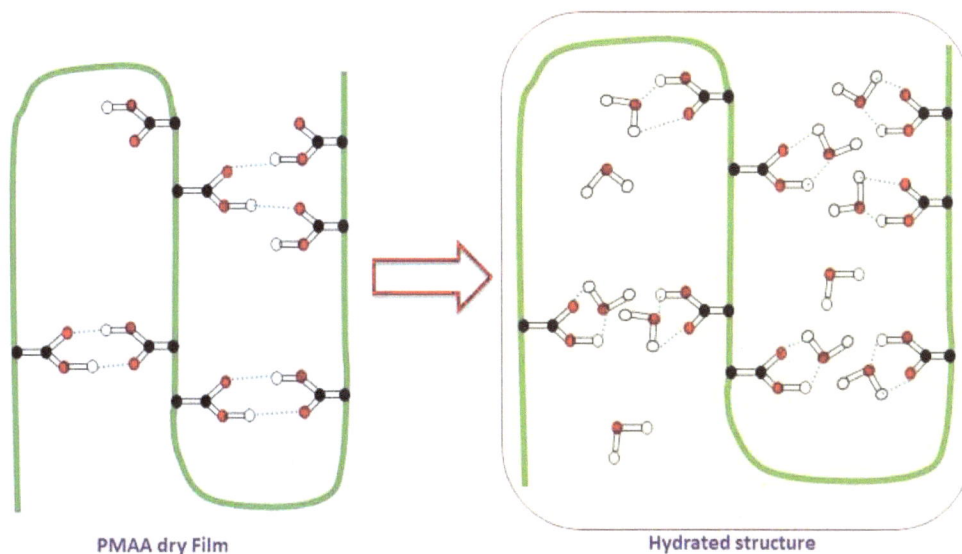

**PMAA dry Film**                                    **Hydrated structure**

**Fig. (2).** Mechanism of hydration of PMAA film.

Static light scattering (SLS) and dynamic light scattering (DLS) techniques are useful experimental tools for the investigation of the structure and dynamics of polyelectrolyte solutions and the characterization of polymer systems in general, with the time resolution limited to a few microseconds [30, 31]. The structure of the polymers is generally probed by static light scattering technique on a certain length scale (typically from 20 nm to several microns) limited mainly by the wavelength of the light used. The accessible structural information, therefore, is usually on the level of a whole polymer chain and the interchain correlations. But, the information on single chain dimensions and interchain separation distances could not be obtained from such studies. Light scattering technique is incompetent to yield direct information on the local structure inside the chain, also referred to as the primary and secondary structure [31]. Charge interactions in polyelectrolyte solutions dominantly influence the structure and dynamics in the above space and time scales, and therefore light scattering provides information on the character of these interactions [32]. Techniques with the time resolution of a few picoseconds are needed to probe dynamics at the single molecular level. Fluorescence spectroscopic measurements offer such resolution, and also allow *in-situ* monitoring of processes that occur within the polymer chains when the pH of the solution is changed [9, 26, 33].

## PHOTOPHYSICS OF FLUOROPHORES IN POLYELECTROLYTES

In order to follow the dynamics and conformations of polyelectrolytes using photophysical techniques, a fluorescent dye molecule is used with the polymer under study as a free probe or covalently attached to the polymer [34 - 37]. Different probes are used to study the conformational transition of PMAA. Pyrene is used to investigate the kinetics of coil expansion and the existence of structural micro–domains within the hyper–coiled conformation at low pH [38]. Other fluorescence probes used to study PMAA include solubilized probes such as coumarin or covalently bound labels such as dansyl, diphenylanthracene, vinylnaphthalene, acridine and phenazine class of dyes [39 - 41].

## PHOTOPHYSICS OF EXCITED STATE FLUOROPHORES

Interaction of visible light with light absorbing organic molecules produces electronically excited state and the dissipation of the excess energy of the excited molecules occurs in various pathways as shown in the Jablonski diagram (Fig. **3**) [42]. The light absorption by the molecules occurs within few femtoseconds which produces the excited singlet state of the molecule S1, S2 and S3. The higher excited singlet states (S2 and S3) dissipate its excess energy in the form of heat and the processes are completed within a few picoseconds. From the lowest singlet excited state, S1, the excited molecule relaxes in two ways: The molecule comes to the ground state by releasing its excess energy in the form of a visible photon or as thermal energy in the nanosecond time region. In other cases, the excited molecule undergoes intersystem crossing to the triplet state and the excess energy is released as phosphorescence from the triplet state. From the triplet state, the molecule absorbs another photon and is promoted to higher triplet states (T2, T3……Tn). The absorption of the molecule from the triplet state to the higher triplet states is termed as triplet–triplet (T–T) absorption and the process takes place in microseconds.

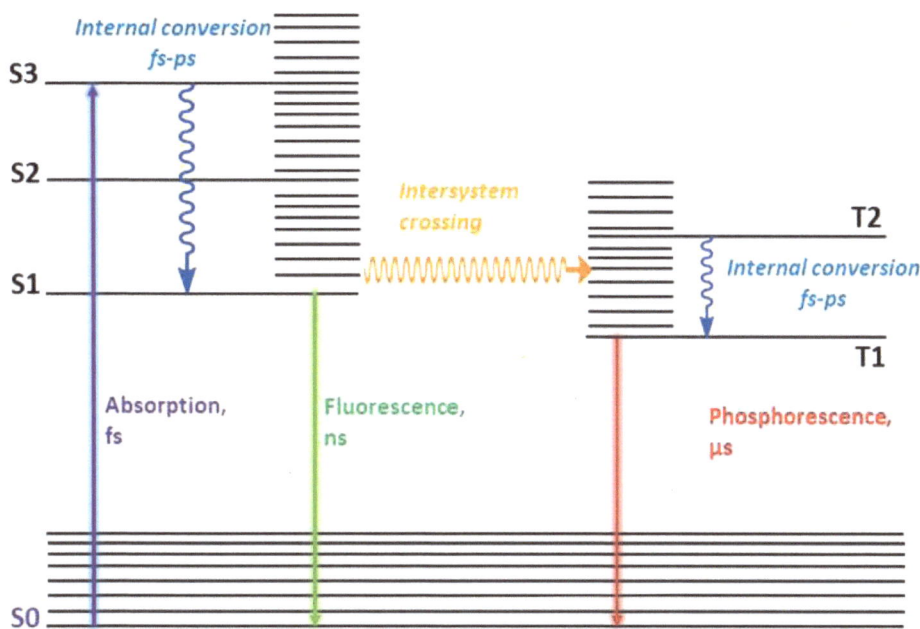

**Fig. (3).** Jablonski diagram and a time scale of photophysical processes for organic molecules.

## IMPORTANCE OF PHOTOPHYSICS AND PHOTOCHEMISTRY OF POLYMERIC MATERIALS

There is considerable interest in the photophysics and photochemistry of water soluble polyelectrolytes functionalized as pendant or copolymerized on the polymer backbone itself [43 - 46]. A promising feature of functionalized polyelectrolyte is that a large variety of chemical modifications based on molecular design is possible as compared to conventional organized assemblies such as micelles and vesicles [47 - 50]. Photoactive polymers have important applications in photoresists, xerography, photocuring of paints and resins, and solar energy conversion systems [51 - 54]. These polymeric systems are broadly classified into two categories: (1) chromophores are directly attached to the backbone of the polymer as pendant and (2) polymer acts as a host to the photosensitizing molecules [55, 56]. Various aspects of photochemical and photophysical processes in polymers are discussed earlier in detail [57, 58].

The photochemistry of polymeric electrolytes with photoactive functional groups is investigated as an artificial model system for photosynthesis. Photosynthetic systems utilize light–absorbing chlorophyll "antenna" moieties to transfer absorbed light energy to nearby energy acceptors, leading to the cascade of processes that result in the production of glucose and $O_2$ from sunlight, $CO_2$ and $H_2O$ (Fig. **4**) [59]. Artificial light harvesting systems capable of converting solar radiation into a useful source of fuel with similar efficiencies are beneficial and are extensively investigated. Recognizing this fact, polymeric assemblies are developed to mimic the natural photosynthetic process, where energy transfer is the mechanism of energy capture and initial routing as shown in Scheme 2 [60].

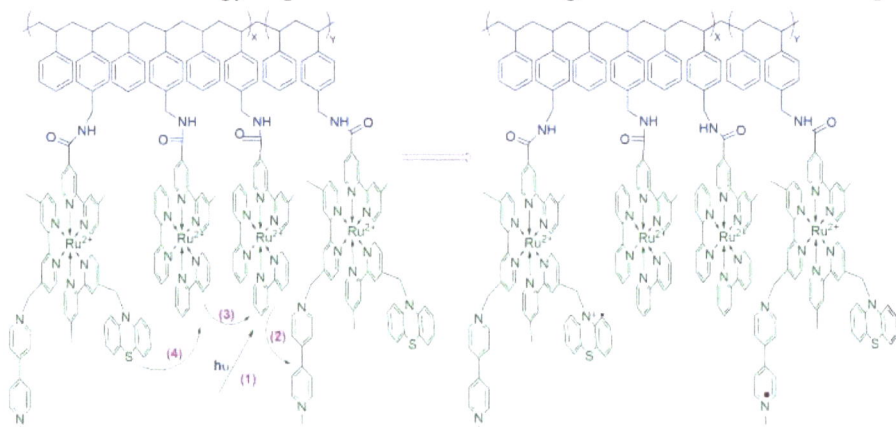

**Fig. (4).** Scheme illustrating the antenna effect in antenna/reaction centre polymer.

Irradiation of the assembly by visible light leads to the excitation of a bound chromophore unit, which acts as an antenna fragment within the polymer. After absorption of a photon and thermal equilibration, the excited state is formed rapidly from which subsequent energy transfer to a neighbour occurs. In this way, the energy migration takes place along the polymer chain until it reaches the reaction centre where the charge separation takes place.

Photopolymerization and photocuring science and technologies are developed particularly with regard to design novel and specific initiators and materials for specialised applications [61]. Novel macroinitiators based on thioxanthone are developed and found to be highly efficient when bound to a polymer chain which are used in conjunction with a free amine co-synergist [62]. The photocrosslinking of polymer materials continues to be attractive for many applications in electronics, electricals, and insulation property enhancements [63,

64]. N–isopropylacrylamide with di–methylmaleinimidoacrylamide are crosslinked to produce thermally sensitive nano–gels [65]. The photocrosslinking of polymers is an attractive process in terms of enhancing the physical and mechanical properties of electronic materials and the development of liquid crystalline materials [66]. The optical properties of polymers remain an active area of research that lead to a continued growth in photochromic and liquid crystalline materials [67]. Chemiluminescence of polymer materials is an active area of interest as an analytical probe for various biological samples. The effectiveness of various commercial antioxidants and their ability to inhibit hydroperoxide formation are determined through their chemiluminescence activity when incorporated into poly(ethylene) [68].

The organic, polymer–based photovoltaic materials are introduced with the potential for obtaining cheap and easy methods to produce useful energy from natural light source [69]. Photoinduced electron transfer from donor–type semiconducting polymers onto acceptor–type polymers or molecules, such as $C_{60}$ was utilized in these organic solar cells [70]. The concept of "double cable" polymers is introduced to have a control on the morphology at the molecular level [71]. Chemically attaching the electron acceptor moieties directly to the donor polymer backbone prevents the phase separation and improves the energy conversion efficiency [72].

## PHOTOPHYSICS AND CONFORMATIONAL DYNAMICS OF POLYELECTROLYTES

In addition to the above mentioned potential applications, the photophysics and photochemistry of polymeric materials is extensively studied to understand the structure and dynamics of parent polymer to which the fluorophores are tagged [73]. In particular, polyelectrolytes of both synthetic and natural origin are investigated by making use of various photophysical processes of excited chromophoric moiety [74].Advantage of using photophysical techniques are summarised below:

- Measurements can be carried out at extremely low chromophoric concentration. Thus, a polymer may be labelled with low concentration of the fluorophore and the label does not disturb the properties of the polymer.
- The emission intensity is extremely sensitive to the local environment of the label.
- The dependence of the emission intensity on the concentration of the quencher may provide the information about heterogeneous distribution of the fluorophore and quencher
- Intramolecular excimer formation between the neighbouring chromophores

labelled on the same polymer backbone may provide information about the rate of conformational transition.

- Fluorescence polarization experiments of the fluorophores bound to the polymer give information on the rotational diffusion pathways of the fluorophore in a restricted environment.
- Above all, the information extracted from the fluorescence studies provides the nature of the macromolecular structure at the molecular level compared to other ensemble measurements.

The photophysical methods used to study the structure and dynamics of polyelectrolytes are

a. Steady state fluorescence studies and fluorescence quenching
b. Excited state energy and electron transfer
c. Excimer and exciplex formation
d. Time–resolved studies and solvation dynamics

## STEADY STATE FLUORESCENCE STUDIES AND FLUORESCENCE QUENCHING

The absorption and emission spectra of fluorophores change with the nature of solvents. The specific solute–solvent interactions such as dipole–dipole interactions, van der Wall's interaction and hydrogen bonding are the major factors governing the electronic spectral properties of fluorophores [75]. Absorption and emission spectral characteristics of the fluorophores due to the polarity of the medium are used to probe the local environment in microheterogeneous systems such as micelles, polymers, vesicles and solid matrices [76 - 80].

Earlier, changes observed in absorption and fluorescence spectra are used to probe the polymerization process using fluorophores [81]. Pyrene bound to acrylic acid and its methyl ester compounds and other molecules with intramolecular charge transfer (ICT) dynamics, such as [p–(N,N–dialkylamino)benzylidene] malononitrile are used as fluorescent probes for the cited purposes [82, 83]. Auramine–O is used to study the polymerization process of methacrylic acid in aqueous solution [84]. The pioneering work of Morawetz initiated the applications of steady state spectroscopy for the conformational kinetics of polyelectrolytes in solution [85 - 87]. The phenomena accompanying the ionization of PMAA in aqueous solution is investigated by monitoring the absorption and emission spectra of dansyl dye bound to the polymer. Morawetz proposed that the structural transition from compact globular form to the stretched linear chain occurs around pH 5.0 in aqueous solution which mimics the helix–globule transition of polypeptides [88].

The formation of aggregates of the fluorophores in aqueous polyelectrolyte solution occurs under favourable circumstances [89]. The possibility of utilizing electrostatic interactions between anionic polyelectrolytes and cationic dye molecules to foster aggregation has garnered a lot of interest in recent years due to its potential biological applications [90]. The absorption maximum of the aggregates varies with the orientation of the molecules and the aggregation number [91]. The clear distinction in the absorption bands of the aggregates helps in the identification of the nature of aggregation and eventually the charge distribution and structure of polyelectrolytes [92]. In homogeneous solution, the excited fluorophore in the presence of a quencher undergoes deactivation of the excited state as a result of diffusional encounter process. The emission intensity of the fluorophore depends upon the quencher concentration as given below

$$\frac{I}{I_0} = 1 + K_{sv}[Q] \qquad (1)$$

Here, I and $I_0$ are the emission intensities in the presence and absence of the quencher, respectively. The above relationship is applicable only when the quencher is homogeneously distributed in solution. Deviation from the above relation is observed for fluorophore bound to the polymer in solution in the presence of a quencher [93]. The unequal distribution of the quencher in the polymeric medium is the reason for the deviation. The microheterogeneity in PMAA is demonstrated with fluorescent uranyl acetate in the presence of $Fe^{2+}$ as a quencher [94].

## EXCITED STATE ENERGY AND ELECTRON TRANSFER

Electronic energy transfer between chromophores attached to a polymer chain is invoked to explain several phenomena in polymer photophysics [95]. Excited state fluorescence energy transfer is a useful tool for the investigation of molecular arrangements and fluctuations thereof on a nanometer scale, as the efficiency of energy transfer is strongly distance dependent in the FRET mechanism [96]. Two dyes located at strategic points of the same or two different molecules may thus provide extensive information on the distance and the relative orientation of molecules to which the dyes are attached [97]. Under certain conditions, (such as distance, orientation and nature of the fluorophore pairs) the excited molecule may transfer its energy nonradiatively as shown below

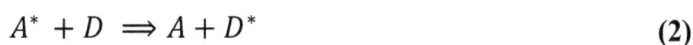

$$A^* + D \implies A + D^* \qquad (2)$$

Winnik and co–workers have extensively studied the end–to–end cyclization of polymers terminated with pyrene and dimethylaniline or benzil moieties [98]. The polysoap–to–polyelectrolyte transition of hydrophobically modified, water soluble poly(sodium maleate–alt–ethylvinyl ethers) is investigated using the energy transfer of covalently bound naphthyl and dansyl choromophores as energy transfer pair in combination with steady state absorption and emission measurements [99]. Recently, Swanson and co–workers studied the dynamics of PMAA in aqueous solution by using fluorescence energy transfer technique from acenaphthalene to anthracene moiety which are covalently tagged to the polymer [100].

Triplet–triplet energy transfer (TTET) between a xanthone donor and a naphthylalanine acceptor group is used to observe intrachain loop formation in unfolded polypeptide chains (Fig. **5**) [101]. The formation of van der Waal's contact between two specific amino acid side chain groups is also studied using TTET technique [102]. With this technique, the chain dynamics in unfolded peptide chains, as well as secondary structure formation and fluctuations in the native state of proteins on the tens of picosecond up to tens of microsecond time scale can be studied.

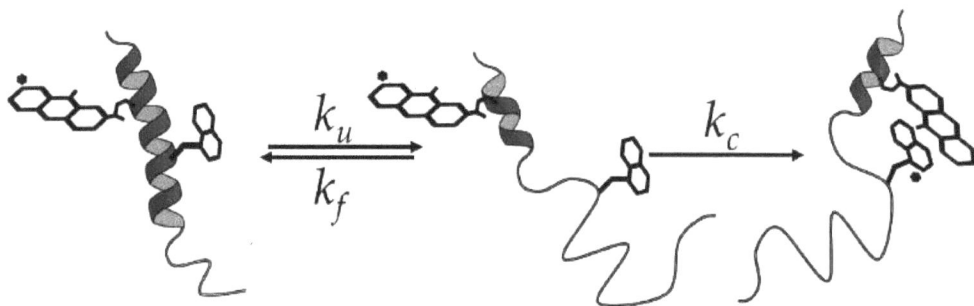

**Fig. (5).** Intrachain loop formation in polypeptides as probed by TTET.

The primary photochemical step in photosynthesis is known to be a one electron transfer process from the excited singlet state of a chlorophyll species to an electron acceptor [103]. The main feature of the primary event is that electron transfer leads to separation of the charged species which undergo further reactions in the photosynthetic sequence [104].Also, photoinduced electron transfer is of prime importance for light energy conversion. Guillet and co–workers studied the intramolecular photoinduced charge transfer process between end–capped porphyrin and quinolone moieties to a polyelectrolyte in aqueous solution. It is shown that the efficiency of electron transfer can be controlled by adjusting the pH of the solution and/or ionic strength of the medium [105].The dependence of electron transfer on the end–to–end distance and the degree of dissociation of polyelectrolyte are also investigated in aqueous solution.

## EXCIMER AND EXCIPLEX FORMATION

Excimer formation between an excited species and the ground state species of the same molecule is represented as:

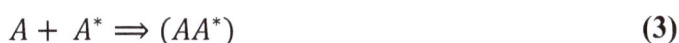

$$A + A^* \Longrightarrow (AA^*) \tag{3}$$

When the interacting molecules are different, then the excited state complex is said to be exciplex. The excimer formation in poly(styrene) investigated earlier shows that the interaction between the neighbouring phenyl rings in poly(styrene) leads to the excimer emission at 355nm (Fig. **6**) [106].In dilute solution, as the chains are far separated from one another, excimer formation between the phenyl groups is largely intramolecular rather than intermolecular. In the bulk state, excimers may involve neighbouring chains as well. The influence of tacticity of the polymer is probed by monitoring the ratio between the intensity of excimer emission and monomer emission ($I_D/I_M$). It is found that the $I_D/I_M$ ratio increases from 10 to 100 on increasing the tacticity of poly(styrene) from atactic to isotactic. The effect of long range polymer motion on the kinetics of excimer formation and exciplex formation in the presence of an external quencher is probed by using pyrene excimer which is randomly labeled on a poly(styrene) backbone and the kinetic scheme is shown below in Fig. (**6**) [107].

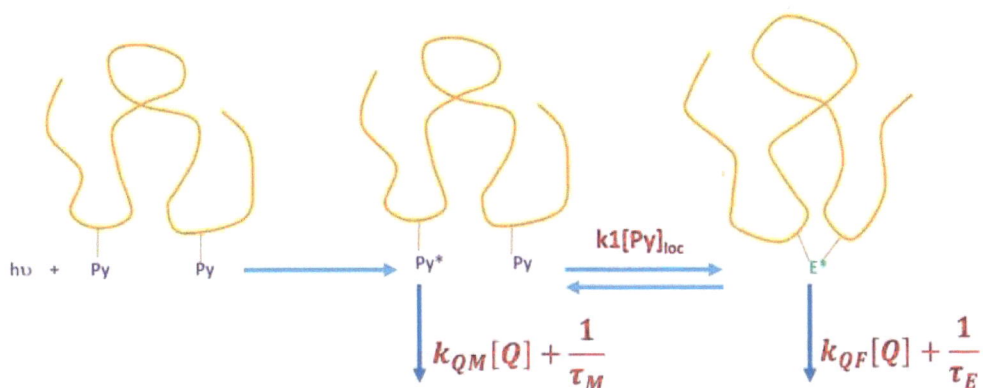

**Fig. (6).** Kinetic scheme showing the rates of excimer formation in pyrene attached to poly(styrene) in the presence of an external quencher.

# TIME–RESOLVED FLUORESCENCE STUDIES AND SOLVATION DYNAMICS

Time–resolved fluorescence spectroscopy is found to be particularly useful in the study of several types of polymers in solution and in the presence of additives. Time–resolved anisotropy decay of a fluorescent probe attached to a polymer chain reflects the rotational relaxation of the probe and the local dynamics of the polymer segments in solution [108]. The dependence of the segmental dynamics as a function of solvent viscosity is studied by time–resolved fluorescence using anthracene attached to poly(butadiene) [109]. The polymer dynamics as well as the structural properties and relaxation processes of the polymer and blends are studied by steady state and dynamic fluorescence using anthranyl and pyrenyl groups [110]. Fluorescence lifetime studies of poly(methacrylic acid), PMAA and poly(acrylic acid), PAA containing aromatic fluorescent probes such as naphthalene, anthracene and pyrene have been carried out by many authors to elucidate the properties of those polyelectrolytes in different solvents, pH, and surfactant addition [111]. Fluorescence anisotropy reflects the microviscosity and the solubilisation dynamics of the probe molecule in solutions of polyelectrolytes. The photophysical properties of thionine and phenosafranine dyes covalently bound to poly(acrylamidoglycolic acid) and poly(methylolacrylamide) are investigated using steady state and time–resolved fluorescence studies [41]. The effect of the induced friction by the polymer chain in the rotational relaxation dynamics of the same dye probe under different conditions, and the relation with the polymer conformation are studied by time–resolved fluorescence anisotropy measurements [35].

Solvation dynamics, the process of solvent shell reorganization after electronic excitation of the solute due to the change in the dipole moment, is one of the most studied processes in chemical dynamics [112]. In most experiments, ultrashort laser pulses excite (chemically inert) solute molecules into a state with an electronic charge distribution substantially different from that in the ground state (Fig. **7**). When small polar solvent molecules surround the solute molecules, following the excitation pulse the solvent molecules will rapidly reorient on account of the electrostatic solute–solvent molecular interactions, such that a dynamic equilibrium in accordance with the new excited–state charge distribution of the solute is established [113]. The time scale for the system to reach the new dynamic equilibrium extends from tens of femtoseconds up to the picosecond time regime. Chemical reactions occurring in solution are strongly affected by the surrounding solvent. For biochemical reactions occurring in DNA, these solvent effects are replaced by interactions with the complex DNA structure surrounding the reaction site [114]. Charge transfer between DNA bases is one example of a reaction where these unusual solvent effects are especially important. This is based on its importance in liquid phase chemistry and in biology, where the solvent shell has an active role in assisting, hindering, or triggering chemical reactions. Since the 1980s, solvation dynamics is a field of intense research using different ultrafast, optical UV–vis and IR laser techniques and with the help of various theoretical models and simulations. In the experimental studies, a short laser pulse excites a dye molecule in a solvent, inducing a sudden dipole moment change, to which the solvent molecules react by minimizing the free energy. This rearrangement is probed by recording the fluorescence, the excited–state absorption, or the stimulated emission of the dye molecule on the femtosecond time scale.

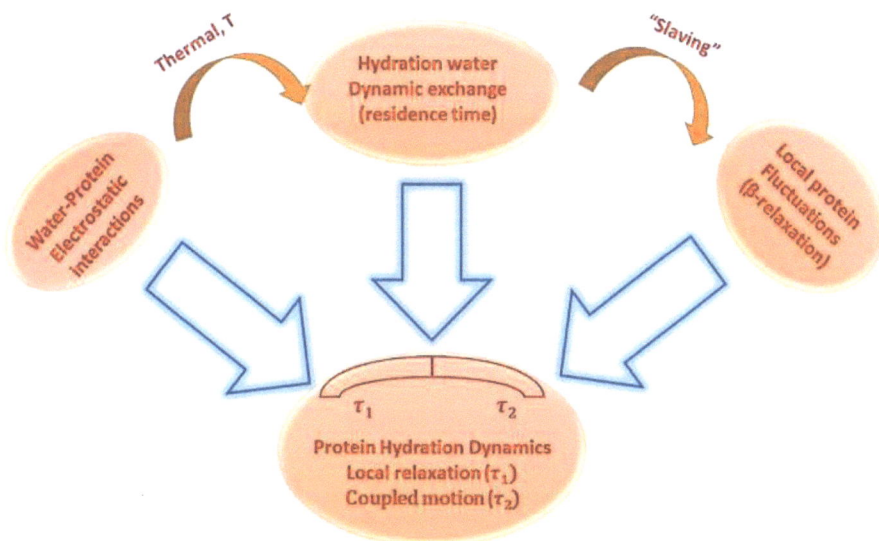

**Fig. (7).** Hydration dynamics of proteins.

In proteins and synthetic polyelectrolytes, the dynamics depend on the probe's position in the microheterogeneous environment. The time–resolved fluorescence Stokes shift (TRFSS) of a probe bound in the minor groove of DNA is measured over five decades in time from 100 fs to 10 ns [115]. The dynamics measured in probe position shows two components, the shorter lifetime component varies from 100 fs to ~100 ps and longer lifetime component varies from 100 ps to 10 ns depending on the position of the probe molecule [116].The processes involved in the hydration dynamics of these natural polyelectrolytes are schematically shown in Fig. (7).

Further, the effect of fluorescent probes with few orders of difference in fluorescence lifetime has been employed to investigate the solvation process between sub-nanosecond to nanosecond time region during the structural transition of PMAA in aqueous solution [36]. The relative amplitudes of the corresponding longer and shorter lifetimes of poly(MAA-Ph) in aqueous solutions increases from $40 \pm 5\%$ to $70 \pm 5\%$ while increasing the pH from ~3.0 to ~5.0 and reaches the minimum of less than $40 \pm 5\%$ with increase in the pH above 5.0; the reverse trend is observed for the short lived component (Fig. **8**). It is shown that there is a structural transition starts at pH 3.0 and continue to expand till basic pH condition. The solvent to structural relaxation occur during the lifetimes of the fluorophore in poly(MAA-Ph) at 930 ps. The solvent relaxation is still observable in the case of longer lifetime of poly(MAA-Ph) at low pH [117].

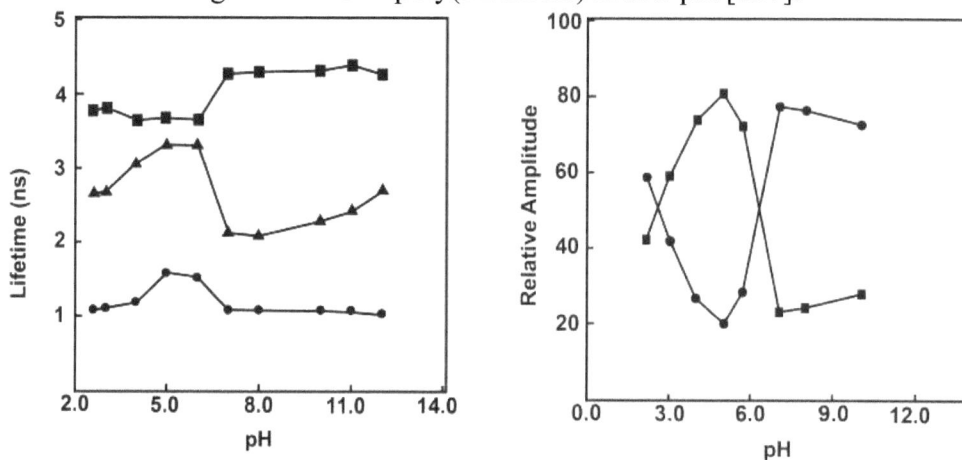

**Fig. (8).** (a) Fluorescence lifetime of poly(MAA-Ph) ($\blacksquare=\tau_2$; $\blacktriangle=<\tau>$; $\bullet=\tau_1$) vs pH and (b) The relative amplitude of poly(MAA-Ph) ($\blacksquare=A_2$; $\bullet=A_1$) vs pH; Polymer concentration = $1.0\times10^{-3}$ M; $\lambda_{ex}$ =295 nm; $\lambda_{em}$ = 590 nm; temperature = $23.0 \pm 0.5°C$. (Reproduced with permission from reference-117).

Only single transition point is noticed in the case of poly(MAA-Th) in aqueous solution at low pH with a longer lifetime as compared to that of thionine in aqueous solution as a result of more hydrophobic environment. These results indicating the reorientation of solvent to polymer structure occurs in the time range of 300–1000 ps (Fig. **9**). The fluorescence lifetimes of both shorter and longer lifetime components of polymer bound thionine at low pH (pH < 4.0) are higher than the lifetime of free thionine in water (~300 ps), which infers that the excited state decay is faster than the solvent relaxation processes.

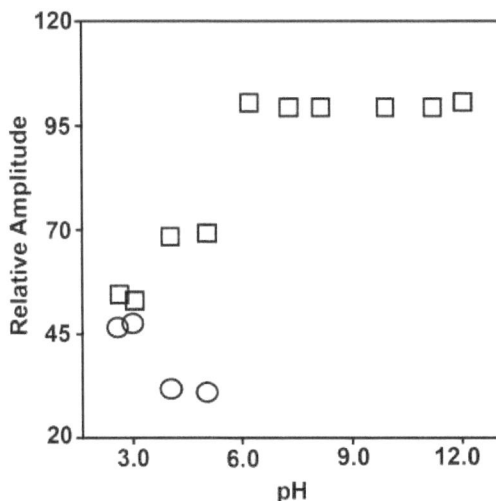

**Fig. (9).** The relative amplitude of poly(MAA-Th) ($\square=A_2$; $\circ= A_1$) vs pH; Polymer concentration = $1.0\times10^{-3}$ M; $\lambda_{ex}$ =295 nm; $\lambda_{em}$ = 622 nm; temperature = $23.0 \pm 0.5°C$. (Reproduced with permission from reference-117).

The dynamics and conformational transition of PMAA in aqueous solution are investigated using the triplet state of the cationic dye covalently attached to the polymer chain. The dye phenosafranine covalently bound to PMAA shows that the triplet-triplet absorption spectrum shifts to red region by 40 nm as compared to that of the free dye in aqueous solution and the triplet state lifetime is enhanced by 20 fold (Fig. **10**) [118]. Laser flash excitation shows that the environment of the triplet state of the dye bound to the polyelectrolyte at pH below 5.5 in aqueous solution is more rigid and less polar resulting in a highly compact globular nature of the polymer.

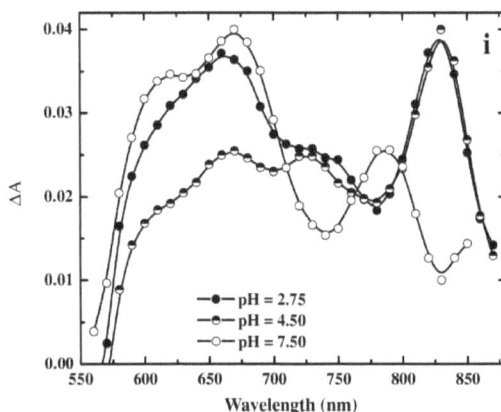

*(Fig. 10) contd.....*

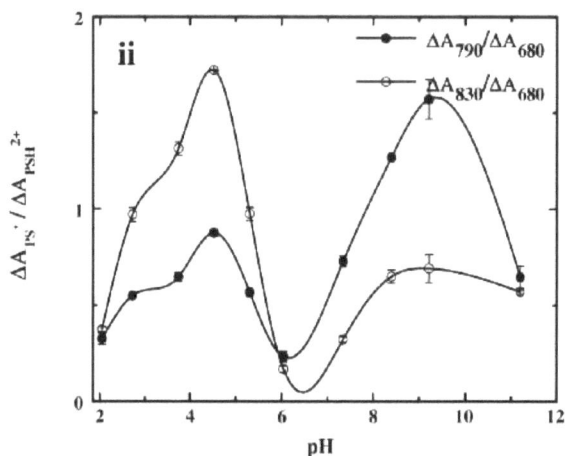

**Fig. (10).** (i) Absorption spectra of the transient plotted 10 1s after laser pulse of PMAA bound phenosafranine in aqueous solution at different pH and (ii) The ratio of monoprotonated form of the triplet in less polar environment (830 nm; solid line) and in polar environment (790 nm; dashed line) to diprotonated form of the triplet at as a function of pH. (Reproduced with permission from reference-118).

The decay of the triplet state of the dye bound to the polymer is attributed to the quenching of the excited state by the carboxylate groups of polyacrylic acids and to the decay process of the triplet in the tightly coiled polymer environment in the pH range 2.0–5.0. The spectra of the triplet dye molecules bound to the polymer at different degree of ionization of the polyelectrolyte suggest that the structural transition from compact globular structure to stretched rod like structure is cooperative involving a series of structural transitions. The observation of diprotonated triplet state of the PMAA bound dye at higher pH (i.e. pH above 7.0) reveals the existence of an intermediate structure akin to a micellar segment in PMAA prior to the formation of elongated linear chain (Fig. 11).

The dynamics and structural characteristics of polymethacrylic acid bound rhodamine-123 (PMAA–R123) and its interpolymer complex formed through hydrogen bonding between the monomeric units with poly(vinylpyrrolidone) were investigated using single molecular fluorescence studies. The time resolved fluorescence anisotropy decay of PMAA-R123 under acidic pH exhibits an associated anisotropy decay behavior characteristic of two different environments experienced by the fluorophore with one shorter and another longer rotational correlation time (Fig. 12) [119]. The anisotropy decay retains normal bi-exponential behavior under neutral pH. Fluorescence correlation spectroscopic investigation reveals that the attached fluorophore undergoes hydrolysis under basic condition which results in the release of the fluorophores from the polymer backbone.

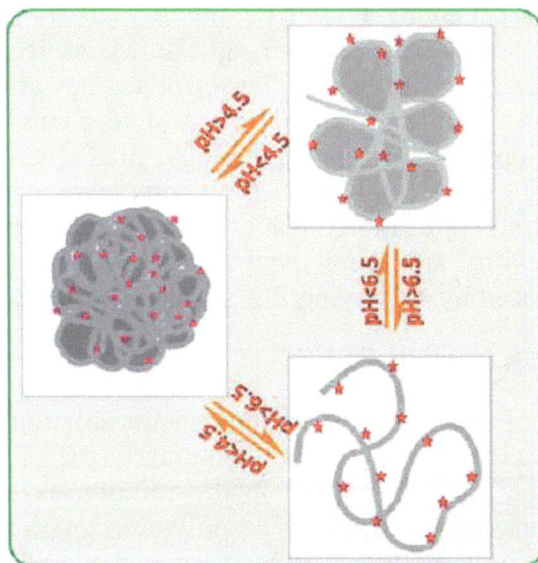

**Fig. (11).** Structural transitions of PMAA as function of pH as proposed by triplet-triplet absorption spectra of covalently bound phenosafranine (Reproduced with permission from reference-118).

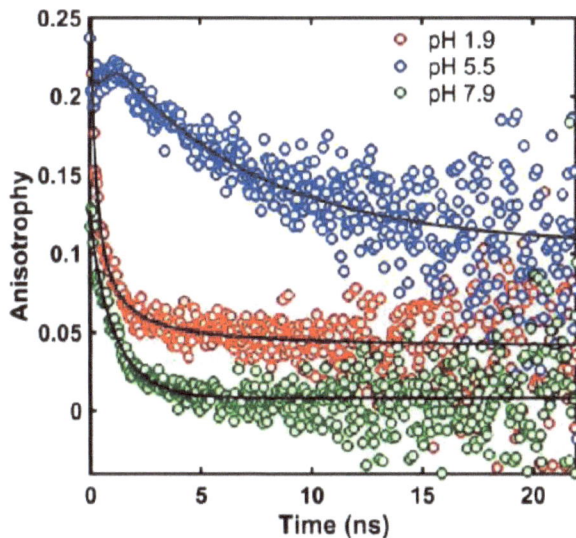

**Fig. (12).** Time resolved fluorescence anisotropy decay of PMAA-R123 in aqueous solution at various pH.(Reproduced with permission from reference -119).

## POLYMER COMPLEXES

In solution, polymers are known to associate due to coulombic interaction, hydrogen bonding and van der Wall's interactions due to the functional units present on polymer backbone. Polymer complexes are formed between a polymer and metal ions, surfactant, polymer and other molecular systems [120 - 122].Complex formation between polyelectrolytes has attracted a great deal of interest, mainly due to the use of polyelectrolyte complexes (PECs) for wide range of applications covering biotechnology, medicine, surface coatings, and papermaking [123 - 125]. Polyelectrolyte complexes have also found use in large–scale industrial applications such as flocculants and binders and it has been found that the stability and rheological properties of colloidal suspensions are significantly enhanced by employing PECs instead of polyelectrolytes.

## INTERPOLYMER COMPLEXES

The associations of two different polymers in a solution formed by the weak interactions are known as interpolymer complexes (IPCs). The polymer which is associated with the polymer present in the solution as a major component is known as complementary polymer. The polymer complexes are classified into four classes based on the interaction between the associating polymers, such as [126]

a. polyelectrolyte complexes
b. hydrogen bonding complexes
c. stereo complexes
d. charge transfer complexes

Polyelectrolyte complexes are formed by mixing oppositely charged polyelectrolytes (polyanions and polycations) due to coulombic forces. During the formation of polyelectrolyte complexes, microions are released quantitatively. Since many of the biopolymers are polyelectrolytes, polyelectrolyte complexes of synthetic macromolecules are studied as models of complicated biological systems [127].

Isotactic and syndiotactic poly(methyl methacrylate) are well–known to exhibit a strong tendency to form intermacromolecular complex through stereospecific interactions in a suitable solvent and bulk. This type of complex formation has not been studied extensively [128]. The formation of such complexes is strongly affected by the solvent [129]. Charge transfer complexes are formed between electron accepting polymer and electron donating polymer through charge transfer interactions. There are only a few studies on charge transfer complexes, since the synthesis of electron accepting polymer is difficult. Sulzberg *et al.* first studied the C–T complexes by successfully synthesizing the following electron donating and electron accepting polymers [130].

# HYDROGEN BONDED INTERPOLYMER COMPLEXES

The fundamental role of hydrogen bond in the structure of DNA and the secondary and tertiary structures of proteins is known [131]. Many interpolymer complexes governed through hydrogen bonds occur in biological systems and synthetic polymers. Examples include complex formation between poly(carboxylic acids) and poly(ethylene oxide), PEO, poly(vinyl alcohol), PVA, poly(vinylpyrrolidone), PVP [126] Polymers containing the base pairs such as uracil, cytosine, thymine, adenine and guanine are also studied as nucleic acid models [132].

The study of these complexes has varied potential applications in the areas of chemical engines, biosensors, separations, drug delivery and space energy supply systems [133 - 136]. One of the most recent and significant advances in the area of IPCs *via* hydrogen–bonding is the design of multilayered materials *via* the layer–by–layer (LBL) approach [137]. It involves an alternating exposure of various solid substrates to solutions of polymers forming insoluble complexes. The LBL approach is successfully employed by Stockton and Rubner and many others for designing multilayered assembly through complexation *via* hydrogen–bonding [138]. The formation of interpolymer complexes through hydrogen bonding is extensively studied between different polymeric pairs both in aqueous and organic solutions as well as at interfaces [123]. Stoichiometric interaction between proton accepting polymer and proton donating polymer in aqueous medium and organic solvents leads to the formation of hydrogen bonded interpolymer complexes (Fig. **13**) [139]. The complex formation is affected by temperature, polymer structure, polymer concentration, solvent and other weak interactions. The interpolymer complexes stabilized by hydrogen bonds are more stable in an aqueous solution than in organic solvents [126].

Interpolymer complex formation involving poly(acrylic acid)s as proton donating polymer, the complexation is strongly affected by the pH of the solution [139]. As a rule, complexes are formed in weakly or strongly acidic media and dissociate upon the increase of pH. pH below which the IPC starts to precipitate is called the critical pH of complexation ($pH_{crit}$). For a given polymer–polymer system $pH_{crit}$ is a specific value that depends on the nature of both polymers, the molecular weight and concentration as well as the presence of various small molecules and ions in the solution [126]. An increase in molecular weight and hydrophobicity of interacting polymers as well as their concentration in solutions leads to higher $pH_{crit}$ values.

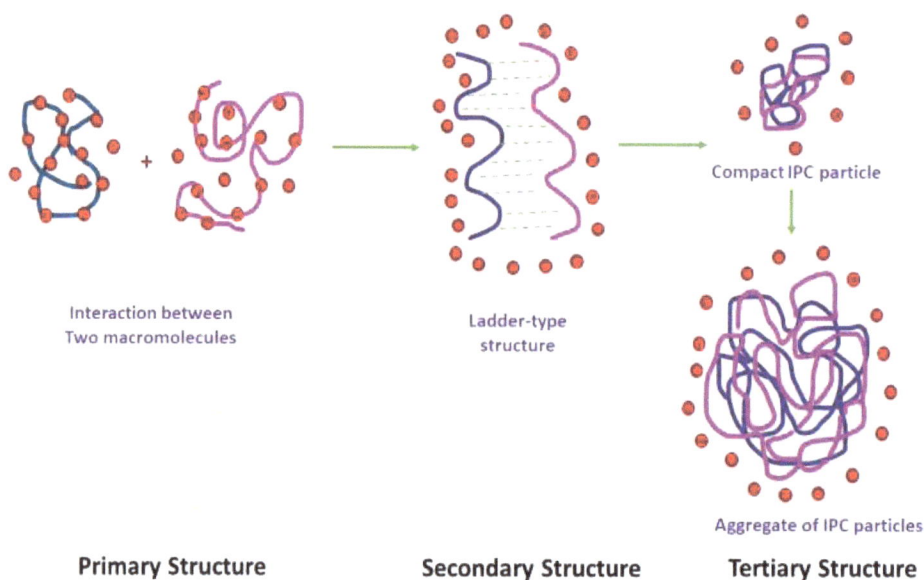

**Primary Structure**      **Secondary Structure**      **Tertiary Structure**

**Fig. (13).** Diagram depicting the multistage formation and structures of hydrogen bonded interpolymer complexes.

Interpolymer association between poly(ethylene oxide) and poly(carboxylic acid)s, such as poly(acrylic acid) and poly(methacrylic acid) by hydrogen bonding leads to the formation of an insoluble complex at low pH at 1:1 molar ratio of the polymers [140]. Changes in the conformation and microstructure of the interpolymer complexes are investigated by small–angle neutron scattering measurements using isotopic substitution [141]. The interpolymer complex formation is investigated earlier by fluorescence spectroscopy using the dansyl labelled poly(acrylic acid) and their complementary polymers [24]. The effect of solvent on the structure and properties of IPCs formed by PAA and PVP both in solutions and at solution–glass interfaces are studied [139]. A series of novel copolymers NVP with more hydrophobic vinyl propyl ether (VPE) is synthesised to look at how the introduction of hydrophobic units into PVP affects the complexation [142].

Interpolymer complex formation between the dansyl labelled poly(acrylic acid) and the complementary polymer leads to the displacement of water molecules from the neighbourhood of the dansyl. As a result, there is a sharp increase in the fluorescence intensity of labelled dansyl on interpolymer complexation with complementary polymers [24]. The preferential location of the fluorophore in the polymer adduct when the fluorophore was covalently attached to the complementary macromolecules was studied using time–resolved measurements [143, 144]. When the fluorophore is attached to poly(acrylic acids), the complex formation with the complementary polymers keeps the fluorophore in hydrophobic environment. The fluorophore experiences a hydrophilic environment when it is bound to the complementary polymer PVP as shown in Fig. (**14**).

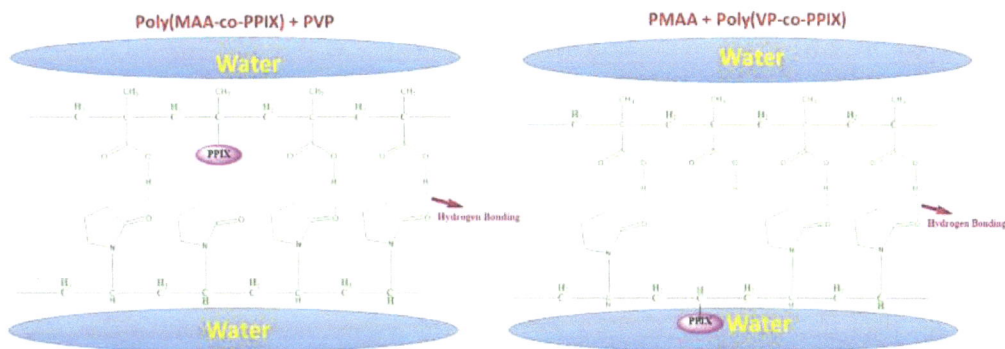

**Fig. (14).** Preferential location of the fluorophores bound to the complementary polymers upon interpolymer adduct formation.

## CONCLUSION

We have described the importance of fluorescently labelled macromolecules. Detailed discussions have been included on various photophysical and photochemical processes of free fluorescent molecules as well as their conjugates with polymeric materials. The applications of steady state and time resolved photophysical techniques to study the nature of interaction between the fluorophores and macromolecular entities have been explained in the subsequent sections. Further, specific applications of photophysical phenomenon to understand the dynamics of polyelectrolytes at various time regions have been explained with examples from our previous publications. Finally, the nature of self-assembly between these polyelectrolytes through hydrogen bonding probed by excited state dynamics of fluorescent labels has been given with relevant examples.

## CONSENT OF PUBLICATION

The author grants the publisher the sole and exclusive license of the full copyright of this material.

## CONFLICT OF INTEREST

The authors declare no conflict of interest, financial or otherwise.

## ACKNOWLEDGMENT

The author D. K thank the financial support from RUSA 2.0 (MHRD-India) grant sanctioned vide Letter No. F. 24-51/2014-U, Policy (TNMulti-Gen), Dept. of Edn. Govt. of India, Dt. October 09, 2018.

## REFERENCES

[1]    Dobrynin, A.; Rubinstein, M. Theory of polyelectrolytes in solutions and at surfaces. *Prog. Polym. Sci.,* **2005**, *30*(11), 1049-1118.
       [http://dx.doi.org/10.1016/j.progpolymsci.2005.07.006]

[2]    Kriwet, B.; Walter, E.; Kissel, T. Synthesis of bioadhesive poly(acrylic acid) nano- and microparticles using an inverse emulsion polymerization method for the entrapment of hydrophilic drug candidates. *J. Control. Release,* **1998**, *56*(1-3), 149-158.
       [http://dx.doi.org/10.1016/S0168-3659(98)00078-9] [PMID: 9801438]

[3]    Decher, G. Fuzzy Nanoassemblies: Toward Layered Polymeric Multicomposites. *Science,* **1997**, *277*(5330), 1232-1237.
       [http://dx.doi.org/10.1126/science.277.5330.1232]

[4]    Fan, Y.; Wang, Y.; Fan, Y.; Ma, J. Preparation of insulin nanoparticles and their encapsulation with biodegradable polyelectrolytes *via* the layer-by-layer adsorption. *Int. J. Pharm.,* **2006**, *324*(2), 158-167.
       [http://dx.doi.org/10.1016/j.ijpharm.2006.05.062] [PMID: 16814967]

[5]    Bajpai, A.K.; Shukla, S.K.; Bhanu, S.; Kankane, S. Responsive polymers in controlled drug delivery. *Prog. Polym. Sci.,* **2008**, *33*(11), 1088-1118.
       [http://dx.doi.org/10.1016/j.progpolymsci.2008.07.005]

[6]    Anspach, W.M.; Marinsky, J.A. Complexing of nickel(II) and cobalt(II) by a polymethacrylic acid gel and its linear polyelectrolyte analog. *J. Phys. Chem.,* **1975**, *79*(5), 433-439.
       [http://dx.doi.org/10.1021/j100572a008]

[7]    Balasubramaniam, E.; Natarajan, P. Photophysical properties of protoporphyrin IX and thionine covalently attached to macromolecules, *J. Photochem. Photobiol. A. Chem,* **1997**, *103*, 201-211.

[8]    Tripathy, S.K.; Kumar, J.; Nalwa, H.S. *Handbook of polyelectrolytes and their applications*; American Scientific Stevenson Ranch CA, **2002**, Vol. 3, .

[9]    Morawetz, H. On the versatility of fluorescence techniques in polymer research. *J. Polym. Sci. A Polym. Chem.,* **1999**, *37*(12), 1725-1735.
       [http://dx.doi.org/10.1002/(SICI)1099-0518(19990615)37:12<1725::AID-POLA1>3.0.CO;2-D]

[10]   Kuhn, W.; Kuhn, H. Statistische und energieelastische Rückstellkraft bei stark auf Dehnung beanspruchten Fadenmolekeln. *Helv. Chim. Acta,* **1946**, *29*(5), 1095-1115.
       [http://dx.doi.org/10.1002/hlca.19460290514]

[11]   Schatzki, T.F. Molecular interpretation of the γ-transition in polyethylene and related compounds. *Journal of Polymer Science Part C: Polymer Symposia,* **1966**, *14*(1), 139-140.
       [http://dx.doi.org/10.1002/polc.5070140114]

[12]   Rudin, A.; Choi, P. *The Elements of Polymer Science and Engineering,* 3rd ed; , **2013**, pp. 1-62.

[13]   Winnik, M.A.; Redpath, T.; Richards, D.H. The dynamics of end-to-end cyclization in polystyrene probed by pyrene excimer formation. *Macromolecules,* **1980**, *13*(2), 328-335.
       [http://dx.doi.org/10.1021/ma60074a023]

[14]   Redpath, A.E.C.; Winnik, M.A. Temperature dependence of excimer formation between pyrenes at the ends of a polymer in a good solvent. Cyclization dynamics of polymers. 9. *J. Am. Chem. Soc.,* **1982**, *104*(21), 5604-5607.

[http://dx.doi.org/10.1021/ja00385a006]

[15]  Mahadevan, L. Polymer science and biology: structure and dynamics at multiple scales. *Faraday Discuss.,* **2008**, *139*, 9-19.
[http://dx.doi.org/10.1039/b809771m] [PMID: 19048987]

[16]  Russel, D.; Lasker, K.; Phillips, J.; Schneidman-Duhovny, D.; Velázquez-Muriel, J.A.; Sali, A. The structural dynamics of macromolecular processes. *Curr. Opin. Cell Biol.,* **2009**, *21*(1), 97-108.
[http://dx.doi.org/10.1016/j.ceb.2009.01.022] [PMID: 19223165]

[17]  Fersht, A.R. On the simulation of protein folding by short time scale molecular dynamics and distributed computing. *Proc. Natl. Acad. Sci. USA,* **2002**, *99*(22), 14122-14125.
[http://dx.doi.org/10.1073/pnas.182542699] [PMID: 12388785]

[18]  Bednar, B.; Trnena, J.; Svoboda, P.; Vajda, S.; Fidler, V.; Prochazka, K. Time-resolved fluorescence study of chain dynamics. 1. Poly(methacrylic acid) in dilute water solutions. *Macromolecules,* **1991**, *24*(8), 2054-2059.
[http://dx.doi.org/10.1021/ma00008a053]

[19]  Costa, T.; Miguel, M.G.; Lindman, B.; Schillén, K.; Lima, J.C.; Seixas de Melo, J. Self-assembly of a hydrophobically modified naphthalene-labeled poly(acrylic acid) polyelectrolyte in water: organic solvent mixtures followed by steady-state and time-resolved fluorescence. *J. Phys. Chem. B,* **2005**, *109*(8), 3243-3251.
[http://dx.doi.org/10.1021/jp046589w] [PMID: 16851348]

[20]  Anufrieva, E.V.; Birshtein, T.M.; Nekrasova, T.N.; Ptitsyn, O.B.; Sheveleva, T.V. The models of the denaturation of globular proteins. II. Hydrophobic interactions and conformational transition in polymethacrylic acid. *Journal of Polymer Science Part C: Polymer Symposia,* **1967**, *16*(6), 3519-3531.
[http://dx.doi.org/10.1002/polc.5070160645]

[21]  Delben, F.; Crescenzi, V.; Quadrifoglio, F. On the enthalpy of dissociation of poly(methacrylic acid) in aqueous solution. *Eur. Polym. J.,* **1972**, *8*(7), 933-935.
[http://dx.doi.org/10.1016/0014-3057(72)90054-7]

[22]  Soutar, I.; Swanson, L. Luminescence studies of polyelectrolyte behavior in solution. 3. Time-resolved fluorescence anisotropy measurements of the conformational behavior of Poly(methacrylic acid) in dilute aqueous solutions. *Macromolecules,* **1994**, *27*(15), 4304-4311.
[http://dx.doi.org/10.1021/ma00093a035]

[23]  Katchalsky, A. Solutions of polyelectrolytes and mechanochemical systems. *Journal of Polymer Science,* **1951**, *7*(4), 393-412.
[http://dx.doi.org/10.1002/pol.1951.120070403]

[24]  Horský, J.; Petrus, V.; Bohdanecký, M. The characteristic ratio of poly(methacrylic acid) in organic solvents. *Makromol. Chem.,* **1986**, *187*(11), 2621-2628.
[http://dx.doi.org/10.1002/macp.1986.021871111]

[25]  Crescenzi, V.; Quadriioglio, F.; Delben, F. Calorimetric investigation of poly(methacrylic acid) and poly(acrylic acid) in aqueous solution. *J. Polym. Sci., Polym. Phys. Ed.,* **1972**, *10*, 357-368.

[26]  Seixas de Melo, J.; Costa, T.; Francisco, A.; Maçanita, A.L.; Gago, S.; Gonçalves, I.S. Dynamics of short as compared with long poly(acrylic acid) chains hydrophobically modified with pyrene, as followed by fluorescence techniques. *Phys. Chem. Chem. Phys.,* **2007**, *9*(11), 1370-1385.
[http://dx.doi.org/10.1039/B613382G] [PMID: 17347710]

[27]  Díez-Peña, E.; Quijada-Garrido, I.; Barrales-Rienda, J.M.; Schnell, I.; Spiess, H.W. Advanced 1H Solid-State NMR Spectroscopy on Hydrogels. *Macromol. Chem. Phys.,* **2004**, *205*, 430-437.

[28]  Ortiz, C.; Hadziioannou, G. Entropic Elasticity of Single Polymer Chains of Poly(methacrylic acid) Measured by Atomic Force Microscopy. *Macromolecules,* **1999**, *32*(3), 780-787.
[http://dx.doi.org/10.1021/ma981245n]

[29]  Tajiri, T.; Morita, S.; Ozaki, Y. Hydration mechanism on a poly(methacrylic acid) film studied by in

situ attenuated total reflection infrared spectroscopy. *Polymer (Guildf.),* **2009**, *50*(24), 5765-5770.
[http://dx.doi.org/10.1016/j.polymer.2009.09.060]

[30]   Hao, J.; Yuan, G.; He, W.; Cheng, H.; Han, C.C.; Wu, C. Interchain Hydrogen-Bonding-Induced
Association of Poly(acrylic acid)- *graft* -poly(ethylene oxide) in Water. *Macromolecules,* **2010**, *43*(4),
2002-2008.
[http://dx.doi.org/10.1021/ma9025515]

[31]   Chen, Y.; Gao, D.; Tian, Y.; Ai, P.; Zhanga, H.; Yu, A. A novel extremophile strategy studied by
Raman spectroscopy, *Spectrochim. Acta. Part A. Mol. Biomol. Spect.,* **2007**, *67*, 1126-1132.
[http://dx.doi.org/10.1016/j.saa.2006.09.032]

[32]   Fernyhough, C.; Ryan, A.J.; Battaglia, G. pH controlled assembly of a polybutadiene–poly
(methacrylic acid) copolymer in water: packing considerations and kinetic limitations. *Soft Matter,*
**2009**, *5*(8), 1674-1682.
[http://dx.doi.org/10.1039/b817218h]

[33]   Raja, C.; Ananthanarayanan, K.; Natarajan, P. Studies on the photophysical characteristics of
poly(carboxylic acid)s bound protoporphyrin IX and metal complexes of protoporphyrin IX. *Eur.
Polym. J.,* **2006**, *42*(3), 495-506.
[http://dx.doi.org/10.1016/j.eurpolymj.2005.09.009]

[34]   Jones, G., II; Oh, C. Photophysical and electron transfer properties of pseudoisocyanine in the
hydrophobic domain of an aqueous polyelectrolyte. *J. Phys. Chem.,* **1994**, *98*(9), 2367-2376.
[http://dx.doi.org/10.1021/j100060a026]

[35]   Oliveira, H.P.M.; Gehlen, M.H. Time resolved fluorescence anisotropy of basic dyes bound to
poly(methacrylic acid) in solution. *J. Braz. Chem. Soc.,* **2003**, *14*(5), 738-743.
[http://dx.doi.org/10.1590/S0103-50532003000500007]

[36]   Natarajan, P.; Raja, C. Studies on the dynamics of poly(carboxylic acids) with covalently bound
thionine and phenosafranine in dilute aqueous solutions. *Eur. Polym. J.,* **2005**, *41*(10), 2496-2504.
[http://dx.doi.org/10.1016/j.eurpolymj.2005.04.031]

[37]   Sukul, D.; Sen, S.; Dutta, P.; Bhattacharyya, K. Isomerization and fluorescence depolarization of
merocyanine 540 in polyacrylic acid. Effect ofpH. *J. Chem. Sci.,* **2002**, *114*(5), 501-511.
[http://dx.doi.org/10.1007/BF02704194]

[38]   Smith, C.K.; Liu, G. Determination of the Rate Constant for Chain Insertion into Poly(methyl
methacrylate)- *block* -poly(methacrylic acid) Micelles by a Fluorescence Method. *Macromolecules,*
**1996**, *29*(6), 2060-2067.
[http://dx.doi.org/10.1021/ma951338u]

[39]   Nabiyan, A.; Max, J.B.; Schacher, F.H. Double hydrophilic copolymers – synthetic approaches,
architectural variety, and current application fields. *Chem. Soc. Rev.,* **2022**, *51*(3), 995-1044.
[http://dx.doi.org/10.1039/D1CS00086A] [PMID: 35005750]

[40]   Bednar, B.; Morawetz, H.; Shafer, J.A. Kinetics of the conformational transition of poly(methacrylic
acid) after changes of its degree of ionization. *Macromolecules,* **1985**, *18*(10), 1940-1944.
[http://dx.doi.org/10.1021/ma00152a024]

[41]   Viswananthan, K.; Natarajan, P. Photophysical properties of thionine and phenosafranine dyes
covalently bound to macromolecules. *J. Photochem. Photobiol A. Chem,* **1996**, *95*, 245-253.

[42]   Rohatgi Mukherjee, K.K. *Fundamentals of Photochemistry*; , **1986**.

[43]   Gopidas, K.R.; Kamat, P.V. Photochemistry in polymers: photoinduced electron transfer between
phenosafranine and triethylamine in perfluorosulfonate membrane. *J. Phys. Chem.,* **1990**, *94*(11),
4723-4727.
[http://dx.doi.org/10.1021/j100374a064]

[44]   Mosnácek, J.; Lukác, I. Reaction mechanism of thymine dimer formation in DNA induced by UV
light. *J. Photochem. Photobiol A. Chem,* **2002**, *151*, 95-101.

[45] Allen, N.S. Polymer Photochemistry. *Photochemistry,* **2007**, *36*, 232-297.
[http://dx.doi.org/10.1039/9781847558572-00232]

[46] Tallent, J.R.; Stuart, J.A.; Song, Q.W.; Schmidt, E.J.; Martin, C.H.; Birge, R.R. Photochemistry in dried polymer films incorporating the deionized blue membrane form of bacteriorhodopsin. *Biophys. J.,* **1998**, *75*(4), 1619-1634.
[http://dx.doi.org/10.1016/S0006-3495(98)77605-2] [PMID: 9746505]

[47] Xu, W.; Lia, T.; Zeng, G.; Zhang, S.; Shang, W.; Wu, Y.; Miyashita, T. Studies on photolithography and photoreaction of copolymer containing naphthyl in ultrathin nanosheets induced by deep UV irradiation. *J. Photochem. Photobiol A. Chem,* **2008**, *194*, 97-104.

[48] Yamashita, K.; Imahashi, S. Photophysics of 2,6-bis(4'-diethylaminobenzylidene)cyclopentanone in a polymer matrix containing carboxyl groups. *J. Photochem. Photobiol A. Chem,* **2000**, *135*, 135-139.

[49] Jaeger, W.; Bohrisch, J.; Laschewsky, A. Synthetic polymers with quaternary nitrogen atoms—Synthesis and structure of the most used type of cationic polyelectrolytes. *Prog. Polym. Sci.,* **2010**, *35*(5), 511-577.
[http://dx.doi.org/10.1016/j.progpolymsci.2010.01.002]

[50] Kreutzer, J.; Demir, K.D.; Yagci, Y. Synthesis and characterization of a double photochromic initiator for cationic polymerization. *Eur. Polym. J.,* **2011**, *47*(4), 792-799.
[http://dx.doi.org/10.1016/j.eurpolymj.2010.09.025]

[51] Kaneko, M.; Yamada, A. Photopotential and photocurrent induced by a tolusafranine ethylenediaminetetraacetic acid system. *J. Phys. Chem.,* **1977**, *81*(12), 1213-1215.
[http://dx.doi.org/10.1021/j100527a020]

[52] Decker, C. UV☐radiation curing chemistry. *Pigm. Resin Technol.,* **2001**, *30*(5), 278-286.
[http://dx.doi.org/10.1108/03699420110404593]

[53] Yagci, Y.; Jockusch, S.; Turro, N.J. Photoinitiated Polymerization: Advances, Challenges, and Opportunities. *Macromolecules,* **2010**, *43*(15), 6245-6260.
[http://dx.doi.org/10.1021/ma1007545]

[54] Tamilarasan, R.; Natarajan, P. Photovoltaic conversion by macromolecular thionine films. *Nature,* **1981**, *292*(5820), 224-225.
[http://dx.doi.org/10.1038/292224a0]

[55] Rousseau, E.; Koetse, M.M.; Van der Auweraer, M.; De Schryver, F.C. Comparison between J-aggregates in a self-assembled multilayer and polymer-bound J-aggregates in solution: a steady-state and time-resolved spectroscopic study. *Photochem. Photobiol. Sci.,* **2002**, *1*(6), 395-406.
[http://dx.doi.org/10.1039/b201690g] [PMID: 12856707]

[56] Fron, E.; Deres, A.; Rocha, S.; Zhou, G.; Müllen, K.; De Schryver, F.C.; Sliwa, M.; Uji-i, H.; Hofkens, J.; Vosch, T. Unraveling excited-state dynamics in a polyfluorene-perylenediimide *copolymer. J. Phys. Chem. B,* **2010**, *114*(3), 1277-1286.
[http://dx.doi.org/10.1021/jp909295h] [PMID: 20050587]

[57] Qin, Y.; Kiburu, I.; Shah, S.; Jäkle, F. Synthesis and Characterization of Organoboron Quinolate Polymers with Tunable Luminescence Properties. *Macromolecules,* **2006**, *39*(26), 9041-9048.
[http://dx.doi.org/10.1021/ma061805f]

[58] Natansohn, A.; Rochon, P. Photoinduced motions in azo-containing polymers. *Chem. Rev.,* **2002**, *102*(11), 4139-4176.
[http://dx.doi.org/10.1021/cr970155y] [PMID: 12428986]

[59] Steinberg-Yfrach, G.; Liddell, P.A.; Hung, S.C.; Moore, A.L.; Gust, D.; Moore, T.A. Conversion of light energy to proton potential in liposomes by artificial photosynthetic reaction centres. *Nature,* **1997**, *385*(6613), 239-241.
[http://dx.doi.org/10.1038/385239a0]

[60] Sykora, M.; Maxwell, K.A.; DeSimone, J.M.; Meyer, T.J. Mimicking the antenna-electron transfer properties of photosynthesis. *Proc. Natl. Acad. Sci. USA,* **2000**, *97*(14), 7687-7691.
[http://dx.doi.org/10.1073/pnas.97.14.7687] [PMID: 10884400]

[61] Corrales, T.; Catalina, F.; Peinado, C.; Allen, N.S.; Rufs, A.M.; Bueno, C.; Encinas, M.V. Photochemical study and photoinitiation activity of macroinitiators based on thioxanthone. *Polymer (Guildf.),* **2002**, *43*(17), 4591-4597.
[http://dx.doi.org/10.1016/S0032-3861(02)00310-5]

[62] Encinas, M.V.; Rufs, A.M.; Corrales, T.; Catalina, F.; Peinado, C.; Schmith, K.; Neumann, M.G.; Allen, N.S. The influence of the photophysics of 2-substituted thioxanthones on their activity as photoinitiators. *Polymer (Guildf.),* **2002**, *43*(14), 3909-3913.
[http://dx.doi.org/10.1016/S0032-3861(02)00211-2]

[63] Tsunooka, M.; Yamamoto, T.; Kurokawa, Y.; Suyama, K.; Shirai, M. Photocuring Systems Using Quaternary Ammonium Thiocyanates. *J. Photopolym. Sci. Technol.,* **2002**, *15*(1), 47-50.
[http://dx.doi.org/10.2494/photopolymer.15.47]

[64] Polo, E.; Barbieri, A.; Sostero, S.; Green, M.L.H. Zirconocenes as Photoinitiators for Free-Radical Polymerisation of Acrylates. *Eur. J. Inorg. Chem.,* **2002**, *2002*(2), 405-409.
[http://dx.doi.org/10.1002/1099-0682(20022)2002:2<405::AID-EJIC405>3.0.CO;2-G]

[65] Kuckling, D.; Vo, C.D.; Wohlrab, S.E. Preparation of Nanogels with Temperature-Responsive Core and pH-Responsive Arms by Photo-Cross-Linking. *Langmuir,* **2002**, *18*(11), 4263-4269.
[http://dx.doi.org/10.1021/la015758q]

[66] Lester, C.L.; Guymon, C.A. Ordering effects on the photopolymerization of a lyotropic liquid crystal. *Polymer (Guildf.),* **2002**, *43*(13), 3707-3715.
[http://dx.doi.org/10.1016/S0032-3861(02)00188-X]

[67] Campidelli, S.; Deschenaux, R.; Swartz, A.; Rahman, G.M.A.; Guldi, D.M.; Milic, D.; Vázquez, E.; Prato, M. A dendritic fullerene–porphyrin dyad. *Photochem. Photobiol. Sci.,* **2006**, *5*(12), 1137-1141.
[http://dx.doi.org/10.1039/B610881D] [PMID: 17136279]

[68] Catalina, F.; Peinado, C.; Allen, N.S.; Corrales, T. Chemiluminescence of polyethylene: The comparative antioxidant effectiveness of phenolic stabilizers in low-density polyethylene. *J. Polym. Sci. A Polym. Chem.,* **2002**, *40*(19), 3312-3326.
[http://dx.doi.org/10.1002/pola.10419]

[69] Cheng, K.W.; Mak, C.S.C.; Chan, W.K.; Ching Ng, A.M.; Djurišić, A.B. Synthesis of conjugated polymers with pendant ruthenium terpyridine trithiocyanato complexes and their applications in heterojunction photovoltaic cells. *J. Polym. Sci. A Polym. Chem.,* **2008**, *46*(4), 1305-1317.
[http://dx.doi.org/10.1002/pola.22471]

[70] Wolffs, M.; Hoeben, F.J.M.; Beckers, E.H.A.; Schenning, A.P.H.J.; Meijer, E.W. Sequential energy and electron transfer in aggregates of tetrakis[oligo(p-phenylene vinylene)] porphyrins and C60 in water. *J. Am. Chem. Soc.,* **2005**, *127*(39), 13484-13485.
[http://dx.doi.org/10.1021/ja054406t] [PMID: 16190697]

[71] Berton, N.; Fabre-Francke, I.; Bourrat, D.; Chandezon, F.; Sadki, S. Poly(bisthiophene-carbazol-fullerene) double-cable polymer as new donor-acceptor material: preparation and electrochemical and spectroscopic characterization. *J. Phys. Chem. B,* **2009**, *113*(43), 14087-14093.
[http://dx.doi.org/10.1021/jp905876h] [PMID: 19813718]

[72] Mohamad, D.K.; Fischereder, A.; Yi, H.; Cadby, A.J.; Lidzey, D.G.; Iraqi, A. A novel 2,7-linked carbazole based "double cable" polymer with pendant perylene diimide functional groups: preparation, spectroscopy and photovoltaic properties. *J. Mater. Chem.,* **2011**, *21*(3), 851-862.
[http://dx.doi.org/10.1039/C0JM02673E]

[73] Košovan, P.; Limpouchová, Z.; Procházka, K. Molecular Dynamics Simulation of Time-Resolved Fluorescence Anisotropy Decays from Labeled Polyelectrolyte Chains. *Macromolecules,* **2006**, *39*(9),

3458-3465.
[http://dx.doi.org/10.1021/ma052557a]

[74]   Anghel, D.F.; Alderson, V.; Winnik, F.M.; Mizusaki, M.; Morishima, Y. Fluorescent dyes as model 'hydrophobic modifiers' of polyelectrolytes: a study of poly(acrylic acid)s labelled with pyrenyl and naphthyl groups. *Polymer (Guildf.)*, **1998**, *39*(14), 3035-3044.
[http://dx.doi.org/10.1016/S0032-3861(97)10126-4]

[75]   Reichardt, C. Solvatochromic Dyes as Solvent Polarity Indicators. *Chem. Rev.,* **1994**, *94*(8), 2319-2358.
[http://dx.doi.org/10.1021/cr00032a005]

[76]   Chakrabarty, A.; Das, P.; Mallick, A.; Chattopadhyay, N. Effect of surfactant chain length on the binding interaction of a biological photosensitizer with cationic micelles. *J. Phys. Chem. B,* **2008**, *112*(12), 3684-3692.
[http://dx.doi.org/10.1021/jp709818d] [PMID: 18307338]

[77]   Wong, K.H.; Chan, M.C.W.; Che, C.M. Modular Cyclometalated Platinum(II) Complexes as Luminescent Molecular Sensors for pH and Hydrophobic Binding Regions. *Chemistry,* **1999**, *5*(10), 2845-2849.
[http://dx.doi.org/10.1002/(SICI)1521-3765(19991001)5:10<2845::AID-CHEM2845>3.0.CO;2-G]

[78]   Sahoo, D.; Chakravorti, S. Dye-surfactant interaction: modulation of photophysics of an ionic styryl dye. *Photochem. Photobiol.,* **2009**, *85*(5), 1103-1109.
[http://dx.doi.org/10.1111/j.1751-1097.2009.00582.x] [PMID: 19558418]

[79]   George, S.; Kumbhakar, M.; Singh, P.K.; Ganguly, R.; Nath, S.; Pal, H. Fluorescence spectroscopic investigation to identify the micelle to gel transition of aqueous triblock copolymer solutions. *J. Phys. Chem. B,* **2009**, *113*(15), 5117-5127.
[http://dx.doi.org/10.1021/jp809826c] [PMID: 19317476]

[80]   Hashimoto, S. Optical Spectroscopy and Microscopy Studies on the Spatial Distribution and Reaction Dynamics in Zeolites. *J. Phys. Chem. Lett.,* **2011**, *2*(5), 509-519.
[http://dx.doi.org/10.1021/jz101572u]

[81]   Pereira, R.V.; Gehlen, M.H. Polymerization and Conformational Transition of Poly(methacrylic Acid) Probed by Electronic Spectroscopy of Aminoacridines. *Macromolecules,* **2007**, *40*(6), 2219-2223.
[http://dx.doi.org/10.1021/ma062020c]

[82]   Pankasem, S.; Biscoglio, M.; Thomas, J.K. Photophysics of Pyrenyl Acrylic Acid and Its Methyl Ester. A Spectroscopic Method to Monitor Polymerization and Surface Properties. *Langmuir,* **2000**, *16*(8), 3620-3625.
[http://dx.doi.org/10.1021/la9911638]

[83]   Loutfy, R.O. High-conversion polymerization of fluorescence probes. 1. Polymerization of methyl methacrylate. *Macromolecules,* **1981**, *14*(2), 270-275.
[http://dx.doi.org/10.1021/ma50003a009]

[84]   Pereira, R.V.; Gehlen, M.H. Spectroscopy of auramine fluorescent probes free and bound to poly(methacrylic acid). *J. Phys. Chem. B,* **2006**, *110*(13), 6537-6542.
[http://dx.doi.org/10.1021/jp054833t] [PMID: 16570951]

[85]   Alfrey, T., Jr; Morawetz, H. Amphoteric Polyelectrolytes. I. 2-Vinylpyridine—Methacrylic Acid Copolymers [1,2]. *J. Am. Chem. Soc.,* **1952**, *74*(2), 436-438.
[http://dx.doi.org/10.1021/ja01122a046]

[86]   Wang, Y.C.; Morawetz, H. Studies of intramolecular excimer formation in dibenzyl ether, dibenzylamine, and its derivatives. *J. Am. Chem. Soc.,* **1976**, *98*(12), 3611-3615.
[http://dx.doi.org/10.1021/ja00428a037]

[87]   Goldenberg, M.; Emert, J.; Morawetz, H. Intramolecular excimer study of rates of conformational transitions. Dependence on molecular structure and the viscosity of the medium. *J. Am. Chem. Soc.,*

**1978**, *100*(23), 7171-7177.
[http://dx.doi.org/10.1021/ja00491a008]

[88]   Pearsall, S.K.; Green, M.M.; Morawetz, H. Titration of Poly(carboxylic acid)s in Methanol Solution. Polymer Chain Extension, Ionization Equilibria, and Conformational Mobility. *Macromolecules,* **2004**, *37*(23), 8773-8777.
[http://dx.doi.org/10.1021/ma0491614]

[89]   Peyratout, C.; Donath, E.; Daehne, L. Electrostatic interactions of cationic dyes with negatively charged polyelectrolytes in aqueous solution. *J. Photochem. Photobiol. A. Chem,* **2001**, *142*, 51-57.

[90]   Sukhorukov, G.; Dähne, L.; Hartmann, J.; Donath, E.; Möhwald, H. Controlled Precipitation of Dyes into Hollow Polyelectrolyte Capsules Based on Colloids and Biocolloids. *Adv. Mater.,* **2000**, *12*(2), 112-115.
[http://dx.doi.org/10.1002/(SICI)1521-4095(200001)12:2<112::AID-ADMA112>3.0.CO;2-P]

[91]   Lai, W.C.; Dixit, N.S.; Mackay, R.A. Formation of H aggregates of thionine dye in water. *J. Phys. Chem.,* **1984**, *88*(22), 5364-5368.
[http://dx.doi.org/10.1021/j150666a051]

[92]   Mirenda, M.; Dicelio, L.E.; San Román, E. Effect of molecular interactions on the photophysics of Rose Bengal in polyelectrolyte solutions and self-assembled thin films. *J. Phys. Chem. B,* **2008**, *112*(39), 12201-12207.
[http://dx.doi.org/10.1021/jp803892g] [PMID: 18774851]

[93]   Horský, J.; Morawetz, H. Kinetics of the conformational transition of poly(methacrylic acid) after a pH jump. Studies of nonradiative energy transfer. *Makromol. Chem.,* **1988**, *189*(10), 2475-2483.
[http://dx.doi.org/10.1002/macp.1988.021891024]

[94]   Morawetz, H.; Taha, I.A.I. Catalysis of ionic reactions by polyelectrolytes. III. Quenching of uranyl ion fluorescence by iron(II) ions in poly(vinylsulfonic acid) solution. *J. Am. Chem. Soc.,* **1971**, *93*(4), 829-833.
[http://dx.doi.org/10.1021/ja00733a005]

[95]   Liu, G.; Guillet, J.E. Application of the "spectroscopic ruler" to studies of the dimensions of flexible macromolecules. 1. Theory. *Macromolecules,* **1990**, *23*(5), 1388-1392.
[http://dx.doi.org/10.1021/ma00207a025]

[96]   Morawetz, H. Characterization of the interpenetration of chain molecules by nonradiative energy transfer.*Photophysical and Photochemical Tools in Polymer Science*; Reidel: Dordrecht, **1986**, pp. 547-559.
[http://dx.doi.org/10.1007/978-94-009-4726-9_24]

[97]   Webber, S.E. Photon-harvesting polymers. *Chem. Rev.,* **1990**, *90*(8), 1469-1482.
[http://dx.doi.org/10.1021/cr00106a005]

[98]   Yang, J.; Roller, R.S.; Winnik, M.A.; Zhang, Y.; Pakula, T. Energy Transfer Study of Symmetric Polyisoprene−Poly(methyl methacrylate) Diblock Copolymers Bearing Dyes at the Junctions: Dye Orientation. *Macromolecules,* **2005**, *38*(4), 1256-1263.
[http://dx.doi.org/10.1021/ma049245c]

[99]   Hu, Y.; Smith, G.L.; Richardson, M.F.; McCormick, C.L. Water Soluble Polymers. 74. pH Responsive Microdomains in Labeled *n* -Octylamide-Substituted Poly(sodium maleate- *alt* -ethyl vinyl ethers): Synthesis, Steady-State Fluorescence, and Nonradiative Energy Transfer Studies. *Macromolecules,* **1997**, *30*(12), 3526-3537.
[http://dx.doi.org/10.1021/ma9613502]

[100]  Ruiz-Pérez, L.; Pryke, A.; Sommer, M.; Battaglia, G.; Soutar, I.; Swanson, L.; Geoghegan, M. Conformation of Poly(methacrylic acid) Chains in Dilute Aqueous Solution. *Macromolecules,* **2008**, *41*(6), 2203-2211.
[http://dx.doi.org/10.1021/ma0709957]

[101] Reiner, A. Triplet-triplet energy transfer studies on conformational dynamics in peptides and a protein. *J. Pept. Sci.,* **2011**, *17*(6), 413-419.
[http://dx.doi.org/10.1002/psc.1353] [PMID: 21360629]

[102] Bieri, O.; Wirz, J.; Hellrung, B.; Schutkowski, M.; Drewello, M.; Kiefhaber, T. The speed limit for protein folding measured by triplet–triplet energy transfer. *Proc. Natl. Acad. Sci. USA,* **1999**, *96*(17), 9597-9601.
[http://dx.doi.org/10.1073/pnas.96.17.9597] [PMID: 10449738]

[103] Krasnovsky, A.A. Photoluminescence of singlet oxygen in pigment solutions. *Photochem. Photobiol.,* **1979**, *29*(1), 29-36.
[http://dx.doi.org/10.1111/j.1751-1097.1979.tb09255.x]

[104] Gust, D.; Moore, T.A.; Moore, A.L. Molecular mimicry of photosynthetic energy and electron transfer. *Acc. Chem. Res.,* **1993**, *26*(4), 198-205.
[http://dx.doi.org/10.1021/ar00028a010]

[105] Nowakowska, M.; Kataoka, F.; Guillet, J.E. Photoinduced Electron Transfer in Porphyrin–Quinone End-Capped Poly(methacrylic acid). 1. Photophysical Studies. *Macromolecules,* **1996**, *29*(5), 1600-1608.
[http://dx.doi.org/10.1021/ma946429q]

[106] Patel, B.; Mendicuti, F.; Mattice, W.L. Dependence on spacer size of the intramolecular excimer emission from model compounds for polyesters derived from isophthalic or terephthalic acid. *Polymer (Guildf.),* **1992**, *33*(2), 239-242.
[http://dx.doi.org/10.1016/0032-3861(92)90978-6]

[107] Duhamel, J. Polymer chain dynamics in solution probed with a fluorescence blob model. *Acc. Chem. Res.,* **2006**, *39*(12), 953-960.
[http://dx.doi.org/10.1021/ar068096a] [PMID: 17176034]

[108] Soutar, I.; Swanson, L.; Annable, T.; Padget, J.C.; Satgurunathan, R. Luminescence techniques and characterization of the morphology of polymer latices. *J. Colloid Interface Sci.,* **2006**, *303*(1), 205-213.
[http://dx.doi.org/10.1016/j.jcis.2006.07.044] [PMID: 16919664]

[109] Bur, A.J.; Lowry, R.E.; Roth, S.C.; Thomas, C.L.; Wang, F.W. Observations of shear-induced molecular orientation in a polymer melt using fluorescence anisotropy measurements. *Macromolecules,* **1991**, *24*(12), 3715-3717.
[http://dx.doi.org/10.1021/ma00012a040]

[110] Sharma, J.; Tleugabulova, D.; Czardybon, W.; Brennan, J.D. Two-site ionic labeling with pyranine: implications for structural dynamics studies of polymers and polypeptides by time-resolved fluorescence anisotropy. *J. Am. Chem. Soc.,* **2006**, *128*(16), 5496-5505.
[http://dx.doi.org/10.1021/ja058707e] [PMID: 16620123]

[111] Hsiao, J.S.; Webber, S.E. Water-soluble polystyrene latexes as photoredox media. 2. Excited-state electron transfer from anthracene covalently bound poly(methacrylic acid). *J. Phys. Chem.,* **1993**, *97*(31), 8296-8303.
[http://dx.doi.org/10.1021/j100133a028]

[112] Majumder, P.; Sarkar, R.; Shaw, A.K.; Chakraborty, A.; Pal, S.K. Ultrafast dynamics in a nanocage of enzymes: Solvation and fluorescence resonance energy transfer in reverse micelles. *J. Colloid Interface Sci.,* **2005**, *290*(2), 462-474.
[http://dx.doi.org/10.1016/j.jcis.2005.04.053] [PMID: 15939425]

[113] Brazard, J.; Usman, A.; Lacombat, F.; Ley, C.; Martin, M.M.; Plaza, P. New insights into the ultrafast photophysics of oxidized and reduced FAD in solution. *J. Phys. Chem. A,* **2011**, *115*(15), 3251-3262.
[http://dx.doi.org/10.1021/jp110741y] [PMID: 21438617]

[114] Fürstenberg, A.; Vauthey, E. Ultrafast excited-state dynamics of oxazole yellow DNA intercalators. *J.*

*Phys. Chem. B,* **2007**, *111*(43), 12610-12620.
[http://dx.doi.org/10.1021/jp073182t] [PMID: 17929857]

[115] Pal, N.; Verma, S.D.; Sen, S. Probe position dependence of DNA dynamics: comparison of the time-resolved Stokes shift of groove-bound to base-stacked probes. *J. Am. Chem. Soc.,* **2010**, *132*(27), 9277-9279.
[http://dx.doi.org/10.1021/ja103387t] [PMID: 20565076]

[116] Samuni, U.; Roche, C.J.; Dantsker, D.; Friedman, J.M. Conformational dependence of hemoglobin reactivity under high viscosity conditions: the role of solvent slaved dynamics. *J. Am. Chem. Soc.,* **2007**, *129*(42), 12756-12764.
[http://dx.doi.org/10.1021/ja072342b] [PMID: 17910446]

[117] Natarajan, P.; Duraimurugan, K.; Kumar, K.S. Photoprocesses of coordination compounds and dyes in solution and nanoporous materials: Evolution from milliseconds to femtoseconds#. *J. Chem. Sci.,* **2011**, *123*(5), 531-553.
[http://dx.doi.org/10.1007/s12039-011-0129-9]

[118] Durai Murugan, K.; Natarajan, P. Studies on the structural transitions and self-organization behavior of polyacrylic acids with complementary polymers in aqueous solution by laser flash photolysis method using the triplet state of covalently bound phenosafranine. *Eur. Polym. J.,* **2011**, *47*(8), 1664-1675.
[http://dx.doi.org/10.1016/j.eurpolymj.2011.05.018]

[119] Kandhasamy, D.M.; Selvaraju, C.; Dharuman, V. Structure and dynamics of poly(methacrylic acid) and its interpolymer complex probed by covalently bound rhodamine-123. *Spectrochim. Acta A Mol. Biomol. Spectrosc.,* **2021**, *248*, 119166.
[http://dx.doi.org/10.1016/j.saa.2020.119166]

[120] Subramanian, R.; Natarajan, P.; Parthasarathi, V. X-ray studies on interpolymer adducts formed between poly(N-vinylpyrrolidone) and poly(acrylic acid)s. *Makromol. Chem., Rapid. Commun.,* **1980**, *1*(1), 47-49.
[http://dx.doi.org/10.1002/marc.1980.030010110]

[121] de Vasconcelos, C.L.; Bezerril, P.M.; Dantas, T.N.C.; Pereira, M.R.; Fonseca, J.L.C. Adsorption of bovine serum albumin on template-polymerized chitosan/poly(methacrylic acid) complexes. *Langmuir,* **2007**, *23*(14), 7687-7694.
[http://dx.doi.org/10.1021/la700537t] [PMID: 17547429]

[122] Bykov, A.G.; Lin, S.Y.; Loglio, G.; Miller, R.; Noskov, B.A. Kinetics of Adsorption Layer Formation in Solutions of Polyacid/Surfactant Complexes. *J. Phys. Chem. C,* **2009**, *113*(14), 5664-5671.
[http://dx.doi.org/10.1021/jp810471y]

[123] Siegel, R.A. Hydrophobic weak polyelectrolyte gels: Studies of swelling equilibria and kinetics. *Adv. Polym. Sci.,* **1993**, *109*, 233-267.
[http://dx.doi.org/10.1007/3-540-56791-7_6]

[124] Demeter, M.; Călina, I.; Vancea, C.; Şen, M.; Kaya, M.G.A.; Mănăilă, E.; Dumitru, M.; Meltzer, V.; Viorica, M. E-beam processing of collagen-poly(N-vinyl-2-pyrrolidone) double-network superabsorbent hydrogels: Structural and rheological investigations. *Macromol. Res.,* **2019**, *27*(3), 255-267.
[http://dx.doi.org/10.1007/s13233-019-7041-4]

[125] Jin, S.; Liu, M.; Zhang, F.; Chen, S.; Niu, A. Synthesis and characterization of pH-sensitivity semi-IPN hydrogel based on hydrogen bond between poly(N-vinylpyrrolidone) and poly(acrylic acid). *Polymer (Guildf.),* **2006**, *47*(5), 1526-1532.
[http://dx.doi.org/10.1016/j.polymer.2006.01.009]

[126] Tsuchida, E.; Abe, K. Interactions between macromolecules in solution and intermacromolecular complexes. *Adv. Polym. Sci.,* **1982**, *45*, 1-119.
[http://dx.doi.org/10.1007/BFb0017549]

[127] Kossel, A. Über die basischen stoffe des Zellkerns. *J. Phys. Chem.,* **1896**, *22*, 178-183.

[128] Lohmeyer, J.H.G.M.; Kransen, G.; Tan, Y.Y.; Challa, G. Stereoassociation between poly(methyl methacrylate) and poly(methacrylic acid). *J. Polym. Sci. Polym. Lett. Ed.,* **1975**, *13*(12), 725-729.
[http://dx.doi.org/10.1002/pol.1975.130131204]

[129] Vorenkamp, E.J.; Bosscher, F.; Challa, G. Association of stereoregular poly(methyl methacrylates): 4. Further study on the composition of the stereocomplex. *Polymer (Guildf.),* **1979**, *20*(1), 59-64.
[http://dx.doi.org/10.1016/0032-3861(79)90043-0]

[130] Sulzberg, T.; Cotter, R.J. Charge Transfer in Donor Polymer-Acceptor Polymer Mixtures. *Macromolecules,* **1968**, *1*(6), 554-555.
[http://dx.doi.org/10.1021/ma60006a020]

[131] Palmer, L.C.; Stupp, S.I. Molecular self-assembly into one-dimensional nanostructures. *Acc. Chem. Res.,* **2008**, *41*(12), 1674-1684.
[http://dx.doi.org/10.1021/ar8000926] [PMID: 18754628]

[132] Pochon, F.; Michelson, A.M.; Polynucleotides, V.I. Polynucleotides. VI. Interaction between polyguanylic acid and polycytidylic acid. *Proc. Natl. Acad. Sci. USA,* **1965**, *53*(6), 1425-1430.
[http://dx.doi.org/10.1073/pnas.53.6.1425] [PMID: 5217645]

[133] Khutoryanskiy, V. Hydrogen-bonded interpolymer complexes as materials for pharmaceutical applications. *Int. J. Pharm.,* **2007**, *334*(1-2), 15-26.
[http://dx.doi.org/10.1016/j.ijpharm.2007.01.037] [PMID: 17320317]

[134] Bekturov, E.A.; Bimendina, L.A. Interpolymer complexes. *Adv. Polym. Sci.,* **1981**, *41*, 99-147.
[http://dx.doi.org/10.1007/3-540-10554-9_11]

[135] Jiang, M.; Li, M.; Xiang, M.; Zhou, H. Interpolymer Complexation and Miscibility Enhancement by Hydrogen Bonding. *Adv. Polym. Sci.,* **1999**, *146*, 121-196.
[http://dx.doi.org/10.1007/3-540-49424-3_3]

[136] Bell, C.L.; Peppas, N.A. Biomedical membranes from hydrogels and interpolymer complexes. *Adv. Polym. Sci.,* **1995**, *122*, 125-175.
[http://dx.doi.org/10.1007/3540587888_15]

[137] Kim, B.S.; Lee, H.; Min, Y.; Poon, Z.; Hammond, P.T. Hydrogen-bonded multilayer of pH-responsive polymeric micelles with tannic acid for surface drug delivery. *Chem. Commun. (Camb.),* **2009**, (28), 4194-4196.
[http://dx.doi.org/10.1039/b908688a] [PMID: 19585018]

[138] Stockton, W.B.; Rubner, M.F. Molecular-Level Processing of Conjugated Polymers. 4. Layer-b--Layer Manipulation of Polyaniline *via* Hydrogen-Bonding Interactions. *Macromolecules,* **1997**, *30*(9), 2717-2725.
[http://dx.doi.org/10.1021/ma9700486]

[139] Zhunuspayev, D.E.; Mun, G.A.; Hole, P.; Khutoryanskiy, V.V. Solvent effects on the formation of nanoparticles and multilayered coatings based on hydrogen-bonded interpolymer complexes of poly(acrylic acid) with homo- and copolymers of N-vinyl pyrrolidone. *Langmuir,* **2008**, *24*(23), 13742-13747.
[http://dx.doi.org/10.1021/la802852h] [PMID: 18980359]

[140] Khutoryanskaya, O.V.; Williams, A.C.; Khutoryanskiy, V.V. pH-Mediated Interactions between Poly(acrylic acid) and Methylcellulose in the Formation of Ultrathin Multilayered Hydrogels and Spherical Nanoparticles. *Macromolecules,* **2007**, *40*(21), 7707-7713.
[http://dx.doi.org/10.1021/ma071644v]

[141] Sotiropoulou, M.; Oberdisse, J.; Staikos, G. Soluble Hydrogen-Bonding Interpolymer Complexes in Water: A Small-Angle Neutron Scattering Study. *Macromolecules,* **2006**, *39*(8), 3065-3070.
[http://dx.doi.org/10.1021/ma052444r]

[142] Nurkeeva, Z.S.; Mun, G.A.; Khutoryanskiy, V.V.; Sergaziyev, A.D. Complex formation of polyvinyl ether of diethylene glycol with polyacrylic acid II. Effect of molecular weight of polyacrylic acid and solvent nature. *Eur. Polym. J.,* **2002**, *38*(2), 313-316.
[http://dx.doi.org/10.1016/S0014-3057(01)00188-4]

[143] Natarajan, P.; Raja, C. Novel features of the interpolymer self-organisation behaviour investigated using covalently linked protoporphyrin IX as fluorescent probe in the macromolecules. *Eur. Polym. J.,* **2001**, *37*(11), 2207-2211.
[http://dx.doi.org/10.1016/S0014-3057(01)00127-6]

[144] Natarajan, P.; Raja, C. Studies on interpolymer self-organisation behaviour of protoporphyrin IX bound poly(carboxylic acid)s with complimentary polymers by means of fluorescence techniques. *Eur. Polym. J.,* **2004**, *40*(10), 2291-2303.
[http://dx.doi.org/10.1016/j.eurpolymj.2004.06.003]

# SUBJECT INDEX

## A

Absorption spectra 153, 169, 197, 198
  triplet-triplet 198
Acid 14, 22, 23, 32, 66, 82, 83, 126, 144, 145,
    148, 162, 183, 189, 193, 197, 200, 201
  acrylamidoglycolic 193
  carboxylic 200, 201
  carminic 14
  chloranilic 148, 162
  gallic 32
  gastric 126
  lewis 82, 83
  methacrylic 183, 189, 193, 201
  oxalic 66
  perfluorooctanoic 22, 23
  picric (PA)144, 145
  polyacrylic 197
  polymethacrylic 197
  sulfuric 83
  trifluoroacetic 83
  trifluoromethanesulfonic 83
Aggregation 6, 94, 96, 98, 99, 103, 140
  induced emission (AIE) 6, 94, 96, 98, 99,
    103
  process 98, 140
AIEgen 94
  behaviour 94
  luminescence 94
AIE turn-on fluorescence behaviours 94
Alkyl chain, dinitrobenzene-containing 96
Alzheimer's disease 146
Amide-functionalized metallacycles 167
Amines 122, 154, 155
Amino acid 18, 123, 124
  chemosensor 123
  enantiomers 18
  residue in peptide sequences 124
  transportation 18
Analysis 116, 149, 162, 163, 166, 171
  absorption titration 163
  emission titration 171

fluorescence spectroscopic 116
fluorescence titration 149, 166
single-crystal X-ray diffraction 162, 163
Analytes 14, 17, 111, 112, 113, 115, 116, 117,
    118, 119, 121, 132
  aliphatic alcohol 14
  amine-based 118
  steroid 17
Angle neutron scattering measurements 201
Anion sensors 43
Anthracene 15, 66, 96, 100, 101, 141, 142,
    149, 164, 165, 166, 193
  bearing Pillar 96
  fluorescent 96
Antibacterial 58, 60, 65, 70, 71
  activity 60, 70
  and antioxidant activity 58, 65, 71
Anticancer activity 42
Antioxidant 58, 65, 70, 71
  activity 58, 65, 70, 71
  agents 58
Anti-tumour agents 70
Applications 58, 60, 65, 70, 71, 84, 86, 100,
    102, 175, 199
  antibacterial 70, 71
  biochemical 175
  catalysis 58
  catalytic 60
  diverse 84, 86, 100, 102
  environmental 60
  industrial 199
  of resorcinarene crowns 65
Architecture 47, 78, 88, 168, 172, 173
  distorted trigonal prismatic 172
  fabricating luminescent 47
  hammock-shaped 168
  novel macrocyclic 78
  rectangular prismatic 173
Arene 42, 46, 68
  metalloporphyrin 68
  oxacyclophane architectures 42
  system 46

**Paulpandian Muthu Mareeswaran, Palaniswamy Suresh and Seenivasan Rajagopal (Eds.)**